Safety Assurance under Uncertainties

From Software to Cyber-Physical/Machine Learning Systems

Editors

Ichiro Hasuo
National Institute of Informatics
and
SOKENDAI
Tokyo, Japan

Fuyuki Ishikawa
National Institute of Informatics
and
SOKENDAI
Tokyo, Japan
and
The University of Electro-Communications
Tokyo, Japan

CRC Press
Taylor & Francis Group
Boca Raton London New York

CRC Press is an imprint of the
Taylor & Francis Group, an **informa** business
A SCIENCE PUBLISHERS BOOK

First edition published 2025
by CRC Press
2385 NW Executive Center Drive, Suite 320, Boca Raton FL 33431

and by CRC Press
4 Park Square, Milton Park, Abingdon, Oxon, OX14 4RN

CRC Press is an imprint of Taylor & Francis Group, LLC

Library of Congress Cataloging-in-Publication Data (applied for)

ISBN: 978-0-367-55401-9 (hbk)
ISBN: 978-0-367-55404-0 (pbk)
ISBN: 978-1-003-09335-0 (ebk)

DOI: 10.1201/9781003093350

Typeset in Times New Roman
by Prime Publishing Services

Preface

Safety assurance of software systems has never been as selected a problem as it is today. Practitioners and researchers who work on the problem face a challenge unique to modern software systems: *uncertainties*. For one, the cyber-physical nature of modern software systems—as exemplified by automated driving systems—mandates environmental uncertainties to be addressed and the resulting hazards to be mitigated. Furthermore, the abundance of statistical machine-learning components—massive numerical computing units for statistical reasoning such as deep neural networks—makes systems hard to explain, understand, analyse, or verify.

Facing the challenge of these physical and statistical uncertainties, no single established method for software safety and reliability would suffice. Rigorous formal verification requires formal modelling of every detail of the target system, which is impossible under uncertainties. Testing suffers from uncertainties, too: notably, it is unclear to what degree of safety assurance a given test result should translate. Therefore, modern efforts towards safe software systems must unite techniques from different disciplines—formal methods, software science, software engineering, control theory, machine learning, etc.—in a way driven by real-world examples and supported by a common theoretical ground.

This book is the first to provide a comprehensive overview of such united and interdisciplinary efforts. Driven by automated driving systems as a leading example, the book describes diverse techniques to specify, model, test, analyse, and verify modern software systems. Coming out of a collaboration between industry and basic academic research, the book covers both practical analysis techniques (readily applicable to existing systems) and more long-range design techniques (that call for new designs but bring a greater degree of assurance). It consists of eleven chapters covering a variety of topics and techniques. Their diversity reflects the diversity of facets of safety assurance of modern software systems. Each chapter is written by experts of the corresponding field. Although there are loose connections between chapters, our intention is that each chapter is self-contained and can be read independent of others.

The book's exposition aims at giving high-level intuitions and use-cases of each technique, rather than technical details, with plenty of pointers for interested readers.

How to Read This Book

The current book consists of eleven chapters covering a variety of topics andtechniques. Their diversity reflects the diversity of facets of safety assurance of modern software systems. Each chapter is written by experts of the corresponding field. Although there are loose connections between chapters, our intention is that each chapter is self-contained and can be read independent of others.

Acknowledgments

The current book originates from *ERATO MMSD–MIRAI eAI Seminars for Automotive Industry*, a seminar series organised by the editors where talks were given by the chapter authors. The target audience was industry practitioners, especially those in the automotive industry. The eleven chapters of this book correspond to the eleven seminars in the series, held in 2020–2021.

The editors would therefore like to express their warm gratitude to the contributors and the participants of the seminar series. The seminar series was organised as a joint activity of two projects: JST ERATO Hasuo Metamathematics for Systems Design project (ERATO MMSD, Grant No. JPMJER1603) and JST MIRAI Engineerable AI project (MIRAI eAI, Grant No. JPMJMI20B8), both funded by Japan Science and Technology Agency. The financial and organisational supports from these projects are gratefully acknowledged.

Tokyo, March 2024

Ichiro Hasuo
Fuyuki Ishikawa

Contents

Chapter Overview

Here is a very high-level overview of the content of each chapter. Readers are encouraged to read it first and choose which chapter to read. As we have previously mentioned, there are no technical dependencies between chapters, therefore each chapter can be read without reading others.

Chapter 1: Optimisation-Based Falsification—Showcasing Black-Box Techniques for System Safety

We start out with an exposition of a concrete technique, namely *optimisation-based falsification*. It is a prototypical formal technique to tackle uncertainties in modern software systems. Notably, it is a safety assurance method that does not require formal modelling (i.e., a white-box system model).

Falsification is a search-based testing task where, given a black-box system model $\mathcal{M}$ and a specification φ on its output, one searches for an input signal σ such that the corresponding output signal $\mathcal{M}(\sigma)$ violates φ. This task is clearly omnipresent in safety assurance efforts.

The emphasis of the chapter is on the crucial role played by *formal logic* as a specification formalism. Falsification relies on formal specification—an important branch of formal methods—and specifications play various roles, from serving as interpretable presentations of objective functions to enhancing search heuristics exploiting logical structures.

Summarising, this chapter: (1) introduces a technique that is widely applicable to various safety assurance problems, and (2) demonstrates how formal methods can broaden its realm to those *uncertain* settings where white-box system models are unavailable.

Chapter 2: Monitoring Temporal Specifications

We continue with another concrete technique that stems out from formal methods but does not require white-box system models. The technique is called *monitoring*: given an execution log σ of a target system and a formal specification φ, monitoring returns if (or in which part) σ satisfies φ. This task too is omnipresent in real-world safety assurance efforts—one generates many execution logs σ and check if nothing is wrong there.

This task of monitoring may not sound too challenging: indeed, in the current industry practice, it is solved mostly by ad-hoc coding specific to individual φ. This is problematic at least for two reasons. For one, such an ad-hoc and repetitive coding effort is often inefficient. For another, such ad-hoc coding usually does not explicate the specification φ, which is undesired in terms of *explainability* and *accountability*.

The chapter discusses techniques that are generic; in particular, the implemented monitor works for any specification φ expressed in a certain formalism. The techniques are derived from *automata theory*—one of the key theoretical backgrounds of formal methods—and the algorithms discussed here exploit latest theoretical developments for the purpose of (1) accommodating more expressive specification formalisms and (2) making the output more informative (e.g., from Boolean verdicts to quantitative ones).

Chapter 3: Formal Specification of Temporal Properties

In Chapter 3, we take a step back, and give a comprehensive introduction of a foundational element that underlies both of Chapters 1 and 2, namely a formalism for expressing various specifications. We are particularity interested in *temporal logics*. Temporal logics extend usual propositional logics (with operators such as ‘and’, ‘or’, ‘not’, etc.) with so-called *temporal operators* (‘some time within 3 seconds’, ‘always henceforth’, etc.); they are important since many modern software systems—especially cyber-physical systems—operate on time-series data.

The exposition here is pedagogical, starting with basic logics and gradually moving on to the logic used in Chapter 1 called *signal temporal logic* (*STL*). In its way, we also emphasise some basic principles of formal logic, such as separation of syntax and semantics, structural induction, etc.

Chapter 4: Testing for Machine Learning-Based Systems

Testing is a principal safety assurance method, for both conventional and modern software systems. Its major advantage is that testing can be conducted if an executable model or code is available: one can press a button and then let the target system run. This is in stark contrast with some formal verification techniques that would require experts spend months or even years for manual efforts on rigorous white-box modeling.

Machine learning-based systems, as one archetype of modern software systems, pose special challenges to the methodology of testing. One example is the absence of *test oracles* that tell whether a test result is pass or fail, for example, we may not know the ''right answer'' for recommendation function. The chapter starts with an overview of (conventional) testing, exhibits some major challenges that they encounter in machine learning-based systems, and then discusses several techniques.

Chapter 5: Safety Standards and Safety Assurance Framework for ADS

This chapter discusses an approach to system safety from a very different direction, namely that of *industry standards* and *social norms*. It in particular focuses on international safety standards and safety assurance frameworks for *automated driving vehicles*.

For one, safety standards in the automotive domain are a well-established field, where a massive amount of technical and organisational effort has resulted in a variety of international standards that are used by a massive industry as guidances. For another, the whole field of automotive safety standards is currently trying to find its new shape, in the rise of automated driving. Therefore, we believe that the trends in the field are instructive when we think more generally about how new kinds of software systems—i.e., new information technologies—should gain social acceptance.

Chapter 6: Uncertainty-wise Testing

This chapter is another testing chapter—after Chapter 4—and its focus is how to deal with *uncertainties*. Uncertainties are a characteristic of modern software systems, especially cyber-physical systems where uncertainties can arise from imperfect sensing, an unexpected variety of physical environments, human interaction, etc.

Such uncertainties pose unique challenges to safety assurance methods. In this chapter, focusing on model-based testing, a framework is introduced in which various uncertainties are explicitly modeled and then tested and monitored so that they are resolved in field tests and operation.

Chapter 7: Decision Making for Automated Driving

In this chapter, we focus on a specific yet representative application domain, namely *automated driving*. Automated driving is an emerging technology with big social and economic impacts, and thus its safety is of wide public interest. Given that automated driving cars operate on public roads where there are a lot of uncertainties, automated driving is a prototypical application domain of the techniques discussed in this book.

The chapter introduces basics on decision making for automated driving. Starting with the usual hierarchy of decision making, it discusses some common techniques—most of them coming from control theory—on motion planning and its safety verification. The chapter ends with an extensive section on the technique called *responsibility-sensitive safety (RSS)*, which is a rigorous framework for safety verification that (1) is based on mathematical proofs and (2) successfully embraces uncertainties (which are often a critical blocker of proof-based approaches).

Chapter 8: Formal Modelling

The chapter discusses *formal modelling*—a crucial "preparation step" for many techniques discussed in the book—where we build a mathematical (*formal*) model of a target system. Such models are typically implemented in a software tool and thus is amenable to automated analysis, such as testing and formal verification.

In the automotive and cyber-physical domains, modelling tools such as Simulink and Modelica are widely used. The "models" built in these tools—together with most *digital twins*—are principally for the purpose of numeric simulation. In contrast, in formal modelling, one seeks *logical* models which expose structural information.

Formal modelling is a costly and laborious task in general; thus a disciplined methodology is needed. The chapter discusses two major approaches. One is *domain-specific modelling*, and the chapter discusses a modelling language called CREST that is tailored for small cyber-physical applications. The other is with *stepwise refinement*, where one starts with a coarse-grained high-level model and gradually adds details that are needed. The chapter uses a specific modelling language called Event-B for illustration.

Chapter 9: Theorem Proving at Work

The chapter discusses another fundamental methodology, namely *theorem proving*. It refers to a body of machine-assisted techniques to write mathematical proofs; in the context of the current book, these proofs are for theorems that state safety of target systems.

The chapter first makes a general introduction to theorem proving. After reviewing the basics of formal logic (i.e., formal languages in which mechanised proofs are written), two major approaches are discussed: *interactive theorem proving* and *automated theorem proving*. The introduction given here is general and goes beyond the scope of this book (namely Safety Assurance under Uncertainties); so it can be read by wider audience who are interested in theorem proving in general.

The chapter proceeds to introduce two specific approaches that are tailored to the book's scope. One is theorem proving for *continuous* and *hybrid systems*, where conventional discrete techniques get combined with ideas from control theory to accommodate continuous dynamics. The other is theorem proving with *model refinement*: continuing the introduction of Event-B in Chapter 8, it is discussed how safety proofs can be established iteratively along gradually refined system models.

Chapter 10: Search-Based Analysis and Engineering

The book takes a turn to the software engineering direction where safety assurance is often conducted by testing (although the use of theorem proving is definitely present there). The chapter in particular focuses on *search-based* techniques, where (1) problems in software engineering and safety assurance get translated to optimisation problems and (2) the latter are solved by advanced optimisation solvers. Objectives in such a scenario include discovering a rare fault, test coverage, etc.

The discussion of the chapter is pinned down by the specific application domain, namely automated driving systems, since it exposes many real-world needs and challenges for search-based testing techniques. The chapter introduces a number of concrete techniques that have been applied to the automated driving domain. It must be clear, however, that those techniques have wider applicability beyond this specific domain.

Chapter 11: Fault Localisation and Understanding

The topic of the final chapter is *fault localisation.* After a problem is found in a target system using the techniques discussed in other chapters (testing, monitoring, formal verification etc.), one needs to *localize* the cause of the problem so that it can be properly addressed. In particular, in complex systems in modern days that consist of numerous components, it is a big challenge to identify which component (line of code, class, etc.) is the source of a problem. Fault localisation is a body of techniques for this purpose, developed in the field of software engineering.

The chapter first introduces the basic ideas of *spectrum-based fault localisation* (*SBFL*), an empirical method to suggest which components are likely to be faulty. SBFL is a well-established technique for software systems expressed as *programs*; however, in modern systems—such as automated driving and machine learning-based systems, i.e., systems of the book's focus—the methodology encounters several issues. The chapter discusses those issues, and exhibits two examples of adaptation of SBFL to those particular domains.

Contributors

Shaukat Ali
Simula Research Laboratory
Oslo, Norway

Paolo Arcaini
National Institute of Informatics
Tokyo, Japan

Rose Bohrer
Worcester Polytechnic Institute
Worcester, USA

Ahmet Cetinkaya
Shibaura Institute of Technology
Tokyo, Japan

Jérémy Dubut
National Institute of Advanced Industrial Science and Technology (AIST)
Tokyo, Japan

Clovis Eberhart
National Institute of Informatics and JFLI
Tokyo, Japan

Ichiro Hasuo
National Institute of Informatics and SOKENDAI
Tokyo, Japan

Fuyuki Ishikawa
National Institute of Informatics, SOKENDAI and The University of Electro-Communications
Tokyo, Japan

Stefan Klikovits
Johannes Kepler University
Linz, Austria

Tsutomu Kobayashi
Japan Aerospace Exploration Agency (JAXA)
Tsukuba, Japan

Sasinee Pruekprasert
National Institute of Advanced Industrial Science and Technology (AIST)
Tokyo, Japan

Kenji Taguchi
Underwriters Laboratories
Tokyo, Japan

Toru Takisaka
University of Electronic Science and Technology of China
Chengdu, China

Masaki Waga
Kyoto University and National Institute of Informatics
Kyoto, Japan

Akihisa Yamada
National Institute of Advanced Industrial Science (AIST)
Tokyo, Japan

Tao Yue
Nanjing University of Aeronautics and Astronautics and Simula Research Laboratory
Nanjing, China

Man Zhang
Kristiania University College
Oslo, Norway

Xiao-Yi Zhang
University of Science and Technology Beijing
Beijing, China

Zhenya Zhang
Kyushu University and National Institute of Informatics
Fukuoka, Japan

Chapter 1

Optimisation-Based Falsification—Showcasing Black-Box Techniques for System Safety

Zhenya Zhang

Kyushu University, Fukuoka, Japan and National Institute of Informatics, Tokyo, Japan

Paolo Arcaini

National Institute of Informatics, Tokyo, Japan

Ichiro Hasuo

National Institute of Informatics and SOKENDAI, Tokyo, Japan

1.1 Introduction

1.1.1 Cyber-Physical Systems

Cyber-Physical Systems (CPS) refer to physical systems with digital control of computers. With the rapid development of computer technologies, especially artificial intelligence in the last decade, the collaboration between physical and digital components of systems is getting much closer than ever, giving birth to many applications of CPS. These applications are facilitating people's daily lives: for example, domestic robots save people's time from household duties; automated driving cars give people easier and safer driving experiences; unmanned aerial vehicles (UAVs) are used for photography, package delivery, or even fire fighting; some advanced medical devices, such as *artificial pancreas*, assist people to fight with diseases. Indeed, it has already been such an era: machines are equipped with "brains", and thus they behave not only mechanically, but also intelligently.

From a mathematical perspective, CPS are also called *hybrid systems*: they combine discrete and continuous dynamics together in a single system. Discrete dynamics exhibits stepwise behaviours, and it is empowered by computer software. Continuous dynamics exhibits continuous behaviours, and it is given by traditional physical systems. In general, computer software is used to monitor and control the behaviours of the physical systems, based on a set of logical rules. The physical systems execute the commands, and then provide feedback information back to the computer system. Therefore, they do not work independently, but closely interact with each other.

1.1.2 Quality Assurance of CPS

Quality assurance of CPS products is a topic of great significance. Firstly, many CPS applications are safety-critical, and their failures can cause unacceptable consequences. For example, many CPS serve as transportation facilities, such as airplanes, trains, cars, etc. System faults of these applications can cause huge financial damages or even loss of human lives. Moreover, software systems themselves are prone to errors; so far, it is still a challenging task to guarantee the quality of software products. In history, there have been many tragedies due to the system failures caused by software errors, such as the well-known Apollo 13 Accident.[1]

Verifying software systems is hard, and it is still one of the major research topics in computer science. CPS, as a family of complicated systems, pose many new challenges in their quality assurance. We list some of the major challenges below:

- **Infinite state space.** Due to the existence of physical components, the system state space of CPS is usually infinite. This feature poses challenges in modelling and reasoning about the system. New formalisms have been invented for modelling hybrid systems, such as hybrid automata [35], Simulink,[2] etc. However, verification of such models is usually undecidable.

[1] https://nssdc.gsfc.nasa.gov/planetary/lunar/ap13acc.html

[2] https://www.mathworks.com/products/simulink.html

- **Complex dynamics.** White-box methods based on the knowledge of system dynamics and rigorous mathematical proofs are considered as a highly reliable approach to system quality assurance. However, the dynamics equations of CPS are typically so complicated that even if they are known, there is still a huge barrier to obtain a tractable method for their verification.
- **Scalability.** Scalability refers to the ability to extend reasoning techniques from small models to systems of much larger magnitude. As real-world CPS products are usually large and complex, how to develop scalable techniques remains a big challenge.
- **Black-box components.** Nowadays, many CPS products are produced by commercialised collaborations, so they contain black-box components. A black-box component indicates that the internal dynamics of the component are not accessible. One reason is that those components include confidential techniques; also, modern CPS are produced by interdisciplinary collaborations, where one individual is not supposed to hold all the domain expertise. As a consequence, the black-box nature gives rise to more uncertainties and makes the system less understandable compared to the white-box models.
- **Environmental uncertainties.** In practice, environmental noises can be introduced into the system execution in many ways, and the uncertainties brought by them are an important issue that needs to be considered. Even worse, they are typically difficult to model or handle. First, few formal models can precisely describe them; second, considering during reasoning is also complicated.

1.1.3 Verification vs. Testing

Classic approaches to quality assurance of software systems include *verification* and *testing*, which exemplify different philosophies. In this section, we introduce these two techniques.

1.1.3.1 Verification

Formal verification is a quality assurance approach that has been widely used in applications such as *cryptographic protocols*, *digital circuits*, etc. It applies mathematical tools to derive rigorous proofs for the properties of interest, and by that, it provides reliable conclusions.

For instance, *model checking* [15] is a basic verification approach to quality assurance of traditional software systems. In general, it models the system under consideration with some formalism (e.g., finite state automata), and transforms the verification problem into a reachability analysis problem (that can be solved by graph search).

As CPS usually have infinite system states, abstraction is inevitable in their modelling. Such abstraction retains the major features of the systems, but ignores some concrete events in the system executions. For instance, *hybrid automata* are a formalism for modelling hybrid systems. See Fig. 1.1 for an example of using a hybrid automaton to describe the dynamics of a bouncing ball [40]. The continuous state transformations are encoded within the node, while the discrete state transformations are represented by the edges.

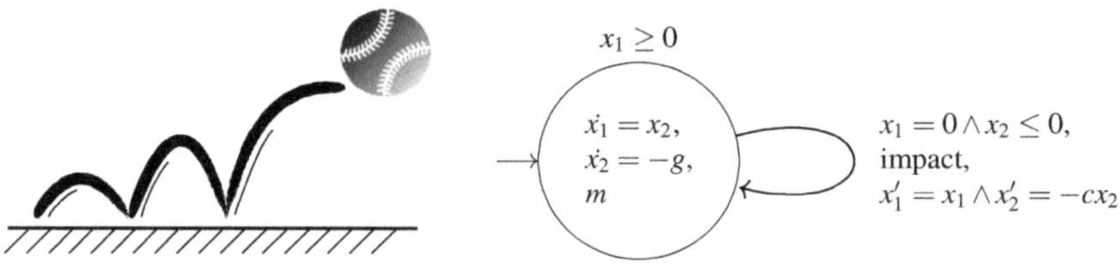

Figure 1.1: An example of hybrid automaton—a bouncing ball (adapted from Example 6 in [40]).

Although verification gives elegant formalization and rigorous guarantees, it also has some fatal weaknesses: first, scalability is a major problem, and due to that, the class of models and specifications that can be handled is limited; second, verification usually relies on the knowledge of system dynamics, and so it is not suited for modern CPS that contain many black-box components. Therefore, the use of formal verification to quality assurance of CPS is highly constrained; more practical methodologies are necessarily needed.

1.1.3.2 Testing

Testing is the dual process of verification. Instead of verifying that a system satisfies a requirement, testing aims to find a counterexample showing that the system violates the requirement. In that way, testing exposes the fact that the system has the potential to behave unexpectedly, and so it requires repairing or redesigning. In some cases when verification is hard to apply testing is potentially a viable solution to understanding the behaviours of systems; indeed, searching for one specific case is usually easier than exploring the whole system thoroughly.

Testing has also been applied to quality assurance of traditional software systems for a long time, and there have been many techniques that proved to be effective, such as search-based testing [34, 44], model-based testing [59], etc. Notably, testing includes methodologies suited for not only the white-box but also the black-box systems. In this regard, testing is a feasible candidate approach to solving the quality assurance problem of CPS.

1.1.4 *Optimisation-Based Falsification as a Practical Quality Assurance Method*

Falsification was born in such a context: when computer scientists found it infeasible to verify CPS via system state exploration, they started to pursue a countermeasure that directly searches for the unexpected system state. Breach [22], a representative falsification tool, was named as the abbreviation of "barbaric reachability" [17]. Indeed, this approach is straightforward, but effective.

A typical falsification tool, e.g., Breach, takes as input a model, a specification, a time budget, and some other configurations, such as the shape and range of input signals. To get the tool run, the user only needs to push a button. After execution, the tool will output a counterexample input signal if it manages to find one that violates the specification, or it reports that no counterexample is found within the time budget. A usage scenario of falsification is shown in Section 1.2.

Falsification has been widely adopted in industry and actively studied in academia. Its application domain covers automotive systems, avionics, etc. In the future, due to its low constraint to the system (black-box suffices), it has the potential to be more widely applied.

1.2 A Usage Scenario: An Automotive System

In this section, we present a usage scenario which engineers may encounter in practice. Suppose an engineer is designing a car in a Simulink development environment (an introduction to Simulink is given in Section 1.3). As Fig. 1.2 shows, the car model[3] is composed of many blocks; it accepts *throttle* and *brake* as inputs, and as outputs *rounds per minute (RPM)*, *speed*, *gear*, etc. Note that this is an automatic transmission system, so *gear* is an output here.

Now a task is given to the engineer. They would like this car to satisfy the following safety property: whenever the *gear* is 4 (a relatively high *gear* value), the *speed* should never be too low, say lower than 35 km/h. This is a reasonable requirement for the car, because, once the car runs at a high *gear* with a low *speed*, it will harm some components, and even threaten the safety of the entire system.

This is not a trivial problem for the engineer, as the car consists of so many blocks that operate according to complicated dynamics. Tentatively, the engineer comes up with the following alternative strategies:

- Figure out the reachable set via reasoning about the dynamics of the system components, to determine whether there exist some states that violate the property;

[3]This Simulink model is from Mathworks (https://www.mathworks.com/help/simulink/slref/modeling-an-automatic-transmission-controller.html).

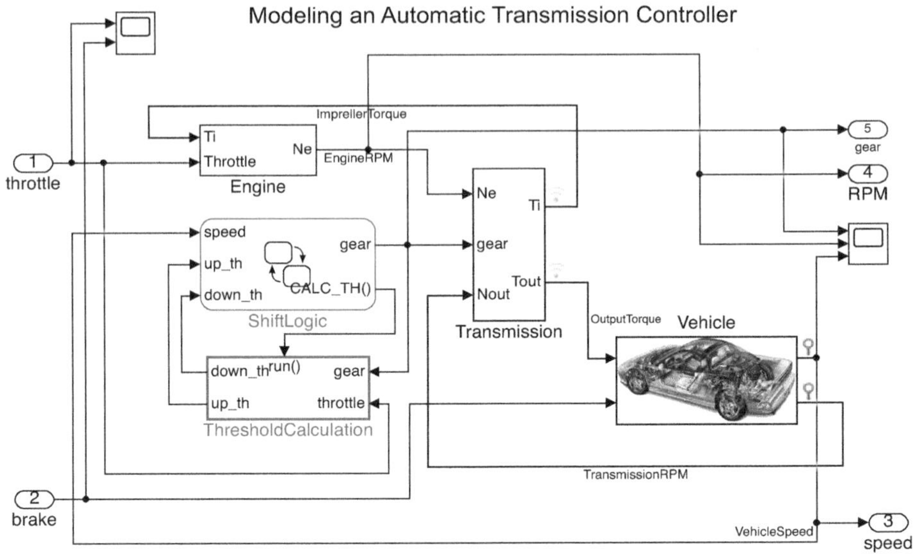

Figure 1.2: Simulink model: automatic transmission.

- Feed the system with some example traces of *throttle* and *brake*, and check whether the output violates the property.

The former refers to verification, which is infeasible to apply to such a complex system; the latter is a more feasible testing approach, however, performing it manually is expensive, and the correctness of those examples is not sufficient. The engineer needs an automatic testing technique, that either says that the system is safe providing different tests as a support of the claim, or reports an unexpected behaviour to show that the system is unsafe.

1.2.1 Counterexamples via Optimisation-Based Falsification

In this section, we show how falsification can help the engineer with that problem.

Let us say the engineer decides to use the falsification tool Breach [22], a representative falsification tool. As shown in Fig. 1.3, Breach takes a model (usually expressed in Simulink), a specification (in *Signal Temporal Logic (STL)*, which we will illustrate later in Section 1.3.1.3), and some other parameters, such as the time budget, the shape of the input signals, the choice of the optimisation algorithm, etc., as inputs; it then performs the falsification workflow shown in Fig. 1.8, and outputs a result—either a success in falsifying the specification with an input signal **u** or a failure to do so.

In an experiment, the engineer gives the automatic transmission model in Fig. 1.2 to Breach as the system model under test, and then expresses the desired property in STL, namely $\Box_{[0,30]}(gear = 4 \rightarrow speed > 35)$; the engineer sets the

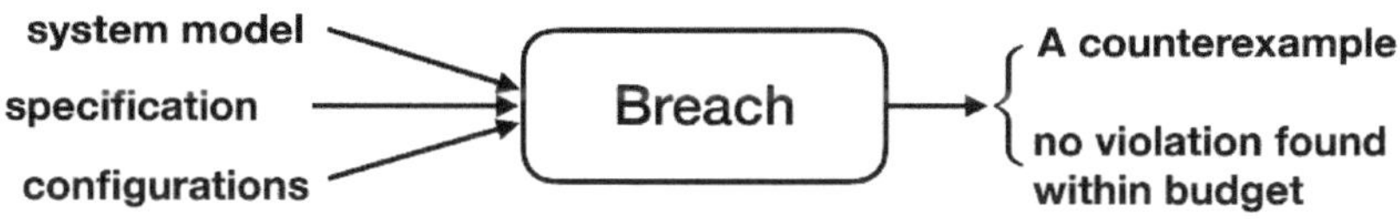

Figure 1.3: Illustration of the tool Breach.

time budget as 100 seconds, and uses *CMA-ES* as the optimiser. The engineer can then start the falsification process, and wait for a while to obtain the results.

```
      ⋮                    ⋮                  ⋮                  ⋮
     198                 51.6        [+1.00000e+00]     (+8.24363e-01)
     199                 51.9        [+1.00000e+00]     (+8.24363e-01)
     200                 52.1        [+1.00000e+00]     (+8.24363e-01)
     201                 52.4        [+6.82844e-01]     (+6.82844e-01)
     202                 52.7        [+1.00000e+00]     (+6.82844e-01)
     203                 52.9        [+2.00000e+00]     (+6.82844e-01)
     204                 53.2        [-1.98605e-01]     (-1.98605e-01)
Falsified with obj = -0.198605

 ---- Best value -0.198605 found with
        brake_u0 = 99.0883
        brake_u1 = 325
        brake_u2 = 86.8103
        brake_u3 = 292.393
        brake_u4 = 316.375
        throttle_u0 = 0
        throttle_u1 = 62.8815
        throttle_u2 = 6.79548
        throttle_u3 = 13.0012
        throttle_u4 = 0
```

Figure 1.4: Screenshot of a trial by Breach.

Breach displays a running instance as reported in Fig. 1.4. On the top of the figure, there are four columns, respectively recording the number of simulations, the total time consumed so far, the robustness value resulting from the current input signal, and the best (i.e., smallest) robustness over the input signal history. It is clear that the best robustness is decreasing, which means that Breach is gradually approaching a falsifying input. After 204 simulations (53.2 seconds), Breach succeeded in finding a falsifying input, shown on the bottom of the figure, with the robustness –0.198605. This result demonstrates that the car designed in Fig. 1.2 is not safe as expected, and the engineer needs to redesign related components to fix some existing problems.

1.3 Optimisation-Based Falsification—An Overview

As we mentioned earlier, falsification is a testing-based approach for quality assurance of CPS. Compared to verification, it does not need the exploration of the whole state space; instead, it pursues one counterexample that refutes the system specification. In this section, we overview the classical workflow of falsification.

1.3.1 Problem Setting

We firstly introduce the basic problem setting of falsification that has been considered in the literature [1, 7, 9, 19, 22, 24, 30, 41, 45, 54, 65, 66, 70, 71]. The inputs to the problem consist of a system model $\mathcal{M}$ and a specification φ. The problem consists in answering the question whether there exists an input signal $\mathbf{u}$ such that the corresponding output signal $\mathcal{M}(\mathbf{u})$ violates the specification φ.

1.3.1.1 System Models

The falsification problem concerns a model, i.e., the system under test. We usually treat the model as a black box, which means that we do not access its internal dynamics, but the only way we evaluate it is by providing input signals and observing the corresponding output signals. This setting makes falsification very practical, in that its usage is not limited to any specific type of model—knowing the inputs and outputs of a model suffices.

In industry, Simulink, a toolbox by MathWorks, is a commonly used development environment for modelling CPS. It provides a visualised interface for fast prototyping. An example of Simulink model is shown in Fig. 1.2. The ports 1 and 2 are input signals for the system, namely *throttle* and *brake*; the ports 3, 4 and 5 are output signals, namely *speed, RPM (rounds per minute)*, and *gear*. It is composed of several blocks, and each of them can have several sub-blocks embedded. When an input signal is provided, the model can be executed and then output signals will be generated.

Since a black-box model suffices, system models for falsification are not limited to Simulink. Recently, have been many studies on applying falsification in the domain of testing of automated driving system [25, 33, 57, 58]. In these works, system models are based on more complicated types of simulators. PSY-TaLiRo [56] is also developed as a falsification tool in *Python* to accommodate the use of falsification in those emerging types of simulators.

1.3.1.2 Input Signals

With the black-box setting, falsification boils down to searching for a specific input signal such that the corresponding output signal violates the system specification. Here, a signal is a time-variant function. As the time domain is continuous, it is impossible to synthesise an input signal by providing its value at each moment. Therefore, we adopt a parametrised representation of input signals as an approximation of them. Commonly-used representations include piecewise constant (see Fig. 1.5a), piecewise linear (see Fig. 1.5b), pulse signals (Fig. 1.5c), etc.

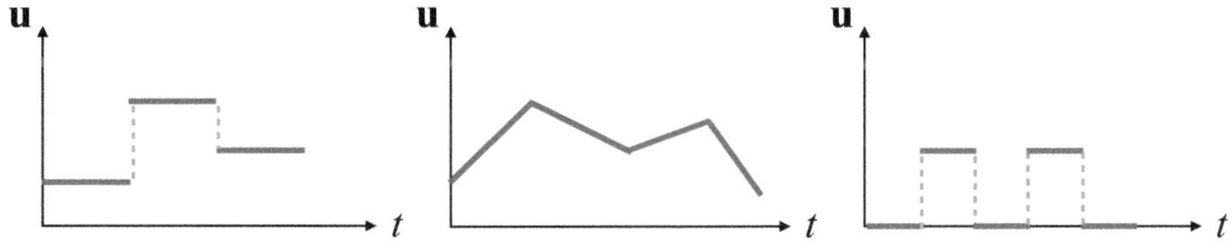

(a) Piecewise constant signal (b) Piecewise linear signal (c) Pulse signal

Figure 1.5: Parametrised representations of input signals.

We elaborate on piecewise constant signals as an example. In Fig. 1.5a, the signal is composed of three constants. In practice, piecewise constant signals can be obtained by sampling the real signals at intervals; as a result, the number of samples, usually called *control points*, decides how closely the piecewise constant signal approximates the real one.

In falsification, after fixing the type of representation and hyperparameters such as the number of control points, the task is reduced to searching for a finite set of parameter values that identify an input signal.

1.3.1.3 Specifications

Specifications are used to describe the requirements to a system, regarding safety, comfort, performance, etc. In some cases, they can be described in a natural language, e.g., when discussing requirements with non-expert stakeholders. However, natural languages usually suffer from the ambiguity issue: the same sentence can be interpreted in a completely different way by different people. Moreover, natural language processing is a non-trivial burden for computer systems, and it is unreasonable to involve this process into the loop of quality assurance. Therefore, formal languages that are easily understandable by computer systems are necessary.

Since the behaviours of CPS are usually constrained by temporal requirements, *temporal logics* are the best choice of formal language to express those requirements. *Linear Temporal Logic (LTL)* [47] is a well-known formalism for reasoning about systems with discrete state transitions. Signal Temporal Logic (STL) [23], as an extension of LTL, is a widely-adopted specification language in most falsification literature. Compared to LTL, STL uses an equality/inequality expression to replace the atomic proposition, and extends the discrete time step to a continuous time domain. The bridges between LTL and STL include *metric temporal logics (MTL)* [39], *metric interval temporal logics (MITL)* [8]. Other formalisms are mostly developed on top of these languages. For instance, AvSTL [5], IA-STL [32], etc.

The feature that enables temporal logics to express temporal properties is the use of temporal operators, namely, *always* (or $\Box$), *eventually* (or $\Diamond$) and *until* (or $\mathcal{U}$). For example, the property in Section 1.2 is expressed in STL as follows:

$$\Box(gear = 4 \rightarrow speed > 35)$$

Here, $\Box$ is the *always* temporal operator, requiring the proposition $gear = 4 \rightarrow speed > 35$ to be true at every moment. Moreover, STL introduces quantitative semantics, which plays an important role in the optimisation-based falsification. In the next section, we will elaborate on the quantitative STL robust semantics.

1.3.2 Quantitative STL Robustness

The Boolean satisfaction of temporal logics is easy to understand. In the context of hybrid systems, it is a relation between a signal $\mathbf{w}$ and an STL formula φ—we write $\mathbf{w} \models \varphi$ if $\mathbf{w}$ satisfies φ, and $\mathbf{w} \not\models \varphi$ otherwise. The quantitative semantics of STL introduces quantities into the relation, which is called *robustness*. We write $[\![\mathbf{w}, \varphi]\!]$ as the robustness of a signal $\mathbf{w}$ to φ; its formal definition is given in Section 3.4.4.

Compared to the Boolean satisfaction, the quantitative semantics does not only say whether $\mathbf{u}$ satisfies φ or not, but also indicates *how robustly* $\mathbf{u}$ satisfies or violates φ. See the examples in Table 1.1.

Table 1.1: Boolean satisfaction $\mathbf{w} \models \varphi$ and quantitative robustness values of three signals of *speed* for the STL formula $\varphi \equiv \Box_{[0,30]}(speed < 120)$

signal **w**	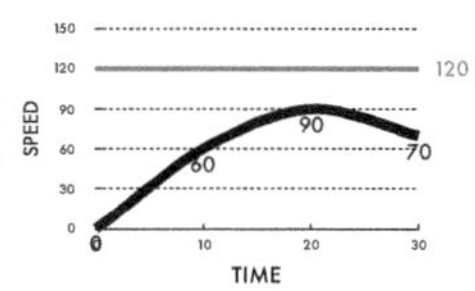	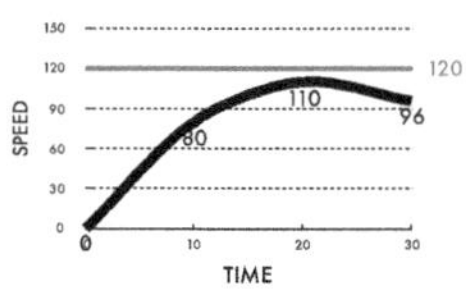	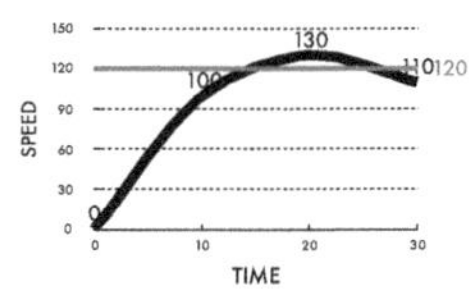
$\mathbf{w} \models \varphi$	True	True	False
$[\![\mathbf{w}, \varphi]\!]$	30	10	-10

There are three signals of *speed*, and we reason about their satisfaction to the temporal property

$$\Box_{[0,30]}(speed < 120)$$

that is, "during the time bound [0, 30], the *speed* is always below 120". It is clear that the first two signals satisfy the property, while the third one violates it, which corresponds to the Boolean satisfaction shown in the second row. Furthermore, although both the first two signals satisfy the property, intuitively the second one is more vulnerable to perturbations than the first one, in the sense that once a small change happens to the signal, it may violate the property. This intuition is embodied by the robustness in the third row. These values are computed from the margin between the peak point of each signal and the threshold 120. The robustness 30 of the first signal shows that this signal is more robust than the second one whose robustness is 10. Moreover, we can see that when the robustness is negative (as the case of the third signal), the property is violated.

This is a simple case where the sub-formula ($speed < 120$) is an atomic proposition. However, in practice, we need to handle specifications that are much more complicated, such as the one in Section 1.3.1.3:

$$\square(gear = 4 \rightarrow speed > 35)$$

where the sub-formula is an implication. Whatever specification it is, STL the robust semantics, given a signal, always returns a real number indicating the quantitative satisfaction or violation of the signal to the specification. This quantity will be utilised as a guidance in the optimisation-based falsification, which will be introduced in the next section.

1.3.3 Stochastic Optimisation-Based Falsification

As we introduced before, the quantitative STL robust semantics returns a quantity that guides the optimisation-based falsification. In this section, we shed a light on this procedure.

Optimisation, in general, is a technique that is used to search for the minimum (or maximum) value of an objective function in an input space. Falsification is formalised as an optimisation problem as follows:

$$\begin{aligned} &\underset{\mathbf{u}}{\textit{minimise}} && [\![\mathcal{M}(\mathbf{u}), \varphi]\!] \\ &\textit{subject to} && \mathbf{u} \in \Omega \end{aligned}$$

where $\mathbf{u}$ stands for an input signal, $\mathcal{M}(\mathbf{u})$ stands for the corresponding output signal, φ is the specification, and Ω stands for a valid range of input signal $\mathbf{u}$. We can see that here the robustness of the output signal $\mathcal{M}(\mathbf{u})$ towards the specification φ is treated as the objective function, and optimisation aims to minimise the robustness as much as possible. Once a negative robustness value is found, which indicates that the system specification is violated, then we are done.

A large number of optimisation algorithms have been proposed in the literature. Many of these algorithms are based on the knowledge of the objective function. For instance, *gradient descent* exploits the gradient information of the objective function to find a direction by which the objective function value can quickly descend. However, in our case, since the system model is black-box, no internal information is available. Therefore, we need to consider another family of optimisation algorithms, which is called *stochastic optimisation*.

1.3.3.1 Stochastic Optimisation

Stochastic optimisation is a family of optimisation techniques that implement diverse metaheuristic strategies. These algorithms introduce randomness into the search process, so that the optimisation is stochastic—different executions of the same algorithm with different random seeds lead to different outcomes. This feature enhances an algorithm's ability of reaching the optimisation goal—even if it fails to reach the goal, one can start a new trial and hope that the different execution can bring a successful result.

Metaheuristics refers to a family of heuristic strategies that can be applied to solve a class of problems, and in our context, these strategies are used to achieve the optimisation goals effectively and efficiently. Compared to random sampling, these algorithms are goal-driven and thus more efficient, i.e., they usually achieve the optimisation goal with less number of samples than random sampling.

Various metaheuristic strategies can be adopted in stochastic optimisation. Specifically, these strategies are usually inspired by biological or natural processes, e.g., *genetic algorithms*, *simulated annealing* [46], etc. Although metaheuristic strategies are often problem-independently applicable, the performances of these strategies can be problem-dependent and, moreover, there is no global best performer. This has been stated by the famous "no free lunch" [60] theorem for optimisation, which indicates that performance gains usually come with costs.

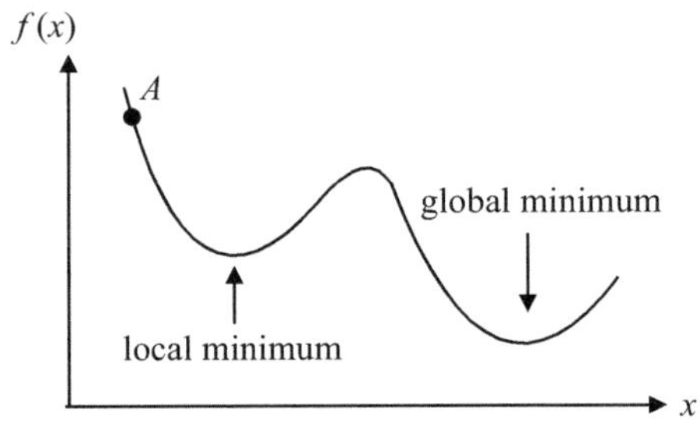

Figure 1.6: "Local optimum".

Let us firstly take a look at how hill climbing works, as depicted in Fig. 1.7. It starts with initial some random samplings and obtains their objective function values; then, it analyses these values and conjectures the direction where the objective function potentially descends; based on that, it comes up with the next sample following that direction. This loop is repeated until a time budget is exhausted; hopefully, by that time, the optimisation goal (i.e., a negative robustness value in our case) can be achieved. Just as the name indicates, this process is like a hill climbing (upwards or downwards).

There are several reasons why these stochastic optimisation algorithms are selected for falsification. Firstly, as we mentioned before, they are applicable to black-box models. Besides that, efficiency and scalability issues are also

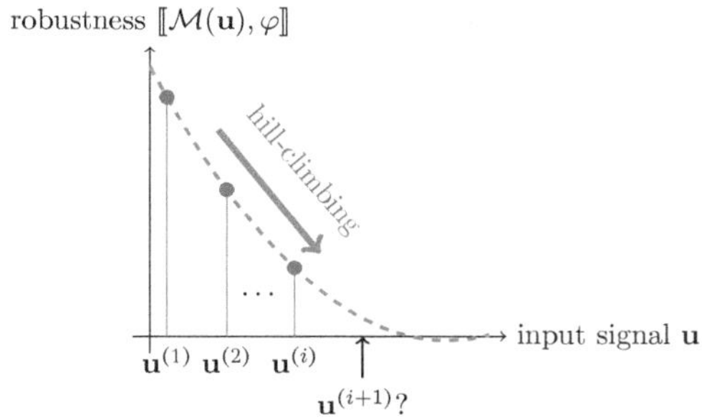

Figure 1.7: Hill-climbing optimisation.

important. Usually the system simulation that, given an input signal, generates the corresponding output signal is computationally expensive, so in practice it is preferred if an algorithm works out efficiently with a small number of simulations. Also, we expect the algorithms to be scalable enough, so that the same workflow can be smoothly applied to systems of larger scale (e.g., with more input signals, system components, etc.). Although stochastic optimisation algorithms also suffer from this scalability issue, the effect is much less severe compared to state exploration-based techniques such as verification.

The details of some commonly used optimisation algorithms are listed below:

- **Global Nelder-Mead** [42] The Nelder-Mead method makes use of the simplex algorithm to perform a local search. It is specialised in searching for a local optimum, but it is unable to jump out to search for the global optimum (see Fig 1.6). Global Nelder-Mead introduces a probabilistic restart mechanism, so that it is possible for the search to be reinitialised elsewhere and thus not trapped in local optima.
- **Simulated Annealing** [46] Simulated annealing is inspired by the physical process of heating a material and then cooling it down to bring it to a stable status. The algorithm starts with random sampling, and accepts all new samples that have lower robustness; meanwhile, it also accepts the samples that have higher robustness, according to a specific mechanism called *temperature*. The temperature is initially high, and decreases as the optimisation goes on. By the temperature mechanism, the algorithm is able to jump out of the local optimum.
- **CMA-ES [11]** CMA-ES is the abbreviation for *Covariance Matrix Adaption Evolutionary Strategy*, belonging to the family of evolutionary strategies that are inspired by *Darwin's theory of evolution*; these techniques evolve a set of possible solutions by mutating and recombining those that have higher fitness values. CMA-ES employs a *multivariate normal distribution* to approximate the objective function, and updates its covariance matrix according to the information learned by performing mutations and crossovers.

1.3.3.2 Evaluation Metrics for Stochastic Optimisation-Based Falsification Algorithms

In the previous section, we have introduced a number of optimisation algorithms. In general, according to the "no free lunch" [60] theorem of optimisation, none of the optimisation algorithms can outperform the others in all cases.

As these optimisation algorithms are stochastic, different falsification trials with different random seeds can produce different results. Taking this feature into account, in order to evaluate an algorithm, we usually run it multiple times and analyse the results statistically.

We consider two metrics, namely *effectiveness* (if an algorithm can find an answer) and *efficiency* (how fast an algorithm manages to find an answer). In general, we prioritise the effectiveness rather than the efficiency of an algorithm, because the ultimate goal of falsification is to find falsifying inputs, as long as the time budget is not run out. These two aspects are embodied by the following two measurements.

- **Falsification success rate:** It refers to the percentage of the trials that succeed in finding falsifying inputs within the budget, over the total number of trials. A high success rate certificates the effectiveness of an algorithm, in the sense that it manages to find falsifying inputs with a high probability. Instead, a low success rate indicates that the algorithm is not effective, or vulnerable to random seeds, thus not stable.
- **Time consumption:** For successful trials, the time consumption is computed as the time when the algorithm returns a falsifying input. Unsuccessful trials usually terminate because the time budget expires; however, there are also cases in which the optimisation algorithm meets other internal termination criteria, e.g., local optimum, and thus quits before the timeout.

Therefore, there are two ways of computing time consumption: one is by computing the average time consumption of only successful trials; the other one is by taking also unsuccessful trials into consideration, and thus computing the average time consumption of both successful and unsuccessful trials, in the latter case taking time budget as their values.

The advantage of the first method is that it reflects the efficiency of algorithms better, e.g., in the case that a "greedy" algorithm is fast but not effective; the second method integrates effectiveness into efficiency, exemplifying the principle that only if one is effective it can be efficient. Both methods are adopted in literature, especially the former one.

Note that these two metrics are not independent—as the success rate metric is subject to the parameter of budget, in many cases, high effectiveness (i.e., a high falsification success rate) can only be achieved when high efficiency is achieved.

1.3.4 Workflow Summary

So far, we have gone through the workflow of the optimisation-based falsification technique, as illustrated in Fig. 1.8. The model accepts an input signal given by the optimiser, and runs simulation to generate an output signal; the robustness that reflects the relation between the output signal and the system specification is computed according to the STL robust semantics; the optimiser then tries to minimise the robustness by proposing new input signals. This loop keeps on iterating until a falsifying input is found, or the pre-specified budget is run up.

1.4 Tools and Applications of Optimisation-Based Falsification

We first describe what tools are currently available, and what applications have been targeted. Scientific observations that underlie these tools, together with topics of future research, are in the next section.

1.4.1 Tools

Over the years, the methodology of falsification has proved to be effective in practice. There have been several tools developed following the workflow depicted in Fig. 1.8, such as Breach [22] and S-TaLiRo [9]. In this section, we give an introduction to these two representative tools.

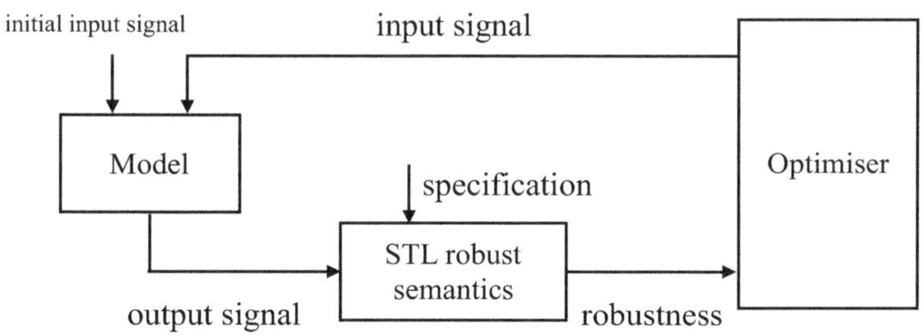

Figure 1.8: Workflow of stochastic optimisation-based falsification.

We present a comparison between Breach and S-TaLiRo in Table 1.2. Since both tools are mainly developed in Matlab, they provide Matlab API and take Simulink models as the basic supported systems under test. However, the type of models they support is not limited to Simulink; with proper interfacing efforts, they theoretically support any type of black-box models. Both tools provide a nice documentation and some example scripts for facilitating the usage.

Table 1.2: A comparison between Breach and S-TaLiRo

	Breach	S-TaLiRo
API language	MATLAB	MATLAB
Supported models	Mainly Simulink, but not limited to that	Mainly Simulink, but not limited to that
Specification language	Signal Temporal Logic (STL)	Metric Interval Temporal Logic (MITL)
Documentation	Tutorial, tool paper, demo, etc.	Tutorial, tool paper, demo, etc.
Demo	Some example scripts are in the distribution	Many example scripts are in the distribution
Difficulty of usage	Clear STL syntax, thus easy to write specifications; easy to specify parameters and run	MITL syntax is based on a template, thus harder to write specifications; easy to specify parameters and run
Availability of optimisation algorithms	CMA-ES, Simulated annealing, global Nelder-Mead, etc.	Simulated annealing, rapidly random tree [24], grey-box methods for NN-based systems [61]

They differ in many implementation details. For example, Breach implements Signal Temporal Logic (STL) as the specification language, while S-TaLiRo adopts Metric Interval Temporal Logic (MITL) [31]. Breach allows users to write specification directly using the STL syntax, while S-TaLiRo provides a template for specifications and users need to transform their specifications in that format. In addition to the syntactic differences, the two tools also adopt different algorithms for (offline) monitoring of specifications, and this can lead to some performance differences in some cases. Moreover, these tools are used for research purposes—their authors include new research outcomes in them, so they also differ in falsification algorithms and they keep evolving over time.

There is an annual friendly competition, held with the workshop of Applied Verification for Continuous and Hybrid Systems (ARCH).[4] In this competition, all the latest falsification tools run experiments on some existing benchmarks, and their performances are compared. See [20, 26–29] for competition results of recent years.

1.4.2 Case Studies

In this section, we introduce some Simulink models commonly-used in the falsification community. Specifically, we select 3 Simulink models from the ARCH competition, and we show their complexity and typical safety requirements.

Automatic Transmission. This is the model that has been shown in Fig. 1.2. It is originally from MathWorks,[5] and firstly used as a falsification benchmark in [36]. It has two input signals, *throttle* and *brake*, and three output signals, *speed*, *gear*

[4]https://cps-vo.org/group/ARCH/FriendlyCompetition

[5]https://www.mathworks.com/help/simulink/slref/modeling-an-automatic-transmission-controller.html

and *RPM*. The number of blocks that constitute the system is listed in Table 1.3a. Research papers, such as [7,65–67], and ARCH competitions [20,26–29] usually reason about the safety requirements on the outputs of the system. For instance, the two examples

$$\square_{[0,30]}(speed < 120) \quad \text{and} \quad \square_{[0,30]}(gear = 4 \rightarrow speed > 35)$$

were mentioned in Section 1.3.2: the former says that the *speed* should always be below 120; the latter says that whenever the *gear* is 4, the *speed* should be higher than 35. In [12, 13, 63, 64], the authors consider a more practical falsification scenario, that is, falsification of the properties under input constraints, e.g., the two system inputs, *throttle* and *brake*, are required not to be positive simultaneously.

Abstract Fuel Control. This is an industry-level Simulink model provided by Toyota in [37]. It models a powertrain controller that aims to improve the system performance in terms of the ratio of *air* and *fuel*. In [37], the authors propose the falsification-based approach to assuring the safety requirements. It accepts two input signals, *engine speed* and *pedal angle*, and gives the ratio of *air* and *fuel* as well as the reference value for the ratio. The basic requirement requires that the ratio does not deviate from the reference value too much. The complexity of the model is shown in Table 1.3b.

Table 1.3: The number of blocks of Simulink models (by the command `sldiagnostics`)

(a) Automatic Transmission

Blocks	Num.
Fcn	2
Gain	6
Inport	16
Integrator	2
Lookup	3
Lookup2D	3
Outport	14
Product	6
Scope	2
Signum	1
Stateflow	1
SubSystem	6
Sum	3
TriggerPort	1
Torque converter	1
Vehicle model	1
total	64

(b) Abstract Fuel Control

Blocks	Num	Blocks	Num.
Clock	1	Mux	1
Constant	25	Outport	28
DataStoreMemory	7	Product	17
DataStoreRead	10	RelationalOperator	3
DataStoreWrite	10	Saturate	4
DataTypeConversion	8	Signum	1
DiscretePulseGenerator	1	Sqrt	1
EnablePort	1	Step	2
Fcn	3	SubSystem	20
From	10	Sum	20
Gain	18	Switch	6
Goto	8	TransferFcn	1
Inport	39	TriggerPort	3
Integrator	5	UnitDelay	6
Logic	6	VariableTransportDelay	1
Lookup_n-D	3		
Math	1		
MinMax	1		
total	271		

(c) Magnetic Levitation

Blocks	Num.
Constant	14
DiscreteStateSpace	1
Fcn	5
Gain	12
Inport	16
Integrator	2
Math	4
Mux	1
Outport	13
Product	5
Saturate	1
Signum	1
SubSystem	11
Sum	11
Switch	4
UnitDelay	1
ZeroOrderHold	2
NARMA-L2 Controller	1
total	105

Magnetic Levitation. This is a model originally from MathWorks[6] as a demo for designing a neural controller. In [26–28], it is used as a falsification benchmark. The model contains a plant and a neural network controller, where the controller levitates a magnet above an electromagnet at a reference position. The model accepts one input signal, that is, the reference value, and gives one output signal, that is, the actual position of the magnet. The specification of the system requires that whenever the reference is changed, the position of the magnet responds fast enough to stabilise around the value. The complexity of the model is reported in Table 1.3c.

1.4.3 Adoption in Industry

The falsification methodology has been adopted by many industrial manufacturers, such as *Toyota* [37], *Bosch* [33], *Airbus* [16], etc. Toyota was one of the earliest manufacturers that applied this methodology. In [37], they present a fuel control system, and list several requirements that the system is supposed to satisfy; they test those requirements with S-TaLiRo, and experiments show that in most cases S-TaLiRo does not find any violation, so they conclude that "the quality of the manual abstractions that we performed vis-à-vis high-level requirements is reasonable". Bosch has also been working on this for several years. They have been trying to apply the technique in testing automated driving systems. In [33], they present their practical experience, in particular they give several interesting tips, such as "simulation is costly", "many properties can be formalised by formulas of simple form", etc., which are helpful for practitioners. Airbus [16] recently collaborated with academia on a project for testing their products. In their work, they apply Monte Carlo Tree Search (MCTS), a technique introduced into falsification by us [65].

1.5 Scientific Challenges and Highlights in Optimisation-Based Falsification

Falsification has proved to be effective in revealing unexpected behaviours of CPS, and it has been adopted by manufacturers such as Toyota [37], Bosch [33], etc. However, the methodology is not perfect yet. In this section, we list and elaborate on the shortcomings, and we also review the existing literature to show the recent progress on those topics.

[6]https://www.mathworks.com/help/deeplearning/ug/design-narma-l2-neural-controller-in-simulink.html

1.5.1 Exploration and Exploitation

The trade-off between *exploration* and *exploitation* is the core of the search techniques. Exploration prefers to cover the search space as broadly as possible, rather than focusing on a specific local area. Exploitation, instead, is the opposite; it eagerly searches for local minima, without considering big jumps in the search space. Given a limited time budget, it is a challenge how to balance between exploration and exploitation.

As introduced in Section 1.3.3.1, falsification employs hill-climbing optimisation as the optimiser, which is known as an algorithm that favours exploitation. This bias affects the falsification performance severely, and so it attracts plenty of research attention. We classify these works according to the methodologies they use, and introduce them in the following sections.

1.5.1.1 Metaheuristics for Exploration

This line of works makes use of metaheuristics to avoid pure exploitation. Many metaheuristic-based approaches implement a mechanism that triggers a jump when the search falls into a local optimum. For example, the strategies in Section 1.3.3.1, namely global Nelder-Mead, Simulated Annealing [46], and CMA-ES, adopt different metaheuristics, respectively. In [10], the authors apply the famous *ant colony optimisation* to falsification, in which the search takes a small probability to start a new exploration. In [19], the authors propose to use *tabu search* for falsification. The algorithm maintains a *tabu list* during the search, which is a memory structure that records the sampling history to guarantee that the same local optimum is not visited twice. *Cross entropy* [49] is an *importance sampling* method that initially explores the search space and then performs biased sampling in the promising area. Its application to falsification is studied in [50]. S-TaLiRo implements the stochastic optimisation with the *adaptive restart* mechanism [43] to avoid trapping into local optima.

1.5.1.2 Coverage-Guided Metrics

Coverage is an important concept in traditional software testing. Originally, it is used to measure how much of a program is executed by a test suite. In the case that no error is found in the program, the larger coverage a test suite achieves, the more reliable the testing result can be considered. Some falsification techniques are developed based on the coverage notion, with the aim of avoiding local optima. These techniques employ coverage as a guidance for the search—the search aims not only at minimizing the robustness, but also at enlarging the coverage of the search space. In this way, the search can avoid pure exploitation that can only lead to a small coverage.

As hybrid systems deal with infinite search spaces, it remains the problem of how to define the coverage metric in this context. Different coverage definitions have been proposed. In [21], coverage is defined based on the internal states of

Simulink models, and it is included as a part of the objective function so that the search is guided by both robustness and coverage. In [24], the coverage is defined by *star discrepancy*, a measure for evaluating how well-distributed a set of samples are. In [3], a coverage notion is defined based on classification of the input space according to robustness values. This work balances well between exploration and exploitation because it considers not only the coverage of input space but also robustness. Moreover, a recent work [69] of ours targets a specific type of hybrid systems that adopts *deep neural networks (DNNs)* as controllers. In [69], we define coverage metrics based on the intermediate outputs of the DNN controllers, and explore the system behavior space guided by the coverage.

1.5.1.3 Machine Learning for Falsification

Machine learning techniques have been applied everywhere, and some researchers have also introduced it into falsification. In general, this line of work learns a model from data; based on that, it proposes new samples with the aim of exploring the search space effectively. Specifically, two machine learning techniques have been studied heavily in falsification, namely, *Bayesian optimisation* and *reinforcement learning*.

Bayesian optimisation makes use of *Gaussian Process Regression* [48] to learn the objective function (i.e., robustness in our context); this results in a Gaussian process over the search space. Gaussian process is a model in which every point in the search space has a Gaussian distribution; in this context, it predicts the probability distribution of the robustness at each point. Bayesian optimisation then samples at some potentially interesting places, where low robustness is likely to locate. The sampling strategy, known as *acquisition function*, guarantees the balance between exploration and exploitation. This technique has been studied in [2,4,6,18,54].

Reinforcement learning is a hot spot in the machine learning community in recent years. It tries to learn the best policy—a sequence of actions—that an agent can take, from the interactions between the agent and the environment. There is also a trade-off between exploration and exploitation in reinforcement learning: when the agent decides which action to take for the next step, the choice is either to explore an action that has not been taken before, or to exploit an already taken action further to get more reliable feedback. The default mechanism of handling the trade-off in reinforcement learning is *ε-greedy*. In recent years, this technique has been applied to falsification [7,38].

1.5.2 Robust Semantics Definition

The definition of robust semantics in our context plays the role of objective functions for the optimisation, and thus it affects the performance of the entire framework significantly. The current widely-used STL robust semantics refers to the spatial robustness (see Section 3.4.4 for the definition and see the example in Table 1.1). There also exists another formalism [23] that captures the temporal robustness; intuitively, it characterises the level of satisfaction of a signal to a property from the perspective of time. These two formalisms are suited for different use cases. In [5], the authors propose a way that takes both perspectives into account, resulting in the *Averaged STL (AvSTL)*. While the computation of spatial or temporal robustness is just a matter of obtaining a vertical or horizontal distance, AvSTL requires to compute the integration of the signal over a certain interval; the authors then design a specific algorithm for that purpose. Their experiments show the strength of that semantics in solving falsification problems.

Another line of work aims to tackle a different problem, that is, the robust semantics for Boolean connectives. In the STL robustness definition, when it comes to the robustness for Boolean connectives, it makes a comparison between the robustness values of different sub-formulas—for conjunctive it takes the minimum, and for disjunctive it takes the maximum (see Section 3.4.4 for details). However, this yields a problem: if the sub-formulas concern with different signals ranging over different scales (e.g., the formula in Section 1.3.2 that involves *gear* and *speed*), then the comparison becomes unfair, because one can always mask the other one. This is the so-called *scale problem*. In [68], we show that a naive signal rescaling (normalization) approach does not always work in solving this problem. We have done a line of work [66,67] solving this problem. Our main idea is to treat the two operands of the connectives as incomparable, and design a reward to decide which one to adopt on the fly. The details will be elaborated in Section 1.6.2.

1.5.3 Input Constraints

In Section 1.3.3.1, we explained that hill-climbing optimisation relies on random sampling at its initial stage. Typically, random sampling works in a search space that is unconstrained (usually a hyperrectangle), and so does hill-climbing optimisation. In this way, falsification can return any input signal that violates the system specification in the search space, once it manages to find one. However, in reality, there usually exist some logical constraints among input signals. One example is that, in an automotive system (e.g., the one in Fig. 1.2), the *throttle* and *brake* cannot be pushed simultaneously. Such constraints also exist in other literature about CPS products; for example in [14], the authors test an assisted driving system under different environment and system parameters; there is a constraint—"when there is no fog, the visibility range is set to maximum"—that restricts the system inputs. Sometimes, the engineers also desire to impose some conditions on the input when testing systems; for example in [37], the authors aim to test a system under the condition that *throttle* increases monotonically. In the presence of such constraints, input signals produced by falsification should be guaranteed to satisfy them; otherwise, those signals would be meaningless.

There have not been many works that address this issue. In [12], Barbot et al. utilise a timed automaton, that implements the input constraints, to generate meaningful words, and then they apply Monte Carlo sampling to produce input signals. In [13], Barbot et al. use a transformation technique so that they can sample in a hyper-box but obtain timed automaton traces that satisfy the input constraints. We also have a line of work solving this problem. In our work [63], we treat the specification violation and the input constraint satisfaction as two objectives, and apply a lexicographic method to highlight the importance of the satisfaction of the input constraints. In another work [64] of us, similar to [13], we tackle this problem via a method called search space transformation, where we map an unconstrained space to the desired space for constrained signals, and then apply the optimisation algorithms in the unconstrained one; we will elaborate on this work in Section 1.6.3. Sato et al. also have a line of works [51, 52] on constrained falsification, but their constraints are imposed on outputs rather than inputs as other works in this section do.

1.5.4 Confidence Estimation of Falsification

As a testing method, falsification has an intrinsic weakness—when a counterexample cannot be detected by a falsification algorithm, no guarantee is provided that the system does not violate the property. In other words, we do not know whether the specification is really not falsifiable, or the falsification algorithm did not explore the search space enough. Indeed, this is a major difference between verification and testing, and it yields another problem—how confident should we be on our falsification results.

In the traditional software testing techniques, people usually use *coverage* as a metric to measure how much the search space has been explored. Commonly-used metrics for program testing include statement coverage, branch coverage, function coverage, MC/DC, etc. The higher coverage is achieved, the more confidence can be obtained. However, these metrics cannot be straightforwardly adapted to hybrid system falsification, because they require the knowledge of system internal structures (recall that falsification usually assumes black-box systems, as mentioned in Section 1.3.1.1). New metrics or methodologies for estimating confidence are needed.

As this is a new research direction, only a few works [55, 62] have addressed this problem. In [62], we leverage *supervised learning* to estimate the falsification confidence. Essentially, our idea is to treat the simulation data in falsification as training data, and train a *Gaussian process* to approximate the function from inputs (signals) to outputs (robustness). Gaussian process is an extension of Gaussian distribution to continuous domains, and, therefore,

Gaussian process is used to model the distribution of a continuous function. The benefit of using Gaussian process in [62] is that each point in the learned Gaussian process is a Gaussian distribution, so it naturally conveys how likely the point is distributed in the negative domain (i.e., it has a negative robustness). Therefore, [62] collects those points whose lower confidence bound is below 0, and computes a probability that there exists a negative value in their joint distribution. Another recent work [55] uses bounding boxes that contain the falsification samplings to characterize the input space covered by the samplings. The sizes of the bounding boxes are dependent on the robustness values of the samples, so a small observed robustness indicates a higher "suspiciousness" of an area, and thus requires more samplings to cover it. Consequently, this technique can also help us understand how much of the space has been covered by the falsification samples, and how much effort is still needed to cover the whole space.

1.6 A Hierarchical Optimisation Framework for Effective Falsification

In this section, we introduce a general hierarchical optimisation framework that can be instantiated to solve some of the problems presented in Section 1.5. The instantiated techniques introduced in the following sections improve the effectiveness of the classic falsification framework in Chapter 1.

The hierarchical optimisation framework is shown in Fig. 1.9. It firstly decomposes the problem into a set of sub-problems, and then uses a hierarchical methodology: the top layer makes a (discrete) decision on selecting one child problem as the next step to proceed, based on the information coming from the bottom layer; the bottom layer performs numerical optimisation on the sub-problem suggested by the top layer, and it returns feedback to the top layer. This process takes place iteratively, until the problem is solved. One feature of this framework is the interaction between the two layers: the top layer makes high-level decisions and advises the bottom layer; the bottom layer performs concrete numerical evaluation and gives feedback to the top layer to facilitate the future decisions.

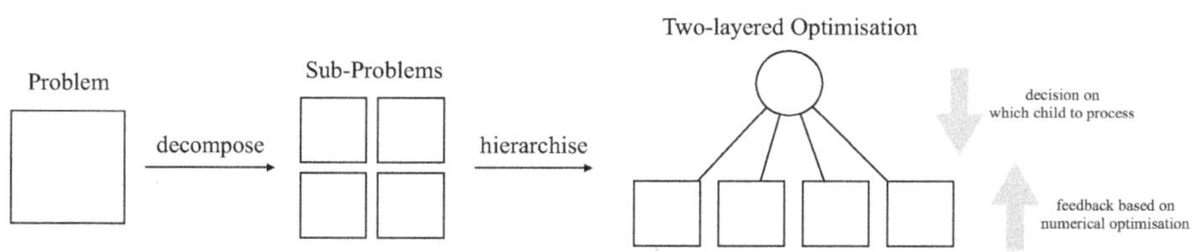

Figure 1.9: Solving problems using a hierarchical optimisation framework.

1.6.1 Monte Carlo Tree Search for Search Space Exploration

Monte Carlo Tree Search (MCTS) is a popular technique in recent years due to its successful application in the computer Go game [53]. This technique is especially effective in handling discrete search problems with large search space. The key of MCTS is that it allocates more budget to the subspace where the search target is more likely to exist, and meanwhile it also explores the less visited areas to take a balance between exploration and exploitation.

In [65], we apply MCTS to solve the "local optima" issue in falsification. As shown in Fig. 1.10a, the search space is organised as a tree by *time staging* and *space discretization*.A key notion of MCTS is *reward*, which is an indicator of the quality of a subspace, and is used for deciding the budget allocation. We connect the definition of reward with robustness—the branches that have witnessed high robustness have low reward, and vice versa. The robustness values are obtained by running hill-climbing optimisation on the subspaces identified by the branches.

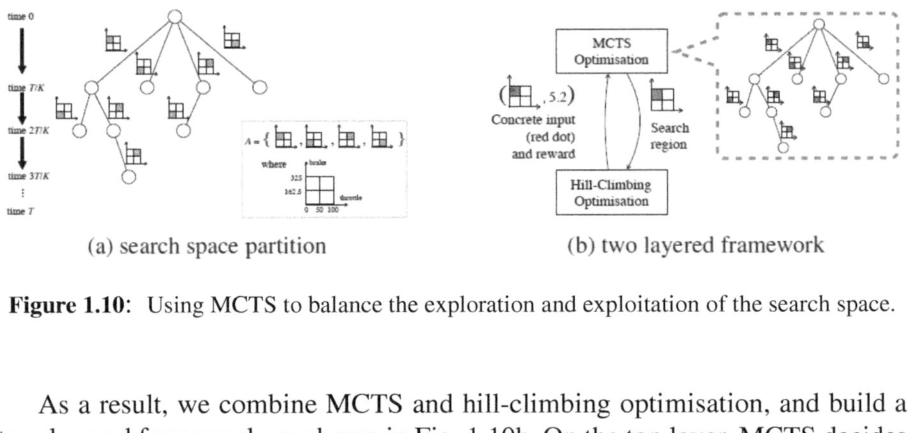

(a) search space partition (b) two layered framework

Figure 1.10: Using MCTS to balance the exploration and exploitation of the search space.

As a result, we combine MCTS and hill-climbing optimisation, and build a two-layered framework, as shown in Fig. 1.10b. On the top layer, MCTS decides which subspace to explore, based on the reward information provided by the low layer; on the low layer, hill-climbing optimisation exploits a specific subspace, and provides feedback to the top layer.

1.6.2 QB-Robustness for the Scale Problem

In this section, we introduce our works for solving the scale problem in the definitions of Boolean connectives (see Section 1.5.2). Our idea consists in exploring the discrete structures of STL formulas to avoid the value comparison between signals with different scales; and this idea, again, results in a two-layered hierarchical optimisation framework that follows Fig. 1.9.

We initiate an early attempt in [66], where we focus on safety properties with Boolean connectives only, namely the formulas in the forms of $\Box(\varphi_1 \wedge \varphi_2)$ and $\Box(\varphi_1 \vee \varphi_2)$. We treat the two sub-formulas φ_1 and φ_2 as two "arms" in the *Multi-Armed Bandit (MAB)* problem, so, during falsification, we can iteratively select one sub-formula by using MAB algorithms, and then try to falsify the selected sub-formula by hill-climbing optimisation. The MAB problem originally models the problem of how to maximise the rewards that a gambler can earn in front of a row of arms of slot machines. Existing research shows that balancing between exploration and exploitation of the machines is necessary to achieve the best rewards. In our work, we associate the "progressiveness" of the falsification process with the rewards of each arm (subformula)—the more progressively the optimisation falsifies a sub-formula, the larger rewards the sub-formula obtains. The selection of sub-formulas is decided by existing algorithms for MAB, such as UCB1, in the way as follows: we allocate more budget to the sub-formula with high reward, and less for the one with low reward. As a result, MAB algorithms work on the top layer, governing the hill-climbing optimisation running on the bottom layer.

The limitation on the target formulas of [66] motivates us to extend this idea to general STL formulas. In [67], we achieve that by formalizing *QB-robustness* and developing an MCTS-based approach. Differently from the classic robustness, QB-robustness introduces another parameter, that declares one of the sub-formulas, in the robustness computation, and gives the robustness value for Boolean connectives in a *case-by-case* manner. For instance, instead of having a real number $[\![\mathbf{w}, \varphi_1 \wedge \varphi_2]\!] \in \mathbb{R}$ as the robustness of $\varphi_1 \wedge \varphi_2$, QB-robustness introduces a parameter that either indicates φ_1 or φ_2: if it is φ_1, then $[\![\mathbf{w}, \varphi_1]\!]$ will be the robustness; otherwise, $[\![\mathbf{w}, \varphi_2]\!]$ will be.

Based on QB-robustness, we develop an MCTS-based algorithm for effective falsification. As shown in Fig. 1.11, we expand the syntax tree of the STL formula, and use MCTS to explore the structure. At each leaf, based on a syntax path, we can compute the QB-robustness, and perform hill-climbing optimisation to obtain the reward (similarly to the "progressiveness" in [66]) of each path. Again, this is a typical hierarchical optimisation framework: the high layer nodes select the best children based on the reward, and the leaf node performs hill-climbing optimisation on the selected child, and back propagates the reward information.

1.6.3 Search Space Transformation for Input Constraints

We have briefly introduced search space transformation in Section 1.5.3, as an approach to solving the input constraint problem in falsification. As illustrated in Fig. 1.12, it allows to sample a point x in a hyperbox Ξ, and then it maps x to a point $y = \mathcal{T}(x)$ in a constrained space Ω_ψ. The robustness $\rho(y)$ of y can be computed by using the STL robust semantics, and then $\rho(y)$ is assigned to $r(x)$ as the fitness of the point x, so that an optimisation can be done that minimises

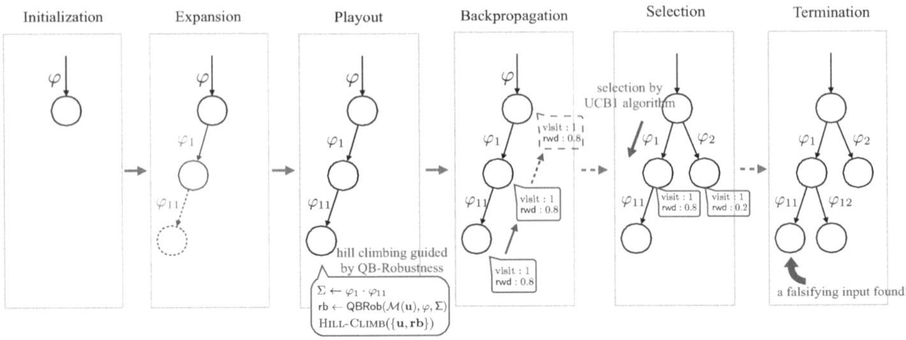

Figure 1.11: An example of falsifying $\Box((\varphi_{11} \wedge \varphi_{12}) \vee \varphi_2)$ using MCTS guided by QB-robustness.

$r(x)$ in space Ξ. As a result, search space transformation turns the constrained optimisation problem into an unconstrained one, and when a counterexample x^* in Ξ is observed, its mapped point $\mathcal{T}(x^*)$ in Ω_ψ can be returned, which is guaranteed to satisfy the constraint.

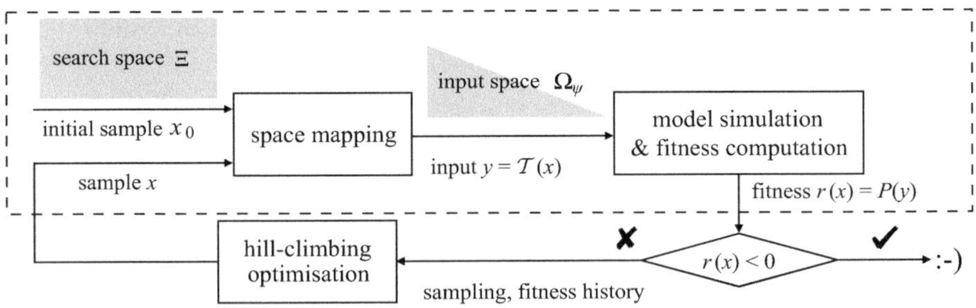

Figure 1.12: Search space transformation

The performance of this framework is subject to a hyperparameter that decides the map function $\mathcal{T}$; more specifically, the hyperparameter is a total order over the dimensions of the search space. In [64], we introduce an MAB-based method to achieve the best performance. The idea is, again, by building a hierarchical optimisation framework: on the top layer, the MAB algorithm selects a total order of dimensions and passes it to the bottom layer; on the bottom layer, the search space transformation applies the total order to falsify a specification with input constraints, and returns feedback to the top layer.

1.7 Conclusions

In this chapter, we give an overview of hybrid system falsification, a technique for quality assurance of CPS via searching for counterexample inputs. We introduce the basic methodologies, tools and applications, scientific challenges and recent advances in the technique. Notably, we present our contributions, i.e., a general hierarchical optimisation framework and its instantiations, that improve the falsification effectiveness.

References

[1] Abbas, H., Hoxha, B., Fainekos, G., Deshmukh, J.V., Kapinski, J., and Ueda, K. 2014. Conformance testing as falsification for cyber-physical systems. arXiv preprint arXiv: 1401.5200.

[2] Abeysirigoonawardena, Y., Shkurti, F., and Dudek, G. 2019. Generating adversarial driving scenarios in high-fidelity simulators. *In*: 2019 International Conference on Robotics and Automation (ICRA), pages 8271–8277. IEEE.

[3] Adimoolam, A., Dang, T., Donzé, A., Kapinski, J., and Jin. X. 2017. Classification and coverage-based falsification for embedded control systems. *In*: Computer Aided Verification, pages 483–503, Cham, 2017. Springer International Publishing.

[4] Akazaki, T. 2016. Falsification of conditional safety properties for cyberphysical systems with gaussian process regression. *In*: Yliès Falcone and César Sánchez, (eds). Runtime Verification—16th International Conference, RV 2016, Madrid, Spain, September 23–30, 2016, Proceeding, volume 10012 of Lecture Notes in Computer Science, pages 439–446. Springer.

[5] Akazaki, T., and Hasuo, I. 2015. Time robustness in MTL and expressivity in hybrid system falsification. *In*: Daniel Kroening and Corina S. Păšareanu, (eds). Computer Aided Verification, pages 356–374, Cham 2015. Springer International Publishing.

[6] Akazaki, T., Kumazawa, Y., and Hasuo, I. 2017. Causality-aided falsification. In Lukas Bulwahn, Maryam Kamali, and Sven Linker, (eds). Proceeding First Workshop on Formal Verification of Autonomous Vehicles, FVAV@iFM 2017, Turin, Italy, 19th September 2017, volume 257 of EPTCS, pages 3–18.

[7] Akazaki, T., Liu, S., Yamagata, Y., Duan, Y., and Hao, J. 2018. Falsification of cyber-physical systems using deep reinforcement learning. *In*: Formal Methods —22nd International Symposium, FM 2018, Held as Part of the Federated Logic Conference, FloC 2018, Oxford, UK, July 15–17, 2018, Proceeding, pages 456–465.

[8] Alur, R., Feder, T., and Henzinger, T.A. 1996. The benefits of relaxing punctuality. Journal of the ACM (JACM). 43(1): 116–146.

[9] Annapureddy, Y., Liu, C., Fainekos, G., and Sankaranarayanan, S. 2011. S-Taliro: A tool for temporal logic falsification for hybrid systems. *In*: Proceeding of the 17th International Conference on Tools and Algorithms for the Construction and Analysis of Systems: Part of the Joint European Conferences on Theory and Practice of Software, TACAS'11/ETAPS'11, pages 254–257, Berlin, Heidelberg. Springer-Verlag.

[10] Singh, Y., Annapureddy, R., and Fainekos, G.E. 2010. Ant colonies for temporal logic falsification of hybrid systems. *In*: IECON 2010–36th Annual Conference on IEEE Industrial Electronics Society, pages 91–96. IEEE.

[11] Auger, A. and Hansen, N. 2005. A restart CMA evolution strategy with increasing population size. *In*: Proceeding of the IEEE Congress on Evolutionary Computation, CEC 2005, pages 1769–1776. IEEE.

[12] Barbot, B., Basset, N., and Dang, T. 2019. Generation of signals under temporal constraints for CPS testing. *In*: NASA Formal Methods, pages 54–70, Cham. Springer International Publishing.

[13] Barbot, B., Basset, N., Dang, T., Donzé, A., Kapinski, J., and Yamaguchi, T. 2020. Falsification of cyber-physical systems with constrained signal spaces. *In*: NASA Formal Methods Symposium, pages 420–439. Springer.

[14] Abdessalem, R.B., Nejati, S., Briand, L.C., and Stifter, T. 2018. Testing vision-based control systems using learnable evolutionary algorithms. *In*: Proceeding of the 40th International Conference on Software Engineering, ICSE '18, pages 1016–1026, New York, NY, USA. ACM.

[15] Clarke Jr, E.M., Grumberg, O., Kroening, D., Peled, D., and Veith, H. 2018. Model checking. MIT press, 2018.

[16] Delmas, R., Loquen, T., Boada-Bauxell, J., and Carton, M. 2019. An evaluation of monte-carlo tree search for property falsification on hybrid flight control laws. *In*: Majid Zamani and Damien Zufferey, (eds). Numerical Software Verification—12th International Workshop, NSV@CAV 2019, New York City, NY, USA, July 13–14, 2019, Proceeding, volume 11652 of Lecture Notes in Computer Science, pages 45–59. Springer.

[17] Deshmukh, J. 2019. Simulation-based and data-driven reasoning for cyber-physical systems. Oded Maler—Barbaric Science from a Captive Poet in 22nd ACM International Conference on Hybrid Systems: Computation and Control, HSCC.

[18] Deshmukh, J.V., Horvat, M., Jin, X., Majumdar, R., and Prabhu, V.S. 2017. Testing cyber-physical systems through bayesian optimization. ACM Trans. Embedded Comput. Syst., 16(5): 170: 1–170:18.

[19] Deshmukh, J., Jin, X., Kapinski, J., and Maler, O. 2015. Stochastic local search for falsification of hybrid systems. *In*: Bernd Finkbeiner, Geguang Pu, and Lijun Zhang, (eds). Automated Technology for Verification and Analysis, pages 500–517, Cham. Springer International Publishing.

[20] Dokhanchi, A., Yaghoubi, S., Hoxha, B., Fainekos, G., Ernst, G., Zhang, Z., et al. 2018. ARCH-COMP18 category report: Results on the falsification benchmarks. *In*: Goran Frehse, (ed). ARCH18. 5th International Workshop

on Applied Verification of Continuous and Hybrid Systems, volume 54 of EPiC Series in Computing, pages 104–109. EasyChair.

[21] Dokhanchi, A., Zutshi, A., Sriniva, R.T., Sankaranarayanan, S., and Fainekos, G.E. 2015. Requirements driven falsification with coverage metrics. *In*: Alain Girault and Nan Guan, (eds). 2015 International Conference on Embedded Software, EMSOFT 2015, Amsterdam, Netherlands, October 4–9, 2015, pages 31–40. IEEE.

[22] Donzé, A. 2010. Breach, A toolbox for verification and parameter synthesis of hybrid systems. *In*: Tayssir Touili, Byron Cook, and Paul B. Jackson, (eds). Computer Aided Verification, 22nd International Conference, CAV 2010, volume 6174 of LNCS, pages 167–170. Springer.

[23] Donzé, A., and Maler, O. 2010.Robust satisfaction of temporal logic over real-valued signals. *In*: Krishnendu Chatterjee and Thomas A. Henzinger, (eds). Formal Modeling and Analysis of Timed Systems—8th International Conference, FORMATS 2010, volume 6246 of LNCS, pages 92–106. Springer.

[24] Dreossi, T., Dang, T., Donzé, A., Kapinski, J., Jin, X., and Deshmukh, J.V. 2015. Efficient guiding strategies for testing of temporal properties of hybrid systems. *In*: Klaus Havelund, Gerard Holzmann, and Rajeev Joshi, (eds). NASA Formal Methods, pages 127–142, Cham. Springer International Publishing.

[25] Dreossi, T., Donzé, A., and Seshia, S.A. 2019. Compositional falsification of cyber-physical systems with machine learning components. Journal of Automated Reasoning, 63(4):1031–1053.

[26] Ernst, G., Arcaini, P., Bennani, I., Chandratre, A., Alexandre Donz, A., Fainekos, G., et al. 2021. ARCH-COMP 2021 category report: Falsification with validation of results. *In*: Goran Frehse and Matthias Althoff, (eds). 8th International Workshop on Applied Verification of Continuous and Hybrid Systems (ARCH21), volume 80 of EPiC Series in Computing, pages 133–152. EasyChair.

[27] Ernst, G., Arcaini, P., Bennani, I., Donzé, A., Fainekos, G., Frehse, G., et al. 2020. ARCH-COMP 2020 category report: Falsification. *In*: Goran Frehse and Matthias Althoff, (eds). ARCH20. 7th International Workshop on Applied Verification of Continuous and Hybrid Systems (ARCH20), volume 74 of EPiC Series in Computing, pages 140–152. EasyChair.

[28] Ernst, G., Arcaini, P., Donzé, A., Fainekos, G., Mathesen, L., Pedrielli, G., et al. 2019. ARCH-COMP 2019 category report: Falsification. *In*: Goran Frehse and Matthias Althoff, (eds). ARCH19. 6th International Workshop on Applied Verification of Continuous and Hybrid Systems, volume 61 of EPiC Series in Computing, pages 129–140. EasyChair.

[29] Ernst, G., Arcaini, P., Fainekos G., Formica, F., Inoue, J., Khandait, T., et al. 2022. ARCH-COMP 2022 category report: Falsification with ubounded resources. *In*: Goran Frehse, Matthias Althoff, Erwin Schoitsch, and Jeremie Guiochet, (eds). Proceeding of 9th International Workshop on Applied Verification of Continuous and Hybrid Systems (ARCH22), volume 90 of EPiC Series in Computing, pages 204–221. EasyChair.

[30] Ernst, G., Sedwards, S., Zhang, Z., and Hasuo, I. 2019. Fast falsification of hybrid systems using probabilistically adaptive input. *In*: Quantitative Evaluation of Systems, pages 165–181, Cham 2019. Springer International Publishing.

[31] Fainekos, G.E. and Pappas, G.J. 2009. Robustness of temporal logic specifications for continuous-time signals. Theory of Computer Science, 410(42): 4262–4291, September 2009.

[32] Ferrère, T., Nickovic, D., Donzé, A., Ito, H., and Kapinski, J. 2019. Interface-aware signal temporal logic. *In*: Proceeding of the 22nd ACM International Conference on Hybrid Systems: Computation and Control, HSCC 2019, Montreal, QC, Canada, April 16–18, 2019, pages 57–66.

[33] Gladisch, C., Heinz, T., Heinzemann, C., Oehlerking, J., von Vietinghoff, A., and Pfitzer, T. 2019. Experience paper: Searchbased testing in automated driving control applications. *In*: Proceeding of the 34th IEEE/ACM International Conference on Automated Software Engineering, ASE '19, pages 26–37. IEEE Press.

[34] Harman, M., Jia, Y., and Yuanyuan Zhang, Y. 2015. Achievements, open problems and challenges for search based software testing. *In*: 2015 IEEE 8th International Conference on Software Testing, Verification and Validation (ICST), pages 1–12.

[35] Henzinger, T.A. 2000. The theory of hybrid automata. *In*: Verification of Digital and Hybrid Systems, pages 265–292. Springer.

[36] Hoxha, B., Abbas, H., and Fainekos, G. 2014. Benchmarks for temporal logic requirements for automotive systems. Proceeding of Applied Verification for Continuous and Hybrid Systems.

[37] Jin, X., Deshmukh, J.V., Kapinski, J., Ueda, K., and Butts, K. 2014. Powertrain control verification benchmark. *In*: Proceeding of the 17th International Conference on Hybrid Systems: Computation and Control, HSCC '14, pages 253–262, NY, USA. ACM.

[38] Kato, K. and Ishikawa, F. 2019. Learning-based falsification for model families of cyber-physical systems. *In*: 2019 IEEE 24th Pacific Rim International Symposium on Dependable Computing (PRDC), pages 236–23609. IEEE.

[39] Koymans, R. 1990. Specifying real-time properties with metric temporal logic. Real-time systems, 2(4): 255–299.

[40] Krishna, S.N. and Trivedi, A. 2015. Hybrid automata for formal modeling and verification of cyber-physical systems. CoRR, abs/1503.04928.

[41] Kŭrátko, J. and Ratschan. S. 2014. Combined global and local search for the falsification of hybrid systems. *In*: Axel Legay and Marius Bozga, (eds). Formal Modeling and Analysis of Timed Systems, pages 146–160, Cham. Springer International Publishing.

[42] Luersen, M.A. and Le Riche, R. 2004. Globalized Nelder–Mead method for engineering optimization. Computers and Structures, 82(23): 2251–2260.

[43] Mathesen, L., Yaghoubi, S., Pedrielli, G., and Fainekos, G. 2019. Falsification of cyber-physical systems with robustness uncertainty quantification through stochastic optimization with adaptive restart. *In*: 2019 IEEE 15th International Conference on Automation Science and Engineering (CASE), pages 991–997. IEEE.

[44] McMinn, P. 2011. Search-based software testing: past, present and future. *In*: Proceeding of the 2011 IEEE Fourth International Conference on Software Testing, Verification and Validation Workshops, ICSTW '11, pages 153–163, USA. IEEE Computer Society.

[45] Nghiem, T., Sankaranarayanan, S., Fainekos, G., Ivanci´c, F., Gupta, A., and Pappas, G.J. 2010. Monte-carlo techniques for falsification of temporal properties of non-linear hybrid systems. *In*: Proceeding of the 13th ACM International Conference on Hybrid Systems: Computation and Control, HSCC '10, pages 211–220, New York, NY, USA. Association for Computing Machinery.

[46] Pincus, M. 1970. Letter to the editor—a Monte Carlo method for the approximate solution of certain types of constrained optimization problems. Operations Research, 18(6): 1225–1228.

[47] Pnueli, A. 1977. The temporal logic of programs. *In*: 18th Annual Symposium on Foundations of Computer Science (SFCS 1977), pages 46–57. IEEE.

[48] Rasmussen, C.E. 2003. Gaussian processes in machine learning. *In*: Summer school on machine learning, pages 63–71. Springer.

[49] Rubinstein, R. 1999. The cross-entropy method for combinatorial and continuous optimization. Methodology and Computing in Applied Probability, 1(2): 127–190.

[50] Sankaranarayanan, S. and Fainekos, G. 2012. Falsification of temporal properties of hybrid systems using the cross-entropy method. *In*:

Proceeding of the 15th ACM International Conference on Hybrid Systems: Computation and Control, pages 125–134.

[51] Sato, S., Saimen, A., Waga, M., Takao, K., and Hasuo, I. 2021. Hybrid system falsification for multiple-constraint parameter synthesis: a gas turbine case study. *In*: Marieke Huisman, Corina Păšareanu, and Naijun Zhan, (eds). Formal Methods, pages 313–329, Cham, 2021. Springer International Publishing.

[52] Sato, S., Waga, M., and Hasuo, I. 2021. Constrained optimization for hybrid system falsification and application to conjunctive synthesis. IFACPapersOnLine, 54(5): 217–222, 2021. 7th IFAC Conference on Analysis and Design of Hybrid Systems ADHS.

[53] Silver, D., Huang, A., Maddison, C.J., Guez, A., Sifre, L., van den Driessche, G., et al. 2015. Mastering the game of Go with deep neural networks and tree search. Nature, 529: 484–489.

[54] Silvetti, S., Policriti, A., and Bortolussi, L. 2017. An active learning approach to the falsification of black box cyber-physical systems. *In*: Nadia Polikarpova and Steve Schneider, (eds). Integrated Formal Methods, pages 3–17, Cham, 2017. Springer International Publishing.

[55] Takisaka, T., Zhang, Z., Arcaini, P., and Hasuo, I. 2023. A robustness-based confidence measure for hybrid system falsification. IEEE Transactions on Computer-Aided Design of Integrated Circuits and Systems, 42(5): 1718–1731.

[56] Thibeault, Q., Anderson, J., Chandratre, A., Pedrielli, G., and Georgios Fainekos, G. 2021. PSY-TaLiRo: A python toolbox for search-based test generation for cyber-physical systems. In Alberto Lluch-Lafuente and Anastasia Mavridou, (eds). Formal Methods for Industrial Critical Systems—26th International Conference, FMICS 2021, Paris, France, August 24–26, 2021, Proceeding, volume 12863 of Lecture Notes in Computer Science, pages 223–231. Springer.

[57] Tuncali, C.E., and Fainekos, G. 2019. Rapidly-exploring random trees for testing automated vehicles. In 2019 IEEE Intelligent Transportation Systems Conference (ITSC), pages 661–666. IEEE.

[58] Tuncali, C.E., Pavlic, T.P., and Fainekos, G. 2016. Utilizing s-taliro as an automatic test generation framework for autonomous vehicles. In 2016 IEEE 19th International Conference on Intelligent Transportation Systems (ITSC), pages 1470–1475. IEEE.

[59] Utting, M., Pretschner, A., and Legeard, B. 2012. A taxonomy of model-based testing approaches. Software testing, verification and reliability, 22(5): 297–312.

[60] Wolpert, D. and Macready, W.G. 1997. No free lunch theorems for optimization. IEEE Trans. Evolutionary Computation, 1(1): 67–82.

[61] Yaghoubi, S. and Fainekos, G. 2019. Gray-box adversarial testing for control systems with machine learning components. In Proceeding of the 22nd ACM International Conference on Hybrid Systems: Computation and Control, pages 179–184.

[62] Zhang, Z. and Arcaini. P. 2021. Gaussian process-based confidence estimation for hybrid system falsification. In Marieke Huisman, Corina Păšareanu, and Naijun Zhan, (eds). Formal Methods, pages 330–348, Cham, 2021. Springer International Publishing.

[63] Zhang, Z., Arcaini, P., and Hasuo, I. 2020. Constraining counterexamples in hybrid system falsification: Penalty-based approaches. In Ritchie Lee, Susmit Jha, and Anastasia Mavridou, (eds). NASA Formal Methods, pages 401–419, Cham, 2020. Springer International Publishing.

[64] Zhang, Z., Arcaini, P., and Hasuo, I. 2020. Hybrid system falsification under (in) equality constraints via search space transformation. IEEE Transactions on Computer-Aided Design of Integrated Circuits and Systems, 39(11): 3674–3685.

[65] Zhang, Z., Ernst, G., Sedwards, S., Arcaini, P., and Hasuo, I. 2018. Two-layered falsification of hybrid systems guided by Monte Carlo tree search. IEEE Transactions on Computer-Aided Design of Integrated Circuits and Systems, 37(11): 2894–2905, Nov 2018.

[66] Zhang, Z., Hasuo, I., and Arcaini, P. 2019. Multi-armed bandits for boolean connectives in hybrid system falsification. *In*: Isil Dillig and Serdar Tasiran, (eds). Computer Aided Verification, pages 401–420, Cham, 2019. Springer International Publishing.

[67] Zhang, Z., Lyu, D., Arcaini, P., Ma, L., Hasuo, I., and Zhao, J. 2021. Effective hybrid system falsification using Monte Carlo tree search guided by QB-robustness. *In*: Alexandra Silva and K. Rustan M. Leino, (eds). Computer Aided Verification, pages 595–618, Cham, 2021. Springer International Publishing.

[68] Zhang, Z., Lyu, D., Arcaini, P., Ma, L., Hasuo, I., and Zhao, J. 2021. On the effectiveness of signal rescaling in hybrid system falsification. *In*: Aaron Dutle, Mariano M. Moscato, Laura Titolo, César A. Muñoz and Ivan Perez, (eds). NASA Formal Methods, pages 392–399, Cham, 2021. Springer International Publishing.

[69] Zhang, Z., Lyu, D., Arcaini, P., Ma, L., Hasuo, I., and Zhao, J. 2023. FalsifAI: Falsification of AI-enabled hybrid control systems guided by time-aware coverage criteria. IEEE Transactions on Software Engineering, 49(4): 1842–1859.

[70] Zutshi, A., Deshmukh, J.V., Sankaranarayanan, S., and Kapinski, J. 2014. Multiple shooting, CEGAR-based falsification for hybrid systems. *In*: Proceeding of the 14th International Conference on Embedded Software, EMSOFT '14, New York, NY, USA. Association for Computing Machinery.

[71] Zutshi, A., Sankaranarayanan, S., Deshmukh, J.V., and Kapinski, J. 2013. A trajectory splicing approach to concretizing counterexamples for hybrid systems. *In*: 52nd IEEE Conference on Decision and Control, pages 3918–3925. IEEE.

Chapter 2

Monitoring Temporal Specifications

Masaki Waga
Kyoto University, Kyoto, Japan and National Institute of Informatics, Tokyo, Japan

Stefan Klikovits
Johannes Kepler University, Linz, Austria

Ichiro Hasuo
National Institute of Informatics and SOKENDAI, Tokyo, Japan

2.1 Introduction

2.1.1 What is Monitoring

Monitoring generally is an activity to observe system behaviour. It is essential to monitor the system for its development and maintenance. For example, in development, engineers modify the system to satisfy certain requirements. To decide the necessary modification, they must monitor the difference between the actual behaviour and the expected behaviour. In maintenance, engineers may have to replace the worn-out components in the system. This decision also requires monitoring because they must detect if the current behaviour deviates from the original behaviour.

The following two examples show concrete usage scenarios of monitoring.

Example 1 (continuous monitoring of a deployed system) *Consider a scenario in which a recently developed factory automation system is deployed. Initially, the system performs effectively, e.g., there is no overshoot, and the system behaves robustly. But with time, the system gradually becomes erroneous. For instance, wearing out of hardware may cause a slight overshoot, which ultimately could stop the system from functioning. Therefore, continuous monitoring of the system is needed to detect and deal with an issue as soon as possible.*

Example 2 (monitoring as a test oracle) *Consider a scenario in which a new engine is being developed from scratch, and there are several prototypes (either as a simulator or a hardware). Owing to the early stage of development, each prototype most likely contains issues, that must be detected by testing. In testing, an oracle is needed to decide if the behaviour of the system under test is erroneous or not; this is also monitoring.*

2.1.2 Monitoring by a Human

One of the most primitive forms of monitoring is by humans: given a specification in a document, an engineer checks if the system observation meets the specification. Manual monitoring is ubiquitous in system development. For example, it is natural to check if the system under development satisfies the given requirement through manual trials.

Despite its ease of application and prevalence in system development, manual monitoring has poor scalability. There tends to be a huge number of observations in cyber-physical system (CPS) development, that requires high scalability of monitoring. For humans, it is extremely challenging to monitor such massive system observations. Manual monitoring is also usually expensive due to the cost of human resources.

2.1.3 Automated Monitoring by a Handcrafted Monitor

To overcome the scalability issue, monitoring can be automated by implementing a dedicated software or hardware. These automated monitors are necessary, especially for continuous monitoring of a running system, e.g., indicators of vehicles or airplanes. However, it is challenging to handcraft the correct monitor for a complicated specification. Developing a monitor that provides more information than just detecting a safety violation, like severity of a safety violation, is also very complex.

Moreover, such ad-hoc monitors are often developed from scratch and are rarely reused, increasing their development cost. If developed by novice developers their performance will also be suboptimal.

Table 2.1: Comparison of monitoring methods

Monitoring Method	Flexibility	Informativeness	Scalability
Human monitoring	very high	very high	very low
Ad-hoc automated monitoring	low	low	high
Formal automated monitoring	high	high	high

2.1.4 Monitoring Based on Formal Specifications

Specification-based monitoring is a monitoring method that uses logics. Given a formal specification (typically of an unsafe behaviour) expressed in some logical formalism, e.g., automata, temporal logic formalism, or regular expressions, specification-based monitoring observes a system execution and evaluates if the observation satisfies the specification. Research has shown that these logical specifications can be automatically turned into an efficient monitor. In other words, there is a specification-generic monitoring algorithm that instantiates to an efficient monitor for different events of interest and requires the user to only provide the formal specification of an event to the algorithm.

Table 2.1 shows a comparison between the three monitoring methods. Human monitoring is highly flexible and informative, but its scalability is extremely limited. Ad-hoc automated monitoring is highly scalable, but its engineering cost is high, especially if the monitored specification is complicated. Moreover, ad-hoc automated monitoring is usually not very informative. The engineering cost of formal automated monitoring is much lower than that of ad-hoc automated monitoring because the user does not have to develop a dedicated program but must only provide a formal specification. As this monitoring is automatically executed, its scalability is high. Research has shown that formal automated monitoring can be more informative than only detecting safety violations, which is presented in the later part of this chapter. We remark that such additional information does not increase the engineering cost of formal automated monitoring because the specification-generic monitoring algorithms can still be used.

2.1.5 About This Chapter

Monitoring-based on formal specifications (also called as "runtime verification") has attracted extensive research efforts, see e.g., [9, 10, 21]. The current chapter is too short to provide a systematic overview of the entire field.

Instead, in this chapter, we will discuss several different problem settings of formal monitoring. We present algorithms and prototype tools for solving those problems—many of which are results of our own—and show that the problems are tractable. The problem formalizations and algorithms are all based on *timed automata* [1], showing the impressive potential of the fundamental construct in theoretical computer science.

The problem settings differ in complexity and tractability—the more complex a setting is, the slower the algorithm. However, the additional complexity is the price for properties of monitoring that are highly desirable in its practical usage scenarios.

2.1.6 Organisation of This Chapter

In Section 2.2, we introduce *timed pattern matching* [22, 26] as the theoretical background of specification-based monitoring. In Sections 2.3 to 2.5, we show the extensions of timed pattern matching by *parametric* specifications (Section 2.3), events equipped with data (Section 2.4), and quantitative monitoring results (Section 2.5). These extensions endow timed pattern matching with extra flexibility, one that is often appreciated in real-world usage scenarios. In Section 2.6, we discuss the methodological difficulty in monitoring continuous behaviour with digital sampling where only discrete observations are available. To tackle this issue, we introduce *model-bounded monitoring*. The key idea is to exploit the prior knowledge of the system under monitoring to interpolate the behaviour between discrete samples. Finally, we conclude in Section 2.7.

2.2 Timed Pattern Matching

2.2.1 Motivation: Locating Unsafe Parts in an Execution Log

There have been various mathematical formulations of specification-based monitoring of CPSs, e.g., Boolean monitoring of discrete-time signals [16], Boolean monitoring of continuous-time signals [20], and quantitative monitoring of continuous-time and continuous-valued signals [13, 14]. Refer to [9] for an example of a survey of CPS monitoring.

The objective of most monitoring problems is to detect a safety violation, i.e., provide a Boolean value showing if the monitored system behaviour satisfies the given specification. This is useful in cases such as raising alarms on a safety violation. However, returning a Boolean value is not always informative enough. For example, when monitoring of a system for debugging, it is helpful to locate part of the log in which an unsafe behaviour is detected, to find which component requires a fix. In such a use case, *locations* of the erroneous part in the log will be needed rather than the Boolean satisfaction value.

2.2.2 Solution: Pattern Matching with the Specification

Timed pattern matching [22,26] is a monitoring problem to locate the satisfaction of the specification in the given log. More precisely, given a system log w and a specification $\mathcal{A}$, timed pattern matching provides the set of the intervals (t,t') for which the restriction $w|_{(t,t')}$ of the log w to the open interval (t,t') satisfies the specification $\mathcal{A}$. Typically, the specification $\mathcal{A}$ represents an unsafe behaviour, e.g., the gear of a vehicle changing too frequently. In this case, timed pattern matching finds when such an unsafe behaviour occurs in the given log.

After the introduction in [22], the timed pattern matching problem has been studied in various settings with technical improvements (e.g., [22, 23, 26, 30]). In [26], the timed pattern matching problem is defined with *timed words* and timed automata (TAs) [1]. A timed word is a sequence of events with timestamps. A TA is a nondeterministic finite automaton (NFA) equipped with clock variables to represent timing constraints on the timed words.

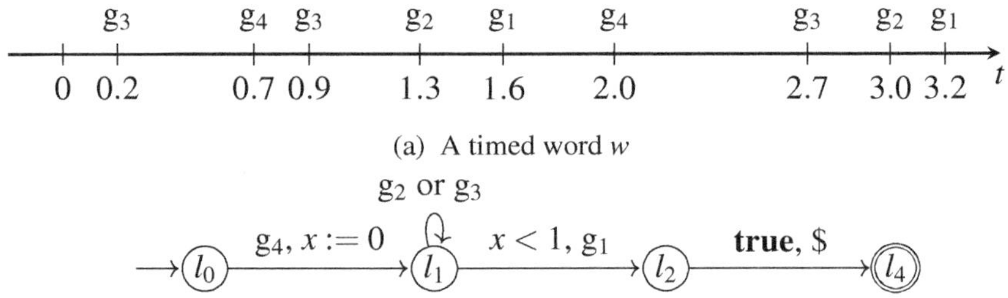

(b) A timed automaton $\mathcal{A}$

Figure 2.1: An example of timed pattern matching.

Example 3 (gear changes of a vehicle) *Consider the example in Fig. 2.1. The timed word w in Fig. 2.1a is the log of the gear changes of a vehicle. The objective of the monitoring with the TA $\mathcal{A}$ in Fig. 2.1b is to find too frequent gear changes. Roughly speaking, $\mathcal{A}$ matches if the gear changes from* g_4 *to* g_1 *within 1 time unit, which is too soon and undesirable. First,* g_4 *is at 0.7 and* g_1 *is at 1.6 in w, $\mathcal{A}$ matches around* $[0.7, 1.6]$. *Next,* g_4 *is at 2.0 and* g_1 *is at 3.2 in w, $\mathcal{A}$ does not match around* $[2.0, 3.2]$ *because the time difference is longer than 1.0.*

Formally, timed pattern matching returns the set of matching intervals $\{(t,t') \mid w|_{(t,t')} \in \mathcal{L}(\mathcal{A})\} = \{(t,t') \mid t \in [0.2, 0.7), t' \in (1.6, 2.0]\}$, *which is the set of open intervals (t,t') such that the restriction $w|_{(t,t')}$ of the timed word w in (t,t') is accepted by the TA $\mathcal{A}$. Here the notion "around* $[0.7, 1.6]$*" is formalised as the open intervals containing* $[0.7, 1.6]$ *but not including any other events, i.e., not including either* 0.2 *or* 2.0. *We note that \$ in the TA $\mathcal{A}$ is the special terminal character representing the end of the matching.*

2.2.3 Algorithms and Implementations

There are several algorithms proposed for the timed pattern matching problem, such as [8,22,23,26,30]. We discuss the one presented in [30].

In [30], we presented an automata-based algorithm for timed pattern matching. Roughly speaking, each event is fed in the given log to the timed automaton to check which subsequence of the given timed word is accepted by the timed automaton. Notably, an interval-based symbolic analysis is needed for the beginning and the end of the matching because the number of the matching is, in general, infinite, and is represented by intervals, as shown in Example 3. The algorithm in [30] is optimised using *skipping* [11,17,19] originally used for string search [15].

Using the algorithm shown in [30], a prototype tool MONAA [31] was implemented.[1] In MONAA, we can provide a specification in *timed regular expressions* [6] as well as timed automata.

Our experimental evaluation shows the efficiency of MONAA. For example, MONAA monitors approximately 250,000 entries per second for a realistic benchmark called GEAR on MacBook Pro 2019 (2.8 GHz Quad-Core Intel Core i7, and 16 GiB RAM). This is as big as the number of the search queries on Google per second and, for example, is fast enough to monitor controller area network (CAN) messages[2] in real time.

2.2.4 Summary

The problem of timed pattern matching, which is the most basic problem setting amongst the ones discussed in the current chapter, is considered in this section. Timed pattern matching is a monitoring problem locating all the unsafe behaviours in the given log. In [30], the problem is formulated with timed automata. There are efficient algorithms and tools for timed pattern matching such as MONAA [31]. Note that the above algorithm achieves "informativeness" in the following sense.

Informativeness All the unsafe behaviours in the monitored log can be located using timed pattern matching. This is more informative than the standard monitoring problem returning Boolean satisfaction, i.e., returning if the log satisfies the specification.

[1]MONAA is distributed at `https://github.com/MasWag/monaa` under GNU General Public License version 3.

[2]CAN bus is a widely used vehicle bus standard. CAN is originally used to connect devices in one vehicle, but it is also used in other application fields, e.g., robotics.

2.3 Parametric Timed Pattern Matching

2.3.1 Motivation: Specification with Unspecified Thresholds

All the occurrences of the unsafe behaviour in the monitored log can be located using timed pattern matching as discussed in Section 2.2. There is an algorithm that uses skipping to optimise this method. Also, the tool MONAA for timed pattern matching is highly scalable.

However, it is often practically challenging to fix a concrete specification. Notably, determining the exact *threshold* in the specification is challenging. For example, for an automatic transmission system of a vehicle, frequent gear changes are not desirable, and the following is a reasonable requirement: "the gear should not change *too soon* after the latest gear change". While providing a specification in a TA, the concrete threshold, that determines what "*too soon*" means for a gear change, needs to be fixed. However, it is not easy to provide a suitable threshold (e.g., 1 s, 300 ms, and so on) as it highly depends on the domain and requires an in-depth domain knowledge of the developed system. This issue can be a significant obstacle to timed pattern matching.

2.3.2 Solution: Pattern Matching with Parametric Specifications

Our solution to the aforementioned issue is to use *parameters for unspecified thresholds*. More precisely, we use *parametric timed automata (PTAs)* [3] for parametric specifications. Figure 2.2 shows an example of a parametric timed automaton (PTA). A PTA is a TA that allows the use of *parameters* and the constant numbers in the timing constraints. For instance, using the parameter p for the unspecified threshold, one can represent "the gear should not change *for* p *seconds* after the latest gear change" in a PTA.

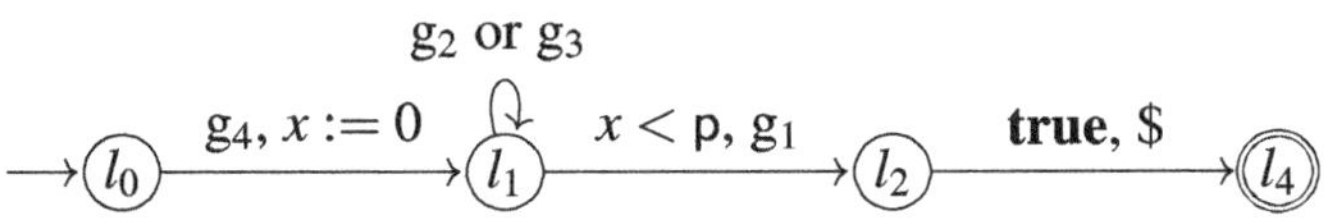

Figure 2.2: The PTA $\mathcal{A}$ such that the timing constraint $x < 1$ in Figure 2.1b is replaced with a parametric timing constraint $x < \mathsf{p}$.

Parametric timed pattern matching [5] is a generalisation of timed pattern matching where the specification is given in a PTA $\mathcal{A}$ rather than a TA. Given a timed word w and a PTA $\mathcal{A}$, it returns the part of the timed word w and the parameter valuation v, assigning a real-value to each parameter, the parametric specification represented by the PTA $\mathcal{A}$ is satisfied.

Example 4 (gear changes of a vehicle) *Consider the timed word w in Fig. 2.1a and the PTA $\mathcal{A}$ in Fig. 2.2. Here, the PTA $\mathcal{A}$ matches if the gear changes from*

g_4 *to* g_1 *within* p *time unit. As we have* g_4 *at 0.7 and* g_1 *at 1.6 in* w*,* $\mathcal{A}$ *matches around* $[0.7, 1.6]$ *for* $p > 0.9$*. As we have* g_4 *at 2.0 and* g_1 *at 3.2 in* w*,* $\mathcal{A}$ *matches around* $[2.0, 3.2]$ *for* $p > 1.2$*.*

Formally, the output of parametric timed pattern matching is as follows, where (t, t') *is a matching interval and* v *is one of the corresponding parameter valuations.*

$$\{(t,t',v) \mid w|_{(t,t')} \in \mathcal{L}(v(\mathcal{A}))\} = \{(t,t',v) \mid t \in [0.2, 0.7), t' \in (1.6, 2.0], v(\mathsf{p}) > 0.9\}$$
$$\cup \{(t,t',v) \mid t \in [1.6, 2.0), t' \in (3.2, \infty), v(\mathsf{p}) > 1.2\}$$

Figure 2.3: An automated cleaning system. Once the camera finds trash, it notifies a robot cleaner to remove the trash.

Example 5 (automated cleaning system) *Consider an automated cleaning system as shown in Fig. 2.3. There are cameras in a public space, and once a camera finds trash, it notifies a robot cleaner to remove the trash. A reasonable specification is that the robot should remove the trash as soon as the camera finds it. However, it is hard to define the exact threshold determining if the robot is too late because it depends on various factors, e.g., the size of the public space, how crowded it is, and the number of the working robots.*

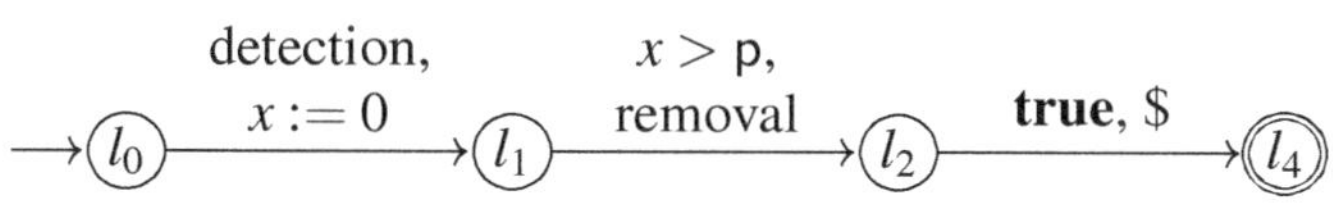

Figure 2.4: A PTA matching too late removal of the trash after its detection.

By using a PTA, this specification can be expressed as in Fig. 2.4. By parametric timed pattern matching, for each trash, the time between detection and removal of trash can be obtained. A global constraint of the parameter p*, to ignore the very short delay between the detection and the removal, can also be provided.*

2.3.3 Algorithms and Implementations

Parametric timed pattern matching is a relatively new problem, and not many algorithms have been proposed. Here we discuss our algorithms, presented in [5,27].

The two algorithms for parametric timed pattern matching have been proposed: an algorithm via model checking and a dedicated algorithm. Both methods are similar in the sense they utilise the finite abstraction of the clock and parameter valuations by *polyhedra*. The latter algorithm is optimised using *skipping* [17] that is originally from string matching. This optimisation is much like the algorithm for (non-parametric) timed pattern matching in [30].

The former algorithm is implemented with IMITATOR [4], which is a model checking tool for PTAs. Meanwhile, the latter is implemented in a prototype tool ParamMONAA.[3] The experimental evaluation in [27] shows the efficiency of these methods, especially of ParamMONAA. For example, ParamMONAA monitors approximately 35,000 entries per second for a realistic benchmark called GEAR and approximately 120,000 entries per second for a realistic benchmark called ACCEL on an Amazon EC2 c4.large instance (2.9 GHz Intel Xeon E5-2666 v3, 2 vCPUs, and 3.75 GiB RAM) that runs Ubuntu 18.04 LTS (64 bit). This is as fast as the Controller Area Network (CAN bus) or hundreds of high-speed sensors.

2.3.4 Summary

Parametric timed pattern matching is a parametric extension of timed pattern matching with parameters in timing constraints introduced in [5]. In [5], this problem is solved via model checking, and a more efficient algorithm is proposed in [28]. Note that the above algorithms achieve “flexibility” and “informativeness” in the following sense.

Flexibility Parameters allows the use of timing constraints with unspecified thresholds in the specification. This allows providing a specification without precisely defining timing constraints, which tends to be practically challenging.

Informativeness With parametric timed pattern matching, how robustly the timing constraints are satisfied (or violated) can be synthesised. This is more informative than returning Boolean satisfaction. The idea is to enlarge the timing constraints using parameters representing the robustness of the satisfaction. For example, by replacing the guard $x < 1$ of Fig. 2.1b with $x < 1 + \mathsf{p_x}$, where $\mathsf{p_x}$ is a parameter, the robust satisfaction degree of $x < 1$ can be synthesised.

[3] ParamMONAA is distributed at `https://github.com/MasWag/ParamMONAA` under GNU General Public License version 3.

2.4 Symbolic Monitoring against Specifications Parametric in Time and Data

2.4.1 Motivation: Beyond Finite Domain Events

By parametric timed pattern matching in Section 2.3, monitoring can be conducted even if the timing constraints in the specification contain unspecified thresholds, represented by parameters. Its essence is the use of PTAs instead of TAs.

In parametric timed pattern matching, the monitored log is a timed word, which is the same as timed pattern matching. A timed word is a sequence of events with timestamps, where the events are from a *finite* domain. However, this finiteness of the events is often too restrictive to represent behaviour of a system with *infinite* or *unbounded* nature.

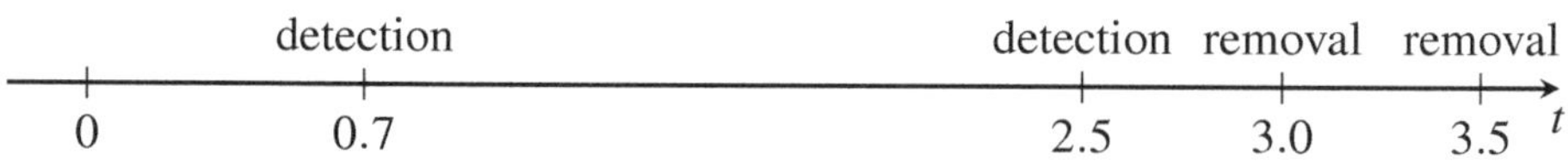

(a) The timed word over the alphabet {detection, removal} modelling the observation in Example 6

(b) The timed data word modelling the observation in Example 6. Each event (detection or removal) is equipped with an identifying string (#1 or #2).

Figure 2.5: The timed word and the timed data word modelling the observation in Example 6 of the automatic cleaning system.

Example 6 (automated cleaning system) *Consider again the scenario in Example 5. In Example 5, the observed actions are abstracted into two events: detection and removal. This abstraction is too coarse when there are multiple pairs of* detection *and* removal. *Consider the following observation.*

- *The camera observed trash (say #1) at 0.7.*
- *The camera observed another trash (say #2) at 2.5.*
- *The cleaning robot removed #2 at 3.0.*
- *The cleaning robot removed #1 at 3.5.*

Figure 2.5a shows the timed word modelling this observation using the events in Example 5. The issue here is that the information on the correspondence between the detection and the removal is abstracted away, and it cannot be monitored in the preferred manner. One possible solution is to equip an identifying string for each event. However, the technical issue is that the number of such strings is unbounded *owing to the unbounded number of observed trash.*

```
@1.5     deposit     100
@3.8     withdraw     50
@4.2     withdraw     30
@6.7     withdraw     80
@8.9     deposit      30
```

Figure 2.6: Example transactions of a bank account.

Example 7 (bank account) *Consider the transactions of a bank account shown in Fig. 2.6. It can be modelled as a timed word by abstracting each transaction to either* withdraw *or* deposit. *However, this abstraction is extremely coarse because the amount is essential in modelling the transactions; without the amount of deposit or withdrawal, it is not possible to monitor, e.g., the extreme decrease in the balance in a short period. The amount of each withdrawal or deposit is a number, which is obtained from an infinite domain.*

Overall, timed words are often not expressive enough, and thus each event needs to be equipped with data from some infinite domain.

2.4.2 Solution: Symbolic Monitoring of Infinite Domain Data

To overcome the above issue of the expressive power of the timed words, *symbolic monitoring* of infinite domain data is introduced in [28]. In the formulation in [28], the log is provided by a *timed data word*, and the specification is provided by a *parametric timed data automaton (PTDA)*.

A timed data word is a timed word such that each event is equipped with infinite domain data as well as a timestamp. For example, Fig. 2.5b shows an example of a timed data word for the automated clearing system; here, #1 and #2 are strings to identify each trash.

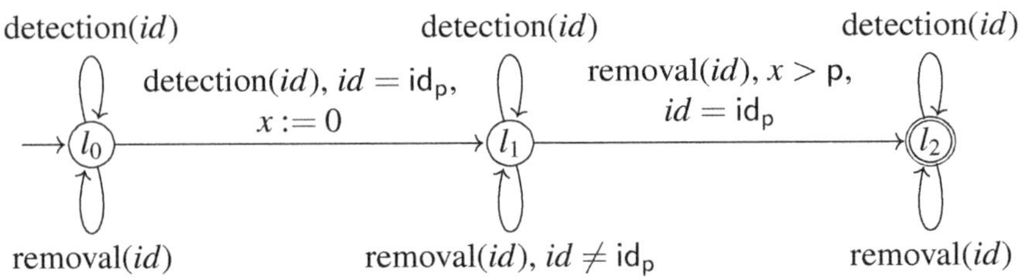

Figure 2.7: A PTDA accepting a timed data word containing a removal of trash that is more than p time units after the detection, where $\mathsf{id_p}$ is the data parameter for the identifying string of the trash, and p is the timing parameter.

PTDA is an extension of PTA with infinite domain data. It can represent constraints on data using data variables, data guards, and variable updates. Parameters can be used for data as well as the timing constraints. For example, Fig. 2.7 shows an example of the PTDA for the automated clearing system with timing parameter p and data parameter $\mathsf{id_p}$. This PTDA accepts a timed data word containing a removal of trash that is more than p time units after the detection.

Given a timed data word and a PTDA, the symbolic monitoring algorithm in [28] returns the set of parameter valuations such that the PTDA accepts the timed data word.

Example 8 (automated cleaning system) *Let us continue the scenario in Example 6. The observation in Example 6 contains two pairs of the detection and the removal of trash; the first trash (say #1) is detected at time 0.7 and removed at time 3.5; the second trash (say #2) is detected at time 2.5 and removed at time 3.0. The removal of the trash* $\mathsf{id_p}$ *more than* p *time unit following the detection is alerted, where* $\mathsf{id_p}$ *is the string-valued data parameter to identify the trash, and* p *is the timing parameter for the timing constraint. For the above observation, an alert exists for* $\mathsf{id_p} = \#1 \land \mathsf{p} < 2.8$ *and* $\mathsf{id_p} = \#2 \land \mathsf{p} < 0.5$.

Formally, these observation and specification are represented by the timed data word w *in Fig. 2.5b and the PTDA* $\mathcal{A}$ *in Fig. 2.7, respectively. The result of the symbolic monitoring for* w *and* $\mathcal{A}$ *is the following set of parameter valuations* v*, where* $v(\mathsf{id_p})$ *is a string and* $v(\mathsf{p})$ *is a rational number.*

$$\{v \mid v(\mathsf{id_p}) = \#1, v(\mathsf{p}) < 2.8)\} \cup \{v \mid v(\mathsf{id_p}) = \#2, v(\mathsf{p}) < 0.5)\}$$

Example 9 (vehicle platooning) *Consider monitoring of a* vehicle platooning *[18] system. Vehicle platooning is a style of self-driving in which a group of vehicles drive in a line. Owing to the inter-vehicle communication, vehicle platooning can reduce the inter-vehicle distance and the brake usage, and thus, improve the efficiency of the road usage and the energy consumption.*

Figure 2.8: Joining process of a vehicle platoon. When a non-member truck sends a joining request (left), the member trucks increase the inter-vehicle distance so that the non-member truck can join. Once the inter-vehicle distance becomes large enough, one of the member truck sends the agreement to the joining truck (center), and the truck joins the platoon (right).

The goal is to monitor a joining process to a platoon as shown in Fig. 2.8 and check if the acceptance of the joining request is too late. In the log in Fig. 2.9, it can be observed that the acceptance of the vehicle #B takes much longer time than #A. The PTDA $\mathcal{A}$ *in Fig. 2.10 can be used to detect such delay in acceptance. The result of the symbolic monitoring is as follows.*

$$\{v \mid (v(\mathsf{p_{id}}), v(\mathsf{delay})) = (\#A, 4.3) \text{ or } (\#B, 11.8)\}$$

```
@0.5      request    #A    #2
@4.8      accept     #A
@8.5      request    #B    #4
@20.3     accept     #B
```

Figure 2.9: Example communication in joining processes of a vehicle platoon. The request event with two strings, *id* and *position* is for the joining request by the vehicle *id* at *position*. The accept event with one string *id* is for the acceptance of the joining by the vehicle *id*. Infinite domain data is needed for identification because the number of the vehicles in a platoon is unbounded.

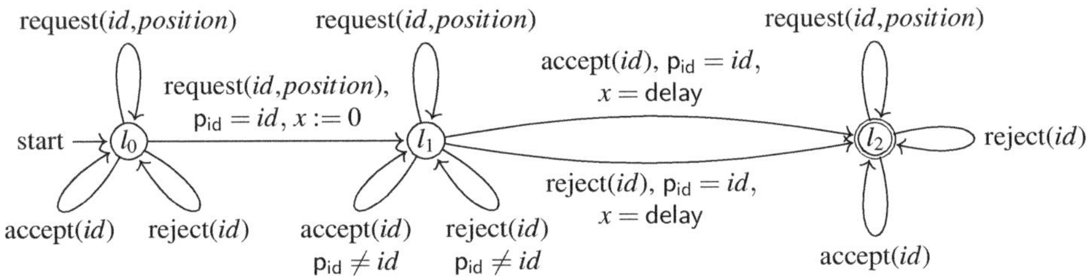

Figure 2.10: A PTDA to detect late acceptance or rejection of the joining request to platoon, where x is the clock variable for the duration after the request, delay is the timing parameter for the delay of the response, and p_{id} is the string-valued data parameter to identify the joining vehicle.

In principle, the symbolic monitoring problem as formulated in [28] does not answer where unsafe behaviour is detected, but it only answers the acceptance with the corresponding parameter valuations. However, it is possible to locate the part of the log in which the unsafe behaviour occurs, using timing parameters. In Example 10, a simpler pattern matching problem than (parametric) timed pattern matching is shown. See [28] for the encoding of parametric timed pattern matching with symbolic monitoring.

Example 10 (bank account) *For the scenario in Example 7, the goal is to locate in which part of the log, the balance decreases by more than 100 in total. In the log in Fig. 2.6, it can be observed that such a decrease happens between 3.8 and 6.7 and between 3.8 and 8.9.*

The PTDA $\mathcal{A}$ in Fig. 2.11 can be used to locate such a decrease. The PTDA $\mathcal{A}$ contains two timing parameters t *and* t' *for the interval* $[\mathsf{t},\mathsf{t}']$ *where such a decrease happens. In $\mathcal{A}$, the summation of the amount of the withdrawals and deposits between* t *and* t' *is considered. The result of the symbolic monitoring is as follows.*

$$\{v \mid (v(\mathsf{t}), v(\mathsf{t}')) = (3.8, 6.7) \text{ or } (3.8, 8.9)\}$$

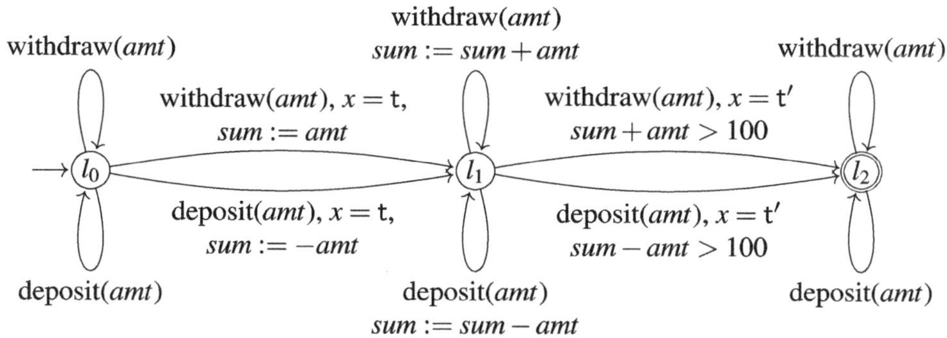

Figure 2.11: A PTDA to locate the interval $[t, t']$ in the log where the balance decreases by more than 100 in total with at least two transactions, where x is the clock variable for the global time, t and t' are timing parameters, and *sum* is the variable to contain how much the balance decreases.

2.4.3 Algorithm and Implementation

In [28], we gave an algorithm to solve the symbolic monitoring problem with infinite domain data. Owing to the finite abstraction of the time and data valuations, e.g., by convex polyhedra for rational-valued data and an ad-hoc data structure based on a list for string-valued data, the main body of the algorithm is a breadth-first search.

SYMON[4] is a prototype tool for symbolic monitoring with infinite domain data. The experimental evaluation in [28] shows the efficiency of SYMON. For example, SYMON monitors approximately 130 entries per second for a benchmark to detect a dominant withdrawal by one user called DOMINANT and approximately 250 entries per second for another benchmark to detect a series of periodic withdrawals called PERIODIC on an Amazon EC2 c4.large instance (2.9 GHz Intel Xeon E5-2666 v3, 2 vCPUs, and 3.75 GiB RAM) that runs Ubuntu 18.04 LTS (64 bit). This is fast enough for online monitoring using 10 sensors with sampling interval of 100 ms.

2.4.4 Summary

The symbolic monitoring problem with infinite domain data [28] is a generalisation of parametric timed pattern matching with data from some infinite domain. In [28], the problem is formulated using timed data words and PTDAs. The experimental results in [28] show the efficacy of the algorithm. Note that the above algorithm achieves "genericity", "flexibility", and "informativeness" in the following sense.

[4]SYMON is distributed at `https://github.com/maswag/symon` under GNU General Public License version 3.

Genericity The symbolic monitoring algorithm in [28] works for any "data" from a domain that offers suitable operations. For example, the algorithm works for both rationals and strings.

Flexibility Data and timing parameters introduced in Section 2.3 can be used. In this sense, the specification in symbolic monitoring is more flexible.

Informativeness Similar to Section 2.3, the safety degree can be synthesised using the parameters. Moreover, timing and data information can be extracted from the log. For example, in Example 10, by monitoring the log, the intervals of the withdrawal were extracted for each range of amounts.

2.5 Online Quantitative Timed Pattern Matching with Semiring-Valued Weighted Automata

2.5.1 Motivation: Online Algorithm for Quantitative Timed Pattern Matching

2.5.1.1 More Informative Timed Pattern Matching

Consider the monitoring of a driving record of an automated driving system (ADS). For example, the goal is to locate where in the driving record the acceleration is too sudden. More precisely, there should be an alert if the velocity v changes from $v < 40$ to $v > 100$ within 3 time units. For the signal of the velocity in Fig. 2.12, such a pattern appears in $[0,2.5)$ and $[6.0,8.7)$. Timed pattern matching in Section 2.2 locates such time intervals.

However, it can be observed that this result is not informative enough to provide the following information: the acceleration is more sudden in $[0,2.5)$ than in $[6.0,8.7)$; the acceleration in $[3.3,5.3)$ is more sudden than in $[9.3,10.8)$, even though both are not too sudden according to the specification. To provide a more fine-grained and *informative* answer, for each area, the *degree of unsafety* is required rather than the *Boolean verdict* of unsafety.

Quantitative timed pattern matching [7] is a quantitative extension of timed pattern matching, which returns quantitative semantics representing the satisfaction degree of the given specification rather than the Boolean satisfaction. For each subsignal, quantitative timed pattern matching returns how robustly the given specification is satisfied or violated. When the specification represents unsafe behaviour, the degree of unsafety for each subsignal is obtained.

Example 11 *For the signal σ in Fig. 2.12, the quantitative semantics $\alpha(\sigma([0,2.5)))$ for the subsignal $\sigma([0,2.5))$ in $[0,2.5)$ is 20 because the*

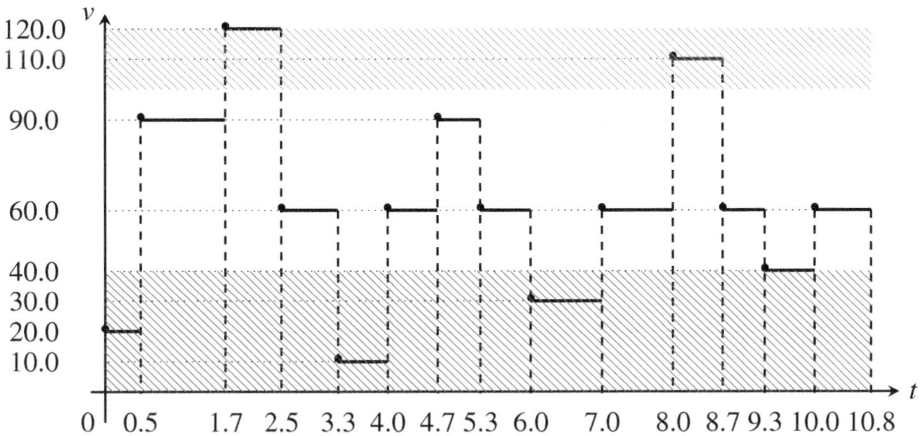

Figure 2.12: A signal σ showing the velocity of a vehicle. A subsignal is unsafe if: (1) it starts in the red area (bottom); (2) it ends in the orange (top) area; (3) its duration is shorter than 3.0 time units.

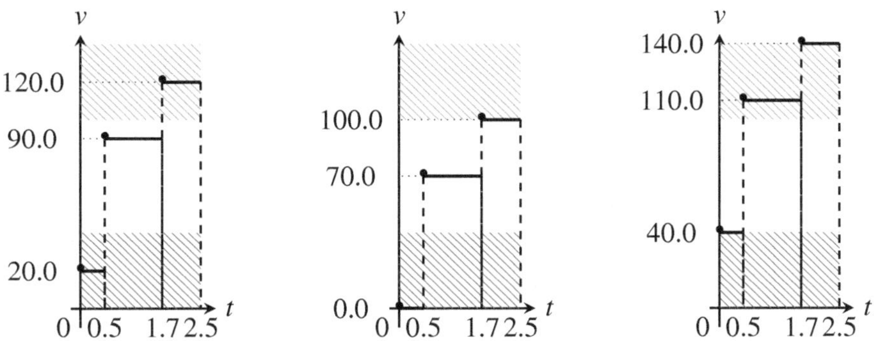

Figure 2.13: An illustration of the quantitative semantics. A signal satisfies the specification if it starts in the red area and ends in the orange area. As this specification represents unsafe behaviour, the quantitative semantics demonstrates an unsafety degree. The quantitative semantics of the specification for the original signal (left) is 20 because it keeps satisfying the specification even after decreasing (centre) or increasing (right) the time-varying values by a gap up to 20.

inequality constraints ($v < 40$ and $v > 100$) are satisfied even if the signal values are shifted up or down, by a gap up to 20. Refer to Fig. 2.13 for illustration. As the quantitative semantics $\alpha(\sigma([6.0, 8.7)))$ for the subsignal $\sigma([6.0, 8.7))$ is 10, $\sigma([0, 2.5))$ satisfies the specification more robustly than $\sigma([6.0, 8.7))$. As the specification represents unsafe behaviour, it can be concluded that σ is safer in $[6.0, 8.7)$ than in $[0, 2.5)$. Similarly, the quantitative semantics $\alpha(\sigma([3.3, 5.3)))$ and $\alpha(\sigma([9.3, 10.8)))$ for the subsignals

$\sigma([3.3,5.3))$ and $\sigma([9.3,10.8))$ are -10 and -40, respectively; therefore, $\sigma([9.3,10.8))$ is safer than $\sigma([3.3,5.3))$.

2.5.1.2 Issue: Online Monitoring Capability

In [7], an *offline* algorithm for quantitative timed pattern matching is introduced. This algorithm requires the entirety of the monitored signal to start monitoring. Such a monitoring algorithm is *offline* monitoring capable for a log of previous system execution, and it can compute the quantitative semantics for each subsignal. However, such a monitoring algorithm is not *online* monitoring capable for a stream of a log of the running system, and it cannot compute the quantitative semantics along with the system execution. This limitation is critical for certain usage scenarios. For example, an offline monitoring algorithm cannot give feedback to a running system using the monitoring result before the execution is complete. The introduction of an online algorithm is valuable because it expands the application area.

2.5.2 Solution: Quantitative Timed Pattern Matching with Weighted Automata

2.5.2.1 Online Algorithm with Automata

In [24], we introduced an online algorithm for quantitative timed pattern matching. The algorithm in [24] utilises *weighted automata*, which is a quantitative extension of NFAs. In general, the use of automata simplifies an online monitoring algorithm.

By combining the concepts from both timed automata and weighted automata, we introduced *timed symbolic weighted automata (TSWAs)*. Fig. 2.14 shows an example of a TSWA. In a TSWA, each location has a constraint on the signal values. The other structure of TSWAs is the same as TAs.

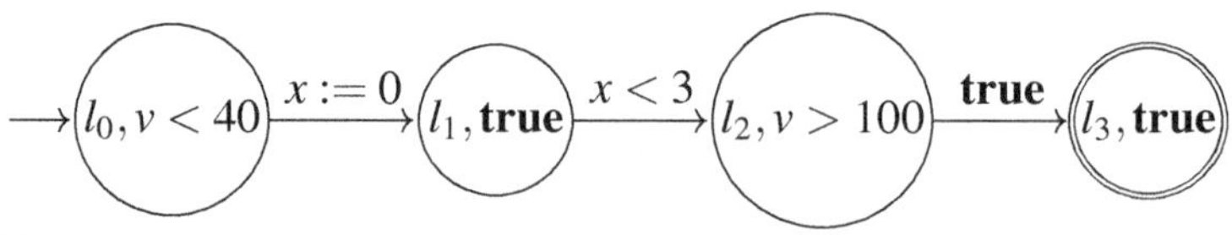

Figure 2.14: The TSWA showing that the velocity v changes from $v < 40$ to $v > 100$ within 3 time units. More precisely, at location l_0 and l_2, the constraints $v < 40$ and $v > 100$ are required to stay at the location, respectively; l_2 should be entered within 3 time units after leaving l_0. The transition from l_2 to l_3 shows that are no constraint on the time elapse in l_2.

Example 12 (velocity monitoring) *Continuing with the scenario in Example 11, let σ be the signal shown in Fig. 2.12. The TSWA in Fig. 2.14* *represents the specification "the velocity v changes from* $v < 40$ *to* $v > 100$ *within 3 time units". Note that the constraints at locations* l_1 *and* l_3 *are* **true**, *and the associated signal values are ignored.*

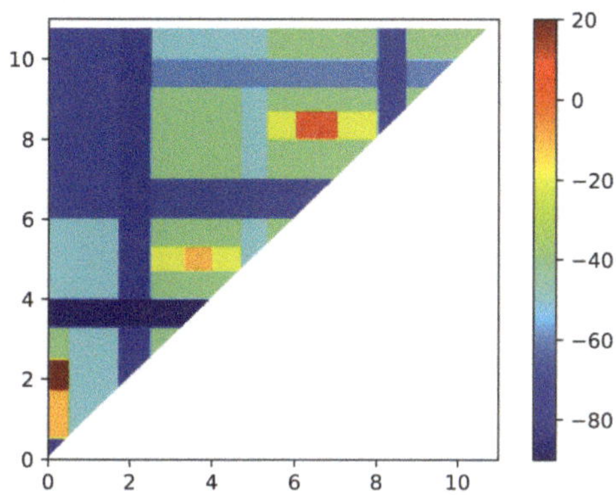

Figure 2.15: The result of the quantitative timed pattern matching. Each point (t,t') shows the quantitative semantics $\alpha(\sigma([t,t')))$ for the subsignal $\sigma([t,t'))$ of the input signal σ in the interval $[t,t')$. For example, the value at $(3.3,5.3)$ is -10, which shows that the quantitative semantics $\alpha(\sigma([3.3,5.3)))$ for the subsignal $\sigma([3.3,5.3))$ in the interval $[3.3,5.3)$ is -10.

For the subsignal $\sigma([0,2.5))$ *of* σ *in* $[0,2.5)$, *the quantitative semantics* $\alpha(\sigma([0,2.5)))$ *is 20 because of Equation* (2.1), *where* $\sigma(0)$ *and* $\sigma(1.7)$ *are the signal values at* 0 *and* 1.7, *respectively, and 40 and 100 are the thresholds in the specification. This follows the intuition in Fig. 2.13.*

$$\alpha(\sigma([0,2.5))) = \min\{(40-\sigma(0)),(\sigma(1.7)-100)\} = \min\{(40-20),(120-100)\} = 20 \tag{2.1}$$

Quantitative timed pattern matching provides such a quantitative semantics for each subsignal. Fig. 2.15 illustrates the result of quantitative timed pattern matching, where the colour at point (t,t') *shows the quantitative semantics* $\alpha(\sigma([t,t')))$ *for the subsignal* $\sigma([t,t'))$.

2.5.2.2 Genericity via Semirings

The monitoring algorithm in [24] is not just an *online* algorithm for quantitative timed pattern matching but also *generic* for various quantitative semantics. We utilise an algebraic structure called semiring in our algorithm, and our algorithm works for various quantitative semantics, i.e., various choices of semirings. For example, the summation of the difference from the threshold, rather than the minimum value as in Equation (2.1), can be considered. Mathematically, this is to use the semantics $\alpha_{\text{sup-+}}$ based on *sup-plus* semiring instead of the semantics α based on the *sup-inf* semiring.

The preference between these semantics depends on the usage scenario. For example, as shown in Example 13, when the quantitative semantics $\alpha_{\text{sup-+}}$ is used with the sup-plus semiring, the "safe" values compensate for the "unsafe" ones. This semantics may be an option, e.g., for cost or energy consumption, but not suitable, e.g., for physical distance, because while the debt can be covered with revenue, very long inter-vehicular distance cannot compensate for a collision of vehicles. See, for example [12] for an example usage of sup-plus semiring to capture the notion of cost. The quantitative semantics by the sup-inf semantics is useful in many settings owing to the intuition in Fig. 2.13.

Example 13 (quantitative semantics by summation) *For the scenario in Example 12, the quantitative semantics $\alpha_{\text{sup-+}}$ defined by sup-plus semiring instead of sup-inf semiring is used. For the subsignals $\sigma([0,2.5))$, $\sigma([3.3,5.3))$, $\sigma([6.0,8.7))$, and $\sigma([9.3,10.8))$, the quantitative semantics are as follows.*

$$\begin{aligned}
\alpha_{\text{sup-+}}(\sigma([0,2.5))) &= (40-20)+(120-100) = 40\\
\alpha_{\text{sup-+}}(\sigma([3.3,5.3))) &= (40-10)+(90-100) = 20\\
\alpha_{\text{sup-+}}(\sigma([6.0,8.7))) &= (40-30)+(110-100) = 20\\
\alpha_{\text{sup-+}}(\sigma([9.3,10.8))) &= (40-40)+(60-100) = -40
\end{aligned}$$

For the quantitative semantics defined by sup-plus semiring, we have $\alpha_{\text{sup-+}}(\sigma([3.3,5.3))) = \alpha_{\text{sup-+}}(\sigma([6.0,8.7)))$ while for the quantitative semantics defined by sup-inf semiring, $\alpha(\sigma([6.0,8.7)))$ is larger than $\alpha(\sigma([3.3,5.3)))$. This inconsistency is because, in the quantitative semantics $\alpha_{\text{sup-+}}$ by sup-plus semiring, small deviation from the specification by the velocity ($v = 10$) at the beginning of $\sigma([6.0,8.7))$ affects the quantitative semantics, while in the quantitative semantics α by sup-inf semiring, large deviation by the velocity ($v = 90$) masks it.

2.5.3 Generic Algorithm and Implementation

In [24], we provided an online algorithm for quantitative timed pattern matching using the shortest distance over (semiring-valued) weighted graphs. It works for any semiring satisfying a certain condition. For example, it captures both quantitative semantics aforementioned. Polyhedra-based abstraction was used when reducing the size of the weighted graph to finite. In signal monitoring, there are continuously many possible timings of the switching in a temporal specification owing to the continuity of the signals. For example, there are continuously many possible durations to stay at l_0 in the automaton in Fig. 2.14. Polyhedra was used to obtain finite abstraction of such possibilities and reduce the size of the weighted graph finite so that the existing shortest distance algorithm can be applied.

The implementation of the algorithm in [24] for online quantitative timed pattern matching is called QTPM. The experimental evaluation in [24] shows the efficiency of QTPM, especially when the specification has a time-bound, i.e., the length of each matching is bounded. For example, for a realistic benchmark called OVERSHOOT with time-bound of length 150 s, when the sampling interval is 10 s, QTPM monitors more than 3,000 entries, i.e., 30,000 s in the simulation time, per second on an Amazon EC2 c4.large instance (2.9 GHz Intel Xeon E5-2666 v3, 2 vCPUs, and 3.75 GiB RAM) that runs Ubuntu 18.04 LTS (64 bit). This assumption on the time-bound is not too restrictive because too long matching is usually not of interest. For instance, for the automaton in Fig. 2.14, although the length of the matching is unbounded, matching longer than 3 time unit is usually not of interest. This is because it is the maximum duration between $v < 40$ and $v > 100$. Therefore, it does not harm the practical usefulness to assume such a bound of the length of the matching.

In addition to the number of the monitored entries, we also measured the shortest feasible sampling interval of QTPM. In signal monitoring, the shortest feasible sampling interval is an important efficiency criterion. This is because by shortening the sampling interval, the monitoring result can be made more accurate. But it does make monitoring more computationally demanding because the algorithm must now handle more information in a certain length of time. For the aforementioned benchmark OVERSHOOT, the shortest feasible sampling interval of QTPM is approximately 450 ms on an Amazon EC2 c4.large instance (2.9 GHz Intel Xeon E5-2666 v3, 2 vCPUs, and 3.75 GiB RAM) that runs Ubuntu 18.04 LTS (64 bit). Although this is not fast enough to monitor the raw sensor values of sampling interval 100 ms, the bound of matching length (150 s) is much longer than the available sampling intervals, and the performance of QTPM may be enough for some applications.

2.5.4 Summary

Quantitative timed pattern matching is a quantitative extension of timed pattern matching, computing the quantitative semantics rather than the Boolean satisfaction. In [24], the quantitative timed pattern matching problem is formulated with TSWAs. The experimental results in [24] show the efficiency of QTPM, which is an implementation of the online algorithm in [24]. We used polyhedra-based symbolic analysis to obtain a finite abstraction of the weighted graph used in the algorithm. Note that the above algorithm achieves "genericity" and "informativeness" in the following sense.

Genericity The quantitative semantics is parameterised with a semiring. The algorithm works for different semirings provided they satisfy some conditions.

Informativeness For each subsignal, quantitative timed pattern matching computes the unsafe degree rather than deciding the Boolean satisfaction.

2.6 Model-Bounded Monitoring of Hybrid Systems

2.6.1 Motivation: Bounding Possible Signals between Samples

2.6.1.1 Signal Monitoring Requires Sampling and Interpolation

In the monitoring of CPSs, the monitored signal is thought of as the system behaviour. For example, in Section 2.5, monitoring the signal σ shown in Fig. 2.12, the quantitative semantics for each subsignal of σ is computed. However, in reality, what is observed is not a *continuous signal* but a series of *discrete sampling points*, which are interpreted as a continuous signal by *interpolation*. For example, in Fig. 2.12, the left end points of the constant pieces of a signal are observed, which are then interpreted as a signal by the piecewise-constant interpolation. Although, the piecewise-constant or piecewise-linear interpolations (see Fig. 2.16) are the commonly used, other interpolations, e.g., with spline curves, can also be used.

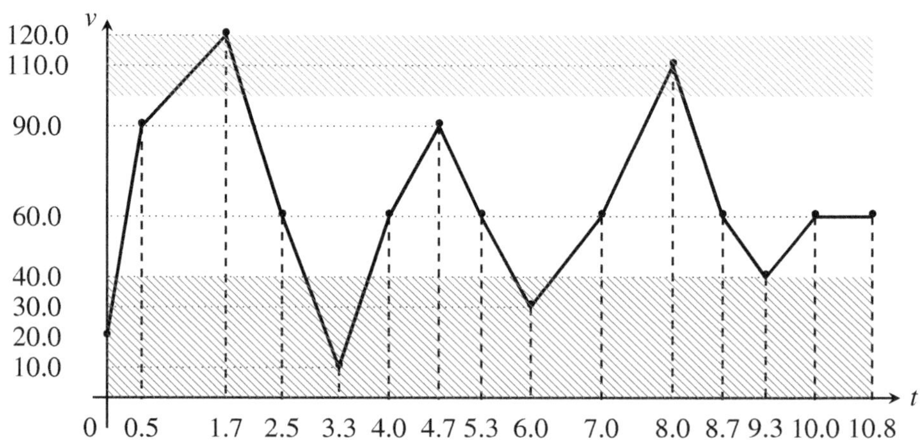

Figure 2.16: The signal constructed by the piecewise-linear interpolation from the samples in Fig. 2.12.

2.6.1.2 Issue: No Guarantee between Samples

In signal monitoring, due to the sampling and interpolation mechanism aforementioned, there is a methodological difficulty in guaranteeing the correctness of the analysis. Namely, the safety of the system between samples cannot be determined.

This issue can be ignored when the sampling intervals are short, implicitly assuming that the system state does not change a lot in a short period. However, this can be a big issue when the sampling interval is longer, which is often the case for an IoT-based monitoring system to reduce power consumption.

2.6.2 Solution: Model-Bounded Monitoring for Flexible Interpolation

2.6.2.1 Idea: Exploit the Knowledge of the Monitored System

Our solution is to bound the unknown behaviour between samples using the prior knowledge of the monitored system. The latter knowledge is expressed by what is called a *bounding model*, which is an abstraction of the monitored system. Utilising the bounding model, the behaviour between samples is bounded and it is determined if there is a possibility of unsafe behaviour between samples. This procedure is called *model-bounded monitoring*.

Linear hybrid automata (LHAs) is used as the formalism for bounding models. The use of LHAs is due to the following observations: While LHAs are expressive, e.g., it allows discrete switching as well as bounding the derivative by the Lipschitz constant; they form a limited subclass of hybrid automata, and their reachability analysis is tractable using convex polyhedra analysis.

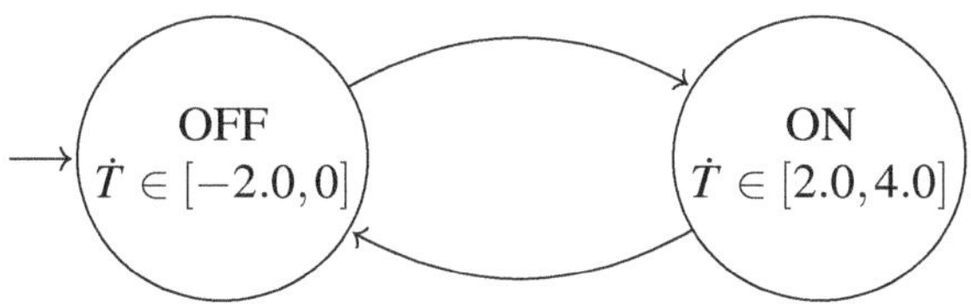

Figure 2.17: A bounding model of the thermostatic chamber in LHA. These intervals can be obtained, e.g., from the Lipschitz constants of the actual dynamics.

Example 14 (thermostatic chamber) *Consider the monitoring of a thermostatic chamber. The thermostatic chamber has two modes, on and off, and at each location, the rate of the temperature change satisfies* $\dot{T} \in [2.0, 4.0]$ *and* $\dot{T} \in [-2.0, 0]$, *respectively. The actual dynamics can be much more complicated, but it is assumed that these intervals bound the actual derivatives of the temperature under the current condition, e.g., the output of the heater, range of the temperature, amount of the water, and room temperature. The LHA in Fig. 2.17 models such a thermostatic chamber.*

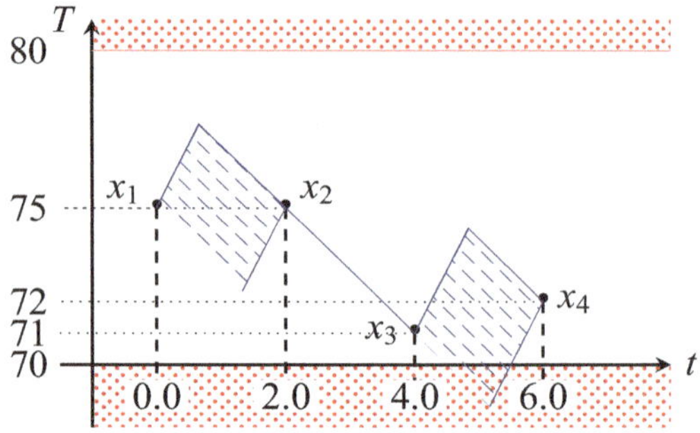

Figure 2.18: A signal of a thermostatic chamber: the black points $(x_1,\ldots,x_4)$ are the samples; the (red) dotted areas are the unsafe areas; the (blue) hatched areas are the reachable areas with the bounding model in FIg. 2.17.

Figure 2.18 illustrates the model-bounded monitoring procedure. The goal is to alert if the temperature enters the unsafe areas, which are shown as red in Fig. 2.18. It is observed that all the samples (the black points $x_1,\ldots,x_4$) are in the safe area. For each pair (x_{i-1},x_i) of samples, the reachable area (the blue areas) is computed and the temperature moves from the previous sample x_{i-1} to the next sample x_i through this. The reachable area is represented by a finite union of convex polyhedra. Here, it is observed that the (blue) hatched and (red) dotted areas intersect in between x_3 and x_4, implying that the unsafe area might be visited. Therefore, there is an alert about a potential safety violation, although no violation is observed in the samples.

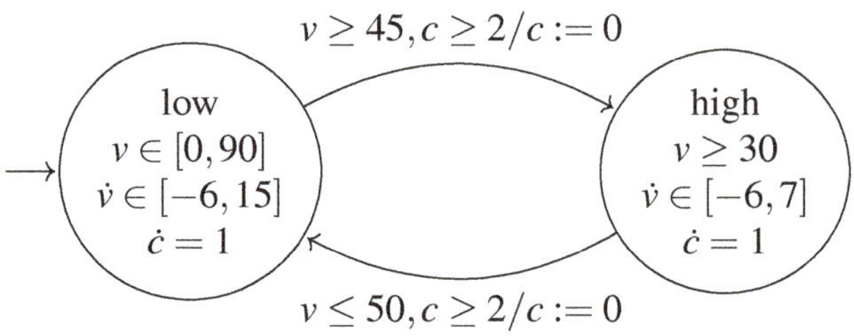

Figure 2.19: A bounding model of an automatic transmission system in LHA. These intervals are e.g., the Lipschitz constants of the actual derivative.

Example 15 (automatic transmission system) *Consider the monitoring of a vehicle. The vehicle has two gear modes, low and high, and at each location, the acceleration of the vehicle satisfies $\dot{v} \in [-6,15]$ and $\dot{v} \in [-6,7]$, respectively. At low and high gear modes, the velocity must be in $v \in [0,90]$ and $v \geq 30$, respectively. The gear can change from low to high and high to low when $v \geq 45$ and $v \leq 50$, respectively. There must be at least 2 time units between each gear change for the comfort. The LHA in Fig. 2.19 models such a vehicle.*

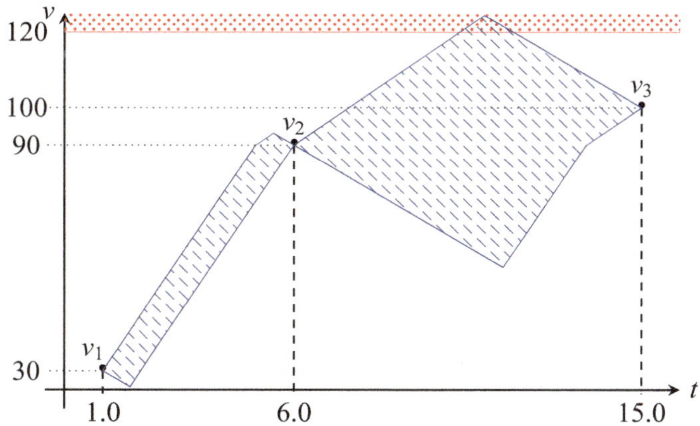

Figure 2.20: A signal showing the velocity of a vehicle: the black points ($v_1, \ldots, v_3$) are the samples; the (red) dotted area is the unsafe area; the (blue) hatched areas are the reachable areas.

Figure 2.20 illustrates the model-bounded monitoring procedure. The goal is to alert if the velocity is too high, i.e., $v \geq 120$*; the red dotted area in Fig. 2.20 shows the unsafe area. It is observed that the reachable area between* v_2 *and* v_3 *is non-convex. This is because the maximum acceleration is different between the low velocity and the high velocity. Nevertheless, the reachable area is represented by a finite union of convex polyhedra, and the symbolic analysis of LHAs with convex polyhedra can be used. Even though all the samples (the black points* $v_1, \ldots, v_3$*) are in the safe area, it is alerted that there is a potential safety violation because an intersection of (red) dotted and (blue) hatched areas is observed between* v_2 *and* v_3*.*

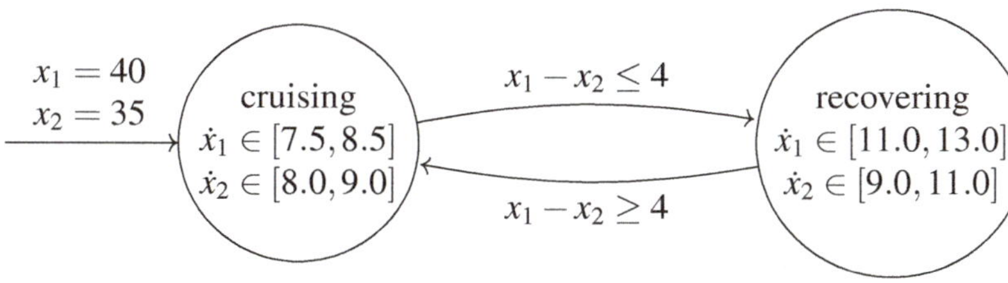

Figure 2.21: A bounding model $\mathcal{A}$ for the platooning example, expressed as an LHA.

Example 16 (vehicle platooning) *Consider a* vehicle platooning *scenario where two vehicles drive one after the other, with small distance between them. As mentioned in Example 9, such vehicle platooning attracts interest as a measure for enhanced road capacity as well as for fuel efficiency.*

The vehicle platooning system has two modes, cruising *and* recovering*. In the normal cruising mode, the relative distance has a drift and the inter-vehicle distance decreases. The recovering mode is used to increase the inter-vehicle*

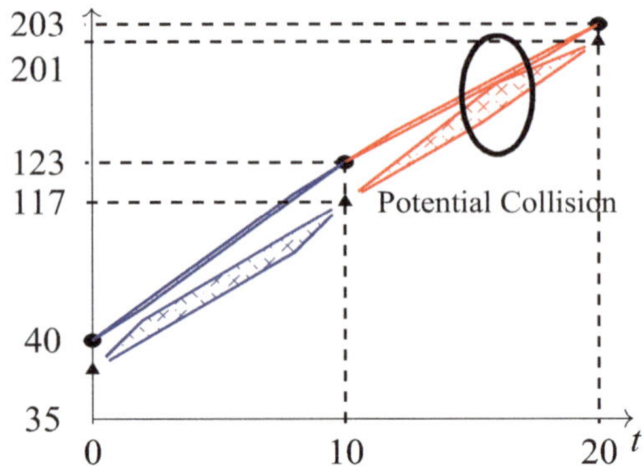

Figure 2.22: A signal of an automotive platooning system: the black circles are the samples of the first vehicle; the black triangles are the samples of the second vehicle; the hatched areas are the reachable areas. There may be a collision in the area where the hatched areas are intersecting.

distance. At each location, the velocity of each vehicle is $\dot{x}_1 \in [7.5, 8.5]$ *and* $\dot{x}_2 \in [8.0, 9.0]$*; and* $\dot{x}_1 \in [11.0, 13.0]$ *and* $\dot{x}_2 \in [9.0, 11.0]$*, respectively. The LHA in Fig. 2.21 models such a vehicle platooning system.*

Figure 2.22 illustrates the model-bounded monitoring procedure. The goal is to alert if the two vehicles could potentially touch each other. Physical contact of the vehicles is not observed in Fig. 2.22, but the same cannot be assured between the sampling instants. The hatched areas show the reachable area i.e., the symbolic interpolation using the bounding model.

Here, in the first interpolation (the blue areas), there is no potential physical contact, whereas in the second interpolation (the red areas), there is a potential physical contact. Therefore, there is an alert about a potential safety violation, even though no violation is observed in the samples.

2.6.2.2 On the Usage of the Model

One might say that the application area of model-bounded monitoring is smaller than the other monitoring methods because it requires a model. It certainly requires an LHA as a bounding model, which is much like a system model for the exhaustive verification of hybrid systems [2]. It is usually challenging to obtain system models, and the use of these usually limits the application of the exhaustive verification.

However, in model-bounded monitoring, the model construction is not a big issue as in exhaustive verification because unlike the system model required for exhaustive verification, the bounding model need not be very precise. The analysis in exhaustive verification is *global*, and the modelling error accumulates over time; thus, even a small modelling error can cause a huge impact on the analysis. In contrast, the reachability analysis in model-bounded monitoring is *local*, and the modelling error is reset at each sample. For example, in Fig. 2.18, it is observed that the reachable area converges to one point at each sample.

2.6.3 Algorithms and Implementations

In [29], we presented two approaches to solve the model-bounded monitoring problem: by a reduction to reachability analysis and by a dedicated algorithm. The former approach is implemented with a model checker PHAVerLite, and the latter is implemented in a prototype tool HAMONI. Both approaches share the technical essence, namely the bounded-time reachability analysis of LHAs. The reachability analysis of PHAVerLite is optimised with recent techniques, while HAMONI is a simple prototype tool using Parma Polyhedra Library for convex polyhedra operations. The experimental results in [29] show the feasibility of model-bounded monitoring, especially by HAMONI. For example, for a benchmark on adaptive cruise control, HAMONI monitors 1,000 samples in about 12 seconds for dimension 5 on an Amazon EC2 c4.large instance (2.9 GHz Intel Xeon E5-2666 v3, 2 vCPUs, and 3.75 GiB RAM) that runs Ubuntu 18.04 LTS (64 bit). Refer to [29] for further experiment results, especially on the number of false alarms (the number was found to be permissible).

2.6.4 Summary

Model-bounded monitoring is a monitoring method guaranteed to detect every unsafe behaviour even if the signal is coarsely sampled. Although there may be false alarms, it is reducible using a more precise bounding model. In [29], the model-bounded monitoring problem is formulated using LHAs with two algorithms. The experimental evaluation suggests the feasibility of the model-bounded monitoring. Note that the above algorithm achieves "flexibility" and "informativeness" in the following sense.

Flexibility Model-bounded monitoring utilises a flexible interpolation of the sampled logs considering the prior knowledge of the system called bounding model. It is guaranteed to detect every unsafe behaviour even if the log is sampled sparsely.

Informativeness Since parametric LHAs is equivalent to LHAs, parameter valuations and unobserved signal values can be synthesised, similar to Sections 2.3 and 2.4. Refer to [25] for the detailed discussion on this point.

2.7 Conclusions

In this chapter, automata-based techniques for monitoring CPSs with formal specifications were reviewed. Formal automated monitoring is much more scalable than human monitoring. Formal automated monitoring is not only more informative but also the associated engineering costs are much lower, in comparison to ad-hoc monitoring. However, it must be noted that human monitoring and ad-hoc automated monitoring can still be viable options in some situations.

Among various techniques for formal automated monitoring, the focus of this chapter was on the timed pattern matching problem (in Section 2.2) and its parametric and quantitative extensions (in Sections 2.3 to 2.5). Timed pattern matching is the most basic and efficient problem as other complex problems take longer time. Nevertheless, the experimental results show that the complex problems, e.g., symbolic monitoring with infinite data domains (Section 2.4) and quantitative timed pattern matching (Section 2.5), are still efficient enough to monitor the values observed by sensors in real-time.

In these monitoring techniques, there is a methodological difficulty. Although the actual behaviour is continuous, the logs can only be observed in discrete time by digital sampling. This approximate observation does not matter if the sampling interval is short enough but can cause an issue when only a sparsely sampled log is available, e.g., due to a network limitation. In Section 2.6, we discussed model-bounded monitoring to flexibly interpolate the behaviour between the discrete samples utilising what is called a *bounding model*. Although model-bounded monitoring is highly complex, the experimental results show that it is still feasible and computationally tractable.

We believe that these formal techniques for system monitoring will play an essential role during the development and after the deployment of safety-critical cyber-physical systems.

References

[1] Alur, R. and Dill, D.L. 1994. A theory of timed automata. Theoretical Computer Science, 126(2):183–235, April 1994.

[2] Alur, R., Henzinger, T.A. and Ho, P-H.1996. Automatic symbolic verification of embedded systems. IEEE Transactions on Software Engineering, 22(3): 181–201.

[3] Alur, R., Henzinger, T.A. and Vardi. M.Y. 1993. Parametric realtime reasoning. *In*: S. Rao Kosaraju, David S. Johnson and Alok Aggarwal, (eds). Proceeding of the 25th Annual ACM Symposium on Theory of computing (STOC 1993), pages 592–601, New York, NY, USA, ACM.

[4] André, E., Fribourg, L., Kühne, U. and Soulat, R. 2012. IMITATOR 2.5: A tool for analyzing robustness in scheduling problems. *In*: Dimitra Giannakopoulou and Dominique Méry, (eds). FM 2012: Formal Methods—18th International Symposium, Paris, France, August 27–31, 2012. Proceeding, volume 7436 of Lecture Notes in Computer Science, pages 33–36. Springer.

[5] André, E., Hasuo, I. and Waga, M. 2018. Offline timed pattern matching under uncertainty. *In*: Anthony Widjaja Lin and Jun Sun, (eds). Proceeding of the 23rd International Conference on Engineering of Complex Computer Systems (ICECCS 2018), pages 10–20. IEEE CPS.

[6] Asarin, E., Caspi, P. and Maler, O. 2002. Timed regular expressions. Journal of the ACM, 49(2): 172–206.

[7] Bakhirkin, A., Ferrère, T., Maler, O. and Ulus, D. 2017. On the quantitative semantics of regular expressions over real-valued signals. *In*: Alessandro Abate and Gilles Geeraerts, (eds). Proceeding of the 15th International Conference on Formal Modeling and Analysis of Timed Systems (FORMATS 2017), volume 10419 of Lecture Notes in Computer Science, pages 189–206. Springer.

[8] Bakhirkin, B., Ferrre, T., Nickovic, D., Maler, O. and Asarin. E. 2018. Online timed pattern matching using automata. *In*: David N. Jansen and Prabhakar Pavithra, (eds). Proceeding of the 16th International Conference on Formal Modeling and Analysis of Timed Systems (FORMATS 2018), volume 11022 of Lecture Notes in Computer Science, pages 215–232. Springer, 2018.

[9] Bartocci, E., Deshmukh, J.V., Donzé, A., Fainekos, G.E., Maler, O., Nickovic, D., et al. 2018. Specification-based monitoring of cyber-physical systems: A survey on theory, tools and applications. *In*: Ezio Bartocci and Yliès Falcone, (eds). Lectures on Runtime Verification—Introductory and Advanced Topics, volume 10457 of Lecture Notes in Computer Science, pages 135–175. Springer.

[10] Bartocci, E. and Falcone, Y. (eds). 2018. Lectures on Runtime Verification—Introductory and Advanced Topics, volume 10457 of Lecture Notes in Computer Science. Springer.

[11] Boyer, R.S. and Moore, J.S. 1977. A fast string searching algorithm. Commun. ACM, 20(10): 762–772.

[12] Daviaud, L. and Johnson, M. 2017. The shortest identities for max-plus automata with two states. *In*: Kim G. Larsen, Hans L. Bodlaender and Jean-François Raskin, (eds). 42nd International Symposium on Mathematical Foundations of Computer Science, MFCS 2017, August 21–25, 2017–Aalborg, Denmark, volume 83 of LIPIcs, pages 48:1–48:13. Schloss Dagstuhl-Leibniz-Zentrum für Informatik.

[13] Donzé, A. and Maler, O. 2010. Robust satisfaction of temporal logic over real-valued signals. *In*: Krishnendu Chatterjee and Thomas A. Henzinger, (eds). Formal Modeling and Analysis of Timed Systems—8th International Conference, FORMATS 2010, Klosterneuburg, Austria, September 8–10, 2010. Proceeding, volume 6246 of Lecture Notes in Computer Science, pages 92–106. Springer.

[14] Fainekos, G.E. and George J. Pappas, G.J. 2009. Robustness of temporal logic specifications for continuous-time signals. Theory of Computer Science, 410(42): 4262–4291.

[15] Faro, S. and Thierry Lecroq, T. 2013. The exact online string matching problem: A review of the most recent results. ACM Computing Surveys, 45(2): 13: 1–13: 42.

[16] Finkbeiner, B. and Kuhtz, L. 2009. Monitor circuits for LTL with bounded and unbounded future. *In*: Saddek Bensalem and Doron A. Peled, (eds). Runtime Verification, 9th International Workshop, RV 2009, Grenoble, France, June 26–28, 2009. Selected Papers, volume 5779 of Lecture Notes in Computer Science, pages 60–75. Springer.

[17] Franek, F., Jennings, C.G. and Smyth, W.F. 2007. A simple fast hybrid pattern-matching algorithm. Journal of Discrete Algoritms, 5(4): 682–695.

[18] Kamali, M., Dennis, L.A., McAree, O., Fisher, M. and Veres, S.M. 2017. Formal verification of autonomous vehicle platooning. Science of Computer Programming, 148:88–106.

[19] Knuth, D.E., Morris Jr., J.H. and Pratt, V.R. 1977. Fast pattern matching in strings. SIAM Journal of Computer, 6(2): 323–350.

[20] Maler, O. and Dejan Nickovic, D. 2004. Monitoring temporal properties of continuous signals. *In*: Yassine Lakhnech and Sergio Yovine, (eds). Formal Techniques, Modelling and Analysis of Timed and Fault-Tolerant Systems, Joint International Conferences on Formal Modelling and Analysis of

Timed Systems, FORMATS 2004 and Formal Techniques in Real-Time and Fault-Tolerant Systems, FTRTFT 2004, Grenoble, France, September 22–24, 2004, Proceeding, volume 3253 of Lecture Notes in Computer Science, pages 152–166. Springer.

[21] Sanchez, C., Schneider, G., Ahrendt, W., Bartocci, E., Bianculli, D., Colombo, C., et al. 2019. A survey of challenges for runtime verification from advanced application domains (beyond software). Formal Methods Syst. Des., 54(3): 279–335.

[22] Ulus, D., Ferrère, T., Asarin, E. and Maler, O. 2014. Timed pattern matching. *In*: Axel Legay and Marius Bozga, (eds). Proceeding of the 12th International Conference on Formal Modeling and Analysis of Timed Systems (FORMATS 2014), Volume 8711 of Lecture Notes in Computer Science, pages 222–236. Springer.

[23] Ulus, D., Ferrère, T., Asarin, E. and Maler, O. 2016. Online timed pattern matching using derivatives. *In*: Marsha Chechik and Jean-François Raskin, (eds). Proceeding of the 22nd International Conference on Tools and Algorithms for the Construction and Analysis of Systems (TACAS 2016), volume 9636 of Lecture Notes in Computer Science, pages 736–751. Springer.

[24] Waga, M. 2019. Online quantitative timed pattern matching with semiringvalued weighted automata. *In*: Proceeding of the 17th International Conference on Formal Modeling and Analysis of Timed Systems (FORMATS 2019), volume 11750 of Lecture Notes in Computer Science, pages 3–22. Springer.

[25] Waga. M. 2020. Empowering Runtime Verification with Polyhedra. PhD thesis, The Graduate University for Advanced Studies, 2020.

[26] Waga, M., Akazaki, T. and Hasuo, I. 2016. A Boyer-Moore type algorithm for timed pattern matching. *In*: Martin Fränzle and Nicolas Markey, (eds). Proceeding of the 14th International Conference on Formal Modeling and Analysis of Timed Systems (FORMATS 2016), volume 9884 of Lecture Notes in Computer Science, pages 121–139. Springer.

[27] Waga, M. and André, E. 2019. Online parametric timed pattern matching with automata-based skipping. *In*: Julia M. Badger and Kristin Yvonne Rozier, (eds). NASA Formal Methods—11th International Symposium, NFM 2019, Houston, TX, USA, May 7–9, 2019, Proceeding, volume 11460 of Lecture Notes in Computer Science, pages 371–389. Springer.

[28] Waga, M., André, E. and Hasuo, I. 2019. Symbolic monitoring against specifications parametric in time and data. *In*: Proceeding of the 31st International Conference on Computer Aided Verification (CAV 2019), Part I, volume 11561 of Lecture Notes in Computer Science, pages 520–539. Springer.

[29] Waga, M., Andre, E. and Hasuo, I. 2021. Model-bounded monitoring of hybrid systems. *In*: Martina Maggio, James Weimer, Mohammad Al Farque and Meeko Oishi, (eds). ICCPS '21: ACM/IEEE 12th International Conference on Cyber-Physical Systems, Nashville, Tennessee, USA, May 19–21, 2021, pages 21–32. ACM.

[30] Waga, M., Hasuo, I. and Suenaga, K. 2017. Efficient online timed pattern matching by automata-based skipping. *In*: Alessandro Abate and Gilles Geeraerts, (eds). Proceeding of the 15th International Conference on Formal Modeling and Analysis of Timed Systems (FORMATS 2017), volume 10419 of Lecture Notes in Computer Science, pages 224–243. Springer.

[31] Waga, M., Hasuo, I. and Suenaga, K. 2018. MONAA: A tool for timed pattern matching with automata-based acceleration. *In*: Proceeding of the 3rd Workshop on Monitoring and Testing of Cyber-Physical Systems (MT@CPSWeek 2018), pages 14–15. IEEE.

Chapter 3

Formal Specification of Temporal Properties

Clovis Eberhart
National Institute of Informatics and JFLI, Tokyo, Japan

Jérémy Dubut
National Institute of Advanced Industrial Science and Technology (AIST), Tokyo, Japan

Ichiro Hasuo
National Institute of Informatics and SOKENDAI, Tokyo, Japan

3.1 Introduction to Formal Specifications

This chapter is a basis for many other chapters as it introduces so-called formal specifications. Formal specifications are necessary whenever one wants to automatically check whether a system behaves as expected on a given input. They are therefore necessary in many testing and verification methods. In particular, in this book, they will be used extensively in Chapter 1, which deals with falsification, an advanced testing technique. They can also be used in relation to monitoring, as explained in Chapter 2.

3.1.1 What is a Formal Specification?

In a nutshell, a formal specification is a description of a property of a system, written in a mathematical language. It is given an unambiguous and precise meaning, unlike specifications written in a natural language.

As such, formal specifications have the advantage that it is possible to reason about them and compute on them automatically, using computers. In particular, they can be used to design (or automatically generate) algorithms for problems such as model-checking, falsification, or monitoring. In such problems, researchers and engineers need to describe what is considered "interesting" without relying on human intervention. The specification therefore has to be completely formalised, hence the need for formal specifications.

On the other hand, they might be perceived as cumbersome to write by non-specialists, especially compared to sentences written in a natural language. The goal of this chapter is to make non-specialist readers familiar with formal specifications, and particularly specifications describing temporal properties.

3.1.2 Challenges for Practitioners

One of the main challenges for practitioners concerning specifications is to choose the right language to describe the particular property they are interested in. Choosing the right language is primarily about striking the right balance between the language's expressiveness and how costly it is to run different algorithms on specifications written in the language. In this regard, the languages we introduce in this chapter are among the most expressive generalistic languages for temporal properties, that is, specifications that describe sequences of events occurring in time, such as *eventually* reaching a target, staying in a safe area *forever*, or not staying out of one's work area *forever*, that possess rather computationally efficient algorithms.

Formal specifications can make things unnecessarily complicated in some cases, in particular when they are meant to be understood by humans. In such cases, natural languages offer better support, as they usually describe high-level concepts that can be subject to interpretation. For example, they are more naturally adapted to write laws, because they must be understood by most people and ambiguity can be broken by jurisprudence.

However, when digital systems, such as computers, are involved, it is necessary to write specifications in an abstract and unambiguous manner. For example, according to Mobileye's Responsibility-Sensitive-Safety philosophy,[1] automated driving systems (ADSs) should verify the following:

You must drive in such a way that you must not be responsible for any crash.

[1] https://www.mobileye.com/responsibility-sensitive-safety/

However, a computerised system cannot understand such a high-level concept, and must instead be given specifications such as the following:

When cruising on a straight line, the distance from the ego vehicle to the leading vehicle must be at least D,
When changing lane between two vehicles, the distance between them must be at least D′,

and so on, where D, D' are given by mathematical expressions depending on the speed of the other vehicles, the response time, the physics of the ego vehicle and so on.

For all these reasons, finding the proper language boils down to balancing intuitive understanding, expressivity, unambiguity, and computational tractability.

3.1.3 Goal of this Chapter

The main goal of this chapter is to introduce the reader to formal specifications, as well as basic mathematical notions needed for them. Numerous examples will be given throughout the chapter to give intuitions on all the concepts, but the reader is encouraged to come up with their own to check their understanding: the end goal is for the reader to be able to write their own specifications, either in the proposed frameworks or in their own framework, adapted to their own problems.

This chapter will serve as a basis for other chapters, particularly Chapter 1 on falsification and Chapter 2 on monitoring, whose problems and algorithms heavily rely on formal specifications, with particular emphasis on *temporal* specifications.

The progression will be as follows. Starting from the basic examples of arithmetic and regular expressions, we will introduce the concepts of syntax and semantics (Section 3.2). The goal of this section is for the reader to understand the distinction between what is written (the syntax), and what is meant (the semantics), but also to understand basic inductive definitions. We will then introduce a syntax and semantics for basic temporal specifications of discrete-time systems: linear temporal logic (LTL) (Section 3.3). The goal for the chapter at this point is to (1) understand basic temporal operators such as until, next, and globally; (2) be able to come up with formal specifications adapted to their own problems; (3) understand the meaning of "a system satisfies a specification" in terms of execution traces. Based on what has been learnt with linear temporal logic (LTL), we will next introduce a more complex framework for temporal specifications of continuous-time systems this time: signal temporal logic (STL) (Section 3.4). The goal in this section is still to be able to write formal specifications, but also to be able to handle a complex, quantitative semantics that answers the question "by how much does a system satisfy a specification?"

3.2 Introduction to Syntax and Semantics

Syntax and semantics are key notions in computer science, formal logic, and elsewhere. Syntax is what allows us to properly write programs, specifications, and sentences by giving symbols to depict notions and rules to structure them, while semantics is what gives a meaning to these objects.

In this section, we will gradually introduce syntax and semantics by examples with arithmetic expressions (Section 3.2.3) and regular expressions (Section 3.2.4), but first, we will explain the intuition (Section 3.2.1) and the purpose of introducing these concepts for formal specifications (Section 3.2.2).

3.2.1 *What are Syntax and Semantics?*

A *syntax* is a collection of symbols that can be combined in a structured manner. A typical example is the syntax of a programming language. This consists in keywords (such as "if", "then", "while" and so on), symbols (such as "+", "=", ";"), variable names, and so on, that can be *written* sequentially following some fixed structural rules (such as typing rules). Another example is the syntax of a natural language consisting of words and punctuation symbols structured by grammar rules. A phrase written using the symbols of the syntax, following the structural rules is called a *term*.

The key is that syntax is *not* interpreted *yet*: these symbols bear no meaning in themselves. This meaning is given by a *semantics*, which consists in a *domain* and an *interpretation*. The domain describes the possible meanings of a given syntax. In the case of a programming language, this domain can be the collection of mathematical functions of a given type, or the collection of assembler functions of a given architecture. The interpretation then associates any valid term of the syntax with an element of the domain. For example, a compiler can be seen as an interpretation of the syntax of a programming language into machine language. It gives a meaning to symbols which are understandable by a machine.

Syntax and semantics fulfil two complementary purposes. The syntax is for describing: it is usually compositional and easy to write, that is, you can write it by block, but its rigidity makes it hard to reason about. Complementarily, it is usually hard to describe the semantics of a given object as it often is infinite by nature (infinite executions traces, functions on integers, and so on), but its mathematical nature makes it easy to formally reason about.

3.2.2 *How are They Related to Formal Specifications?*

In what follows, we introduce some syntaxes and semantics to describe and give a precise meaning to specifications.

The syntax will be given by *formulas* composed of basic blocks (here called *operators*) that would describe specifications with logical connectives (such as conjunctions, implications), and temporal properties (such as the happening of an event at a given time in the future). The domain of the semantics of those formulas will be given by runs (or executions) of a system.

In the remainder of this section, we will give concrete examples of syntaxes and semantics not directly related to specifications. The purpose is to kindly introduce crucial concepts, necessary for the description and the understanding of syntax and semantics on basic examples.

3.2.3 Example: Arithmetic Expressions

Let us start with the example of integer arithmetic. In this example, the syntax is given by the usual symbols for addition, multiplication as well as constant integers. The domain of its semantics is then given by the set of integers and the interpretation by the result of the computation.

3.2.3.1 Syntax

It is usual to describe the syntax by a *grammar*. In the case of the integer arithmetic, the grammar is the following:

$$E \quad ::= \quad 0 \mid 1 \mid -E \mid E+E \mid E*E. \tag{3.1}$$

This means that terms of this syntax are all the object you can *inductively* write using symbols in $\{0,1,-,+,*\}$ following the rules that you must have exactly one expression after a symbol $-$, and exactly one before and one after a symbol $+$ or $*$. Here, *inductively* means that you can iterate the process to construct more and more complex terms. Precisely, it means the following:

- 0 and 1 are arithmetic expressions. They are the basic blocks of the construction of an arithmetic expression.
- If E is an arithmetic expression that one has already constructed, then one can obtain another expression by appending a $-$ symbol in front of E. Namely, $-E$ is another expression.
- If E and F are two already constructed arithmetic expressions, then one can obtain two other expressions by appending a $+$ or a $*$ symbol in-between. Namely, $E+F$ and $E*F$ are two other expressions.

Example 17 *It is for example possible to write*

$$1+0$$

as an arithmetic expression. Indeed, 0 *and* 1 *are arithmetic expressions as basic blocks. Then* $1+0$ *is obtained by appending a* $+$ *symbol between the expressions*

1 *and* 0. *It should be understood that, at this point, the expression* $1+0$ *is not equal to the expression* 1. *Indeed, no meaning has been given to the symbols of the syntax. Consequently,* $1+0$ *is just a different sequence of symbols than* 1. *Later, after introducing the semantics,* $1+0$ *and* 1 *will have the same semantics, namely, they will both be interpreted (or computed in this case) as the integer* 1. *Likewise,* $1+1$ *is an arithmetic expression obtained by appending a* $+$ *symbol between two copies of the expression* 1.

Confusingly enough, it is usual to denote this expression by 2. *At this point, this only means that the symbol* 2 *is a shorthand for the expression* $1+1$. *This process of adding symbols to the syntax to represent bigger terms is usual: it allows to describe a term by blocks which makes the syntax modular. Such shorthands are called* syntactic sugar. *In a similar, manner we can write* 3 *for the expression*

$$(1+1)+1$$

(which is the same as the expression $2+1$*), obtained by appending a* $+$ *symbol between the expressions* $1+1$ *and* 1*;* 4 *for the expression* $((1+1)+1)+1$*; and so on.*

Observe that, in those few last expressions, we added parentheses (*and*). *In this book, those symbols are not part of the syntax, they just help to delimit the scope of each symbol: for example, writing* $1+1*0$ *would be ambiguous between* $(1+1)*0$ *and* $1+(1*0)$. *Another syntactic sugar is given by* E^2, *which is a shorthand for the expression* $E*E$ *for any expression* E. *Using all those pieces, we can write bigger expressions, such as,*

$$1+(3*2^2),$$

which is a shorthand for the expression

$$1+(((1+1)+1)*((1+1)*(1+1))).$$

Remark that the latter is much harder to read than the former, which explains the interest of using syntactic sugar.

3.2.3.2 Semantics

As said, the domain of the semantics of arithmetic expressions is the set $\mathbb{Z}$ of integers. The interpretation of an expression is then the integer obtained after unfolding the computation. For example, the semantics of the expression $1+0$ is the integer 1 (that is the integer 1 in $\mathbb{Z}$, not to be confused with the *expression* 1, that is an expression E built inductively by the grammar (3.1)). Precisely, the semantics of a term is defined by induction on its construction, that is, all symbols of the syntax are assigned a meaning, and the meaning of a term is obtained by unfolding its definition, using the meaning of each encountered symbol. We denote by $[\![E]\!]$ the semantics of an expression.

In the case of integer arithmetic, the interpretation is defined as follows.

- The semantics $[\![0]\!]$ of the expression 0 is the integer 0 (again, not to be confused with the expression 0).
- The semantics $[\![1]\!]$ of the expression 1 is the integer 1.
- The semantics of an expression of the form $-E$ is obtained by taking the opposite of the semantic of E, namely,

$$[\![-E]\!] = -[\![E]\!].$$

  In this equality above, both $-$ are different. The left one is just a symbol without meaning, while the one on the right is the function from $\mathbb{Z}$ to $\mathbb{Z}$ mapping an integer n to $-n$.
- The semantics of an expression of the form $E+F$ is given by the sum of the semantics of E and F, that is,

$$[\![E+F]\!] = [\![E]\!] + [\![F]\!].$$

- The semantics of an expression of the form $E*F$ is given by the product of the semantics of E and F, that is,

$$[\![E*F]\!] = [\![E]\!] * [\![F]\!].$$

Example 18 *Unfolding this process, we can explicitly give the semantics of expressions given in Example 17.*

- $[\![1+0]\!] = [\![1]\!] + [\![0]\!] = 1+0 = 1 = [\![1]\!]$;
- $[\![2]\!] = [\![1+1]\!] = [\![1]\!] + [\![1]\!] = 1+1 = 2$;
- $[\![3]\!] = [\![1+2]\!] = [\![1]\!] + [\![2]\!] = 1+2 = 3$;
- $[\![E^2]\!] = [\![E*E]\!] = [\![E]\!] * [\![E]\!] = [\![E]\!]^2$;
- $[\![1+(3*2^2)]\!] = [\![1]\!] + [\![3*2^2]\!] = 1 + ([\![3]\!] * [\![2^2]\!]) = 1 + (3+2^2) = 8.$

Again, the reader must be careful in understanding that what is inside the $[\![\ ]\!]$ is different from elements of the domain, even though they are written in the same way. This confusion where a symbol and its meaning are written the same way is useful in practice, as this allows us to avoid introducing new symbols. However, this is only a peculiarity of arithmetic expressions, and we will see another example next where this is not the case.

3.2.4 Example: Regular Expressions

In this section, let us overview another example of syntax and semantics for regular expressions, such as in the POSIX standard. Note that the regular expressions presented here are a theoretical notion, and that they are simpler than the regular expressions of the POSIX standard.

3.2.4.1 Syntax

The syntax describes some patterns, using basic constructors such as concatenation or Kleene star, that describe some sets of words. It is given by the following grammar:

$$R \quad ::= \quad \varnothing \mid \varepsilon \mid a \mid R \cdot R \mid R+R \mid R^*. \tag{3.2}$$

Here a ranges over a fixed set Σ called the *alphabet*. In the case of the POSIX standard, Σ contains alphanumeric characters, punctuation signs, spaces, digits, and so on. The other symbols are $\cdot$ the *concatenation*, $+$ the *alternation*, and $*$ the *Kleene star*.

Similarly to the case of arithmetic expressions, this means that regular expressions can be constructed as follows. For each symbol, we also provide the intention behind them in terms of pattern recognition (again, no meaning is given to the symbols at this point).

- Every symbol a in the alphabet Σ is a regular expression. This is a basic block of regular expressions. The intention is to match every character a in a string.

- The symbols $\varnothing$ and ε are regular expressions. These two items are also basic blocks of regular expressions. The intention of the former is to match no strings, while the latter matches only the (zero-length) string with no character.

- If R and S are two regular expressions, then $R \cdot S$ and $R+S$ are two other regular expressions. The former intends to match every string that is formed as a string matched by R followed by a string matched by S. The intention of the latter is to match strings that are matched by R or by S.

- If R is a regular expression, then R^* is another regular expression. The intention here is to match all the strings that are obtained by concatenating a finite (possibly empty) sequence of strings matched by R.

Example 19 *In Example 17, we described some arithmetic expressions in a* bottom-up *manner, that is, we started from basic blocks and combined them to construct bigger and bigger expressions. Here, let us use a* top-down *manner, namely, starting from the outermost symbol, we dig deeper into the expression until we reach the basic blocks. Let us look at the following regular expression:*

$$((a+c) \cdot b)^*,$$

where a, b, and c are symbols in Σ, and let us analyse how it is generated using the grammar. The outermost symbol of this expression is a Kleene star $*$, applied to $(a+c) \cdot b$, so the expression is of the form R^* (again, the parentheses are not part of the syntax and are just for disambiguation). The outermost symbol *of the inner part of this expression is a concatenation $\cdot$, separating $a+c$ and b. Consequently, the term is of the form $(R \cdot R)^*$. Next, the left-hand side of the expression is $a+c$ whose outermost symbol is an alternation, so the expression is of the form $((R+R) \cdot R)^*$, and so on. Pictorially, we can write*

$$\begin{aligned} R \quad \rightsquigarrow \quad R^* \quad \rightsquigarrow \quad (R \cdot R)^* \quad \rightsquigarrow \quad ((R+R) \cdot R)^* \quad \rightsquigarrow \quad ((a+R) \cdot R)^* \\ \rightsquigarrow \quad ((a+c) \cdot R)^* \quad \rightsquigarrow \quad ((a+c) \cdot b)^* \end{aligned}$$

to describe the previous explanation, where $\rightsquigarrow$ represents one step of replacing a missing part of the expression using some rule of the grammar.

In practice, it is usual to use both bottom-up and top-down manners at the same time while writing an expression: it often is clear what basic blocks to use (which is important for bottom-up) but also what outermost symbol to use (which is good for top-down), and the difficulty is to bridge the gap. The importance of outermost symbols will be further illustrated in Section 3.3.2.5.

As for arithmetic expressions, the POSIX standard has a variety of syntactic sugar, of which we introduce some pieces now, together with their intended meanings.

- $[A-Z]$ is a shorthand for the expression $A+(B+(...(Y+Z)))$. The intention of this syntactic sugar is to match all capitalised alphabetic letters.
- R^+ is a shorthand for $R \cdot R^*$. This expression is called the *Kleene plus* and its intention is the same as the Kleene star, except that the empty sequence is forbidden.
- $R?$ is a shorthand for $\varepsilon + R$. This is called the *question mark* and its intention is to match all the strings matched by R or empty. A typical usage is to match British and American English spellings at the same time. For example, behavio $\cdot$ u? $\cdot$ r matches both "behavior" and "behaviour".

3.2.4.2 Semantics

Regular expressions are used to describe patterns of words that one would like to detect in a text for example. Consequently, their meaning can naturally be described in terms of sets of words. Precisely, we call a *word* any finite sequence of symbols in the alphabet Σ. We denote the set of words by Σ^*. The semantics $[\![R]\!]$ of a regular expression will then be a set of words, that is, $[\![R]\!] \subseteq \Sigma^*$. As for the case of arithmetic expressions, the semantics of regular expressions is given by induction on the construction of the expression, as follows.

- The semantics $[\![\varnothing]\!]$ of the expression $\varnothing$ is the empty set of words

$$[\![\varnothing]\!] = \varnothing.$$

- The semantics $[\![\varepsilon]\!]$ of the expression ε is the set consisting of exactly one element, which is the empty word (usually also denoted by ε)

$$[\![\varepsilon]\!] = \{\varepsilon\}.$$

- For a given symbol a of the alphabet Σ, the semantics of $[\![a]\!]$ of the expression a is the set consisting of exactly one element, which is the 1-letter word a

$$[\![a]\!] = \{a\}.$$

- If R and S are two regular expressions whose semantics are $[\![R]\!]$ and $[\![S]\!]$ respectively, then the semantics $[\![R \cdot S]\!]$ of their concatenation is the set of words consisting of concatenations uv of a word u in $[\![R]\!]$ followed by a word v in $[\![S]\!]$. Concretely,

$$[\![R \cdot S]\!] = \{uv \mid u \in [\![R]\!] \text{ and } v \in [\![S]\!]\}.$$

- If R and S are two regular expressions whose semantics are $[\![R]\!]$ and $[\![S]\!]$ respectively, then the semantics $[\![R+S]\!]$ of their alternation is the union of their semantics, that is,

$$[\![R+S]\!] = [\![R]\!] \cup [\![S]\!].$$

- If R is a regular expression whose semantics is $[\![R]\!]$, then the semantics $[\![R^*]\!]$ of its Kleene star is the set of words that are formed by concatenating a finite number of words in $[\![R]\!]$, which means

$$[\![R^*]\!] = \{w \mid \text{there is } n \geq 0 \text{ and } w_1, \ldots, w_n \in [\![R]\!] \text{ such that } w = w_1 \ldots w_n\}.$$

Example 20 *Let us continue Example 19 by unfolding the definition of its semantics.*

$$\begin{aligned}
[\![((a+c) \cdot b)^*]\!] &= \{w_1 \ldots w_n \mid w_1, \ldots, w_n \in [\![(a+c) \cdot b]\!]\} && \text{(semantics of } *\text{)} \\
&= \{u_1 v_1 \ldots u_n v_n \mid u_1, \ldots, u_n \in [\![a+c]\!] \text{ and } v_1, \ldots, v_n \in [\![b]\!]\} \\
& && \text{(semantics of } \cdot\text{)} \\
&= \{u_1 b \ldots u_n b \mid u_1, \ldots, u_n \in [\![a+c]\!]\} && \text{(semantics of } b\text{)} \\
&= \{u_1 b \ldots u_n b \mid u_1, \ldots, u_n \in [\![a]\!] \cup [\![c]\!]\} && \text{(semantics of } +\text{)} \\
&= \{u_1 b \ldots u_n b \mid \text{for all } i.\, u_i \in \{a, c\}\} && \text{(semantics of } a \text{ and } c\text{)}
\end{aligned}$$

Concretely, $((a+c) \cdot b)^*$ represents the set of words obtained by taking any finite sequence of a's and c's, appending the letter b after each symbol. For example, the empty word ε, ab, and cbab are in this set, while a, bc, aba, and cbb are not.

Remark 3.1 The semantics we just gave is useful to give precise meaning to functions such as UNIX's grep command-line: the command shows the lines in a text such that part of the line matches the pattern given by the regular expression, that is, this part belongs to the set of words given by the regular expression's semantics. This type of semantics (which is the only type we are considering in this chapter), which interprets the syntax purely mathematically, is called *denotational semantics*. In contrast, there exist other types of semantics that are more targeted at computation: they are called *operational semantics*. The domain of those semantics are often given by a set of abstract machines that compute the meaning of the syntax. Such an example for regular expressions is given by finite automata, which can computationally check whether a given word is matched by a given regular expression. A reader interested in the rich world of semantics may check [21], and [20] if interested in the connection between regular expressions and automata. ■

3.3 Linear Temporal Logic

In this section, we define LTL, study its properties, and illustrate its usefulness on some examples. In Section 3.3.2 we define LTL's syntax, which is necessary for readers who want to learn to write their own specifications. In Section 3.3.3, we discuss LTL's semantics, which makes formal the intuitions about formulas and can help check whether a specification has the intended meaning. Section 3.3.4 is dedicated to existing tools based on LTL, and Section 3.3.5 to references for readers who want to learn more about LTL.

3.3.1 Introduction to Linear Temporal Logic

Linear temporal logic (LTL) is a logic that describes temporal properties of discrete-time systems. Let us explain the different words in the previous sentence. Since LTL is a *logic*, its main ingredient is *formulas*, which are formal objects (the *syntax*) that represent specifications, together with what these formulas mean (their *semantics*). LTL describes *temporal* properties, by which we mean properties that refer to a system's evolution over time. For example, if we are considering ADSs, properties such as

the ego vehicle will ***never*** *crash*,

the ego vehicle will ***eventually*** *reach its destination*, or

the taxi will arrive at destination ***after*** *picking up its passenger*

are temporal properties, since they deal with the different states of the system over time. LTL specifying properties of *discrete-time* systems means that the behaviours it can specify have been discretised: either because the system itself works in discrete time (such as synchronous software systems), or because systems that work in continuous time are simplified through their modelling as discrete-time systems. Finally, LTL describes *linear* temporal properties, which means that the properties it describes consider each execution of the system separately (as a single linear execution in time), rather than all the possible executions together (which have a tree or graph structure).

LTL is a rather simple logic among temporal logics and therefore a lot of interesting problems are decidable for it. In particular, the *model-checking problem*, which consists in answering whether a (finite-state) system satisfies a given specification, is decidable: there exist algorithms that answer true when a system verifies a given specification, and can give a counter-example otherwise. This approach, known as *verification*, is complementary to testing, and used extensively for example in security-critical software systems. While we do not discuss it in this book, the interested reader will find more in the references in Section 3.3.5.

The main goal of this section is to bring the reader to a level of familiarity with LTL high enough that they will be able to come up with LTL formulas that represent simple specifications on their own. The reader should become familiar enough with temporal logic in general that the definitions and examples of signal temporal logic (STL) in the next section will seem natural to them.

3.3.2 Syntax

In this section, following the basic examples of Sections 3.2.3.1 (for arithmetic expressions) and 3.2.4.1 (for regular expressions), we introduce the syntax of LTL. We define the syntax of LTL formulas in Section 3.3.2.1 and give intuitive meanings to the different constructs in Section 3.3.2.2. These are necessary to be able to understand specifications written in LTL and come up with one's own formal specifications. We define syntactic sugar for common constructions in Section 3.3.2.3 and explain how to extend the modalities with timing constraints in Section 3.3.2.4, both of which are useful shorthands when manipulating LTL specifications. Finally, Section 3.3.2.5 describes several classes of usual specifications expressible in LTL, which will help the reader come up with their own specifications more easily by following predefined templates.

3.3.2.1 Grammar

We fix a set $\mathbb{P}$ of *atomic propositions*, that is, a set of basic properties of interest for the system under study. Let us give a few examples of such atomic propositions that are relevant for ADSs. A first property of interest is whether the ego vehicle's left blinker is on, which can be denoted by an atomic proposition leftBlinker. We may want to be more precise than whether a component is on or off by asking whether that component is in a certain mode, for example whether the vehicle is in third gear, which can be denoted by the atomic proposition gearEq3. Finally, we may also be interested in the value of a particular physical quantity by comparing it to a given value, for example that the vehicle's speed is greater than 20m/s, which can be written as speedGt20.

Definition 3.1 (LTL formulas) The set of LTL formulas is generated by the following grammar:

$$\varphi \quad ::= \quad p \mid \top \mid \neg\varphi \mid \varphi \wedge \varphi \mid \bigcirc\, \varphi \mid \varphi\, \mathcal{U}\, \varphi, \tag{3.3}$$

where p is an atomic proposition in $\mathbb{P}$.

Example 21 *Given atomic propositions* leftBlinker *and* steeringGt0.2, *it is possible to write the following example:*

$$\neg(\top\, \mathcal{U}\, ((\neg \text{leftBlinker}) \wedge \text{steeringGt0.2})), \tag{3.4}$$

which is generated in a top-down manner by the grammar as follows:

$$\begin{aligned}\varphi \quad &\rightsquigarrow \quad \neg\varphi \quad \rightsquigarrow \quad \neg(\varphi\, \mathcal{U}\, \varphi) \quad \rightsquigarrow \quad \neg(\top\, \mathcal{U}\, \varphi) \quad \rightsquigarrow \quad \neg(\top\, \mathcal{U}\, (\varphi \wedge \varphi)) \\ &\rightsquigarrow \quad \neg(\top\, \mathcal{U}\, ((\neg\varphi) \wedge \varphi)) \quad \rightsquigarrow \quad \neg(\top\, \mathcal{U}\, ((\neg \text{leftBlinker}) \wedge \varphi)) \\ &\rightsquigarrow \quad \neg(\top\, \mathcal{U}\, ((\neg \text{leftBlinker}) \wedge \text{steeringGt0.2})).\end{aligned}$$

Throughout the rest of this chapter, we will use *formula* and *specification* interchangeably. From a formal point of view, both terms have the same meaning, but *formula* is more appropriate when discussing syntax and semantics, while *specification* is more natural when discussing desired properties of systems.

3.3.2.2 Intuitive Explanation of the Formulas

The intuitive meaning of the constructors of LTL formulas is as follows.

- An atomic proposition is a basic property that can be either true or false at a given time. Typical examples are whether the velocity of a vehicle is above some threshold (as in speedGt20) or whether it is in a particular gear (as in gearEq3). For more details, see the beginning of Section 3.3.2.1 above.
- The *truth* $\top$ represents a property that is true at all times.
- The *negation* $\neg\varphi$ represents the property that is true whenever φ is false. For example, if leftBlinker represents a proposition that is true whenever the left blinker is on, then $\neg$leftBlinker is true whenever the left blinker is off.
- The *conjunction* $\varphi \wedge \psi$ is a property that is true whenever both φ and ψ are true. For example, if gearEq1 and speedGt30 have their intuitive meaning, then gearEq1 $\wedge$ speedGt30 is true whenever the vehicle is in gear 1 and its speed is greater than 30m/s at the same time.

- The *next* $\bigcirc \varphi$ represents the property that is true whenever φ is true at the next time step. The formula $\bigcirc$ gearEq3 is true at some point in time if the vehicle is in third gear at the next time step.
- The *until* $\varphi \,\mathcal{U}\, \psi$ represents the property that is true whenever ψ becomes true at some point in time, and φ has stayed true at all times until then. For example, the formula speedLt25 $\mathcal{U}$ gearEq4 is true if the vehicle is in fourth gear at some point in time, and its speed has stayed under 25m/s until then.

These intuitions will be made precise when we describe the semantics of these formulas.

Example 22 *Combining those intuitions, the formula* (3.4) *represents the property that is true whenever there is no time in the future where the steering angle is bigger than 0.2 radians (i.e., the vehicle is turning left) and the left blinker is not on. This can be rephrased as "whenever the vehicle turns left, its left blinker is on". Notice that the part of the intuition of the until about all intermediate times can be dropped because it is applied on the truth* $\top$*, which is true at all times.*

3.3.2.3 Syntactic Sugar

It is often convenient to add some notations to usually used formula constructs.

Definition 3.2 Here are some notations that will be used in the rest of this chapter.

- We will denote the *falsehood* $\neg\top$ by $\bot$. Following the intuitive explanation above, this is an expression representing the property that is true whenever the truth $\top$ is false, that is, $\bot$ represents the property that is always false.

- We will denote the *disjunction* $\neg(\neg\varphi \wedge \neg\psi)$ by $\varphi \vee \psi$. This represents the property that is true whenever at least one of φ or ψ is true.

- We will denote the *implication* $\neg\varphi \vee \psi$ by $\varphi \rightarrow \psi$. This represents the property that is true whenever ψ is true or φ is false (this is the usual notion of implication in classical logic).

- We will denote the *finally* (also called *eventually*) $\top \,\mathcal{U}\, \varphi$ by $\Diamond\, \varphi$. This formula represents the property that is true if there is a time in the future when φ is true.

- Lastly, we will denote the *globally* (also called *always*) $\neg(\Diamond\, \neg\varphi)$ by $\square\, \varphi$. It represents the property that is true if at all times in the future φ is true.

Note also that a double negation $\neg\neg\varphi$ represents the same property as φ. This fact will be made formal with the semantics: $\neg\neg\varphi$ and φ have the same semantics. Since we are interested in the specifications represented by the formulas, i.e., in the semantics rather than the syntax, it is sometimes useful to replace a formula with another one with the same semantics to make the semantics clearer from the syntax, as illustrated by the example below.

Example 23 *Using this syntactic sugar and the previous remark, we can shorten the formula* (3.4). *First, we can observe* $\top\ \mathcal{U}\ \varphi$ *which can be shortened as* $\Diamond\ \varphi$. *The formula then becomes* $\neg(\Diamond(\neg\text{leftBlinker}) \wedge \text{steeringGt0.2})$. *Using the syntactic sugar* $\Box$ *together with the double negation, the formula can be rewritten as* $\Box\ \neg((\neg\text{leftBlinker}) \wedge \text{steeringGt0.2})$. *Introducing a disjunction, this becomes* $\Box\ (\text{leftBlinker} \vee \neg\text{steeringGt0.2})$. *Finally, recognising an implication, the formula is shortened as*

$$\Box\ (\text{steeringGt0.2} \rightarrow \text{leftBlinker}). \tag{3.5}$$

Remark that the intuition of this formula following the explanation of this section is slightly different than the one from Section 3.3.2.2: this formula represents the property that is true whenever for all times in the future if the vehicle is turning left then it must have its left blinker on. However, those two formulations are semantically equivalent.

There exist other syntactic sugar notations that are commonly used in LTL, but which we will not use throughout this book. We list them here nonetheless, as they can be useful to the reader to express some specifications.

Definition 3.3 The following notations are common syntactic sugar for LTL.

- The *weak until* $\varphi\ \mathcal{W}\ \psi$ denotes the formula $(\varphi\ \mathcal{U}\ \psi) \vee \Box\ \varphi$. It is true whenever φ stays true until ψ becomes true, but φ has to stay true at all times if ψ does not hold true at any point in time. For example, if systemSafe and protocolEnd are atomic propositions that respectively describe that the system is safe and that a protocol has reached its end, then systemSafe $\mathcal{W}$ protocolEnd describes the fact that the system is safe until the end of the protocol. Note that this is different from systemSafe $\mathcal{U}$ protocolEnd because the latter states that the protocol *will* end at some point (which may or may not be desired in the specification).
- The *release* $\varphi\ \mathcal{R}\ \psi$ denotes the formula $\psi\ \mathcal{W}\ (\varphi \wedge \psi)$. It is true whenever φ is true at some point, and ψ has stayed true at all times until then (including the point in time when φ is true), or if ψ stays true at all points in time. While the meaning is essentially the same as *weak until*, its intended meaning is closer to the intuition that ψ turning true releases the constraint on φ to stay true.

3.3.2.4 Adding Timing Constraints

The operators described until now – except for the Next operator $\bigcirc$ – do not give any precise information on the time an event occurs. For example, the Finally operator $\Diamond$ states that there is a time in the future when a property is true, without giving any explicit information on this time. We can overcome this problem by iterating the Next operator. Let us define some syntactic sugar $\bigcirc^i \varphi$ by induction on $i \geq 0$:

- if $i = 0$, $\bigcirc^0 \varphi$ is a notation for φ itself,
- if $i > 0$, $\bigcirc^i \varphi$ is a shorthand for $\bigcirc \left(\bigcirc^{i-1} \varphi\right)$.

The intuition is that $\bigcirc^i \varphi$ is true whenever φ is true in i time steps in the future.

Using this syntactic sugar, we can define a *timed finally* $\Diamond_{[a,b]} \varphi$ (where $0 \leq a \leq b$) as a shorthand for

$$\bigvee_{i=a}^{b} \bigcirc^i \varphi = \bigcirc^a \varphi \vee \bigcirc^{a+1} \varphi \vee \ldots \vee \bigcirc^b \varphi.$$

The intuition is still that $\Diamond_{[a,b]} \varphi$ is true whenever there is a time in the future where φ is true, but furthermore this time is required to be *between a and b*. Similarly, we can define a *timed globally* $\Box_{[a,b]} \varphi$ as

$$\bigwedge_{i=a}^{b} \bigcirc^i \varphi = \bigcirc^a \varphi \wedge \bigcirc^{a+1} \varphi \wedge \ldots \wedge \bigcirc^b \varphi$$

that is true whenever φ is true at all time in the future *between a and b*, and a *timed until* $\varphi\, \mathcal{U}_{[a,b]}\, \psi$ as

$$\bigvee_{i=a}^{b} \left(\bigcirc^i \psi \wedge \bigwedge_{k=0}^{i-1} \bigcirc^k \varphi \right)$$

that is true whenever there is a time in the future *between a and b* when ψ is true and φ is true at all times before that.

3.3.2.5 Examples of Classes of Property

Here, we describe some recipes to express common specifications using the syntax above.

Safety is a class of properties describing that the system does not show any undesirable behaviours, such as bugs, crashes, dangerous driving, data race, etc. This can be described using LTL syntax in the following equivalent ways: $\neg\Diamond$somethingBadHappens (there is no time in the future when something bad happens) or $\Box\neg$somethingBadHappens (at all time in the future, something bad does not happen). Here somethingBadHappens is a formula formally describing what it means for the system to do something bad. An example can be given by an atomic proposition longDistUnsafe which would be true whenever the longitudinal distance of the ego vehicle with another vehicle is smaller than a certain safe distance.

Guarantee is a class of properties that ensures that some fixed event must happen at some point of the execution, e.g., reaching a target, the end of a protocol, etc. It can simply be described as a finally $\Diamond$desiredEvent, where desiredEvent is a formal description as an LTL formula of the event we want to guarantee. For example, if a system satisfies $\Diamond(\text{speedEq0} \wedge \text{targetReached})$, then it is guaranteed that it will reach its target and stop there at some point, if speedEq0 and targetReached have their intuitive meanings.

Readiness is a class of properties describing that the system is doing some action only when it is ready. This can be formulated as

$$\Box\,(\text{somethingHappens} \rightarrow \text{isReady}),$$

where somethingHappens and isReady are formulas. For example, if the former is an atomic proposition steeringGt0.2 that is true when the vehicle is turning left, and if the latter is another atomic proposition leftBlinker that is true the left blinker is on, then the readiness formula means that the vehicle will only ever turn left when the left blinker is on (note that this is formula (3.5)).

Fairness is a class of properties describing that if all the conditions for an event to happen are satisfied, then this event must happen in the future. This can be described as $\Box\,(\text{conditions} \rightarrow \Diamond\text{eventHappens})$. For example, if the condition formula is given by the conjunction $\text{lowSpeedLaneEmpty} \wedge \text{speedLt25}$ where lowSpeedLaneEmpty and speedLt25 are atomic proposition, and if eventHappens is given by the atomic proposition takeover, then fairness means that if the low-speed lane is empty and if the vehicle has a low speed, then it must change to the low-speed lane at some point in the future, in compliance with driving laws.

Liveness is a class of properties guaranteeing that the system never gets blocked. This can be expressed as $\Box\Diamond$somethingHappens. For example, if greenLight is an atomic proposition that is true whenever a given traffic light is green, then $\Box\Diamond$greenLight is true if the traffic light is green at infinitely many times in the future. Note that this is equivalent to saying that there is no point in the future when the light will never be green again. This will be made formal when discussing semantics, since $\Box\Diamond$greenLight will have the same semantics as $\neg\Diamond\Box\neg$greenLight (by double negation).

Finally, it is also useful to note that these classes of properties can be combined to obtain richer specifications. For example, the formula

$$\Box\,((\Box\Diamond\text{greenLight}) \rightarrow \Diamond\text{targetReached})$$

combines fairness (about traffic lights), guarantee (about reaching the target), and readiness (the overall structure of the formula) properties to obtain a specification that describes the fact that, at any point in time, if the traffic light turns green infinitely often in the future, then the ego vehicle will reach its target at some point, which is a realistic specification.

3.3.3 Semantics

In this section, following the basic examples of Sections 3.2.3.2 (for arithmetic expressions) and 3.2.4.2 (for regular expressions), we introduce the semantics of LTL. The section is organised as follows. We link the semantics of LTL formulas to system executions through the notion of *trace* (Sections 3.3.3.1), inductively define the semantics of LTL formulas (Section 3.3.3.2), and describe how to derive the semantics of the syntactic sugar (Section 3.3.3.3).

3.3.3.1 Systems and Execution Traces

An *execution trace* represents what can be observed of the execution of a system. More precisely, the system state changes over time throughout its execution, and the trace of that execution represents which atomic propositions are true at each point in time. For example, an ADS contains both software and hardware, and we may only be interested in the physical position and velocity of the hardware, while the software variables are of no interest to us. In that case, the system state at some point in time comprises both the physical values of interest to us, and the value of all variables in the software, which we do not care about. On the other hand, since we are only interested in the position and velocity of the vehicle, the execution trace of the system would only contain boolean values derived from those quantities.

Definition 3.4 (traces) An *execution trace of a system* is an infinite sequence of sets of atomic propositions, that is, a sequence of the form $S_0,\ldots,S_n,\ldots$ where each S_n is a subset of $\mathbb{P}$ and is the set of the atomic propositions that are true at time i. More generally, we call *trace* (or execution trace) any sequence as above where all S_n's are subsets of $\mathbb{P}$, even if it does not stem from the execution of a system.

For example, if we are only interested in the two atomic propositions turnLeft and turnRight (which describe that the ego vehicle is turning left and right respectively), then the following is a trace:

$$S_0 = \{\text{turnLeft}\}, S_1 = \{\text{turnLeft}, \text{turnRight}\}, S_2 = \emptyset, \ldots, S_n = \emptyset, \ldots,$$

even though this does not correspond to the execution of a system, since S_1 does not make sense (a vehicle cannot be turning left and right at the same time). We denote by $\mathbb{T}$ the set of all traces.

Both the semantics of an LTL formula and that of a system are given by a subset of $\mathbb{T}$: the semantics $[\![\varphi]\!]$ of a formula φ is the set of traces that satisfy the specification represented by the formula; the semantics $[\![\mathcal{S}]\!]$ of a system $\mathcal{S}$ is the set of all possible execution traces of the system.

This correspondence allows us to formally compare a system with a specification. We say that *a system $\mathcal{S}$ satisfies a specification φ* if

$$[\![\mathcal{S}]\!] \subseteq [\![\varphi]\!].$$

The *model-checking problem* is then to decide whether a given system satisfies a given specification.

3.3.3.2 Inductive Definition of the Semantics

As explained above in Section 3.3.3.1, the semantics of a formula φ is the set of execution traces that satisfy that formula. This is made formal by the following definition.

Definition 3.5 (LTL semantics) The *semantics* of a given LTL formula φ, denoted by $[\![\varphi]\!]$, is a set of execution traces defined by induction on the definition of the formula.

- When φ is an atomic proposition p, its semantics is given by the set of execution traces of the form $S_0, S_1, \ldots$ with $p \in S_0$:

$$[\![p]\!] = \{S_0, S_1, \ldots \mid p \in S_0\}.$$

- When φ is the basic formula $\top$, its semantics is given by the set of all possible traces:

$$[\![\top]\!] = \mathbb{T}.$$

- When φ is formed as a negation $\neg\psi$, its semantics is given by the set of execution traces that are not in the semantics of ψ, that is, $[\![\neg\psi]\!]$ is the complement of $[\![\psi]\!]$:

$$[\![\neg\psi]\!] = \mathbb{T} \setminus [\![\psi]\!].$$

- When φ is formed as a conjunction $\psi_1 \wedge \psi_2$, its semantics is given by the set of execution traces that are both in the semantics of ψ_1 and ψ_2, that is, $[\![\psi_1 \wedge \psi_2]\!]$ is the intersection of $[\![\psi_1]\!]$ and $[\![\psi_2]\!]$:

$$[\![\psi_1 \wedge \psi_2]\!] = [\![\psi_1]\!] \cap [\![\psi_2]\!].$$

- When φ is formed as a next $\bigcirc\, \psi$, its semantics is given by the set of execution traces of the form $S_0, S_1, \ldots$ where $S_1, S_2, \ldots$ is in the semantics of ψ and S_0 is arbitrary:

$$[\![\bigcirc\, \psi]\!] = \{S_0, S_1, \ldots \mid S_1, S_2, \ldots \in [\![\psi]\!] \text{ and } S_0 \subseteq \mathbb{P}\}.$$

- When φ is formed as an until $\psi_1 \,\mathcal{U}\, \psi_2$, its semantics is given by the set of execution traces $S_0, S_1, \ldots$ such that there exists a time $i \geq 0$ where
  - $S_i, S_{i+1}, \ldots$ is in the semantics of ψ_2 and
  - for every $0 \leq k < i$, $S_k, S_{k+1}, \ldots$ is in the semantics of ψ_1.

$$\llbracket \psi_1 \,\mathcal{U}\, \psi_2 \rrbracket = \{S_0, S_1, \ldots \mid \exists i \geq 0.\, S_i, S_{i+1}, \ldots \in \llbracket \psi_2 \rrbracket \text{ and } \\ \forall 0 \leq k < i.\, S_k, S_{k+1}, \ldots \in \llbracket \psi_1 \rrbracket\}$$

In Section 3.3.2.3, we argued that the double negation $\neg\neg\varphi$ represents the same property as φ. This can be made formal using the semantics:

$$\llbracket \neg\neg\varphi \rrbracket = \mathbb{T} \setminus \llbracket \neg\varphi \rrbracket = \mathbb{T} \setminus (\mathbb{T} \setminus \llbracket \varphi \rrbracket) = \llbracket \varphi \rrbracket.$$

Example 24 *Let us unfold this inductive definition on the formula* (3.4).

$$\begin{aligned}
&\llbracket \neg(\top \,\mathcal{U}\, ((\neg \text{leftBlinker}) \wedge \text{steeringGt0.2})) \rrbracket \\
&= \mathbb{T} \setminus \llbracket \top \,\mathcal{U}\, ((\neg \text{leftBlinker}) \wedge \text{steeringGt0.2}) \rrbracket && \text{(semantics of } \neg) \\
&= \mathbb{T} \setminus \{S_0, \ldots, S_n, \ldots \mid \exists i \geq 0.\, S_i, S_{i+1}, \ldots \in \llbracket (\neg \text{leftBlinker}) \wedge \text{steeringGt0.2} \rrbracket \text{ and} \\
&\qquad \forall 0 \leq k < i.\, S_k, S_{k+1}, \ldots \in \llbracket \top \rrbracket\} && \text{(semantics of } \mathcal{U}) \\
&= \mathbb{T} \setminus \{S_0, \ldots, S_n, \ldots \mid \exists i \geq 0.\, S_i, S_{i+1}, \ldots \in \llbracket (\neg \text{leftBlinker}) \wedge \text{steeringGt0.2} \rrbracket\} \\
&&& \text{(semantics of } \top) \\
&= \mathbb{T} \setminus \{S_0, \ldots, S_n, \ldots \mid \exists i \geq 0.\, S_i, S_{i+1}, \ldots \in \llbracket \neg \text{leftBlinker} \rrbracket \text{ and} \\
&\qquad S_i, S_{i+1}, \ldots \in \llbracket \text{steeringGt0.2} \rrbracket\} && \text{(semantics of } \wedge) \\
&= \mathbb{T} \setminus \{S_0, \ldots, S_n, \ldots \mid \exists i \geq 0.\, S_i, S_{i+1}, \ldots \notin \llbracket \text{leftBlinker} \rrbracket \text{ and} \\
&\qquad S_i, S_{i+1}, \ldots \in \llbracket \text{steeringGt0.2} \rrbracket\} && \text{(semantics of } \neg) \\
&= \mathbb{T} \setminus \{S_0, \ldots, S_n, \ldots \mid \exists i \geq 0.\, \text{leftBlinker} \notin S_i \text{ and steeringGt0.2} \in S_i\} \\
&&& \text{(semantics for atomic proposition)}
\end{aligned}$$

As claimed in Example 22, the semantics of this formula can be described as the set of traces such that there is no time in the future where the steering angle is bigger than 0.2 radians (i.e., the vehicle is turning left) and the left blinker is not on.

Furthermore, using basic set theory, this set can also be written as

$$\{S_0, \ldots, S_n, \ldots \mid \forall i \geq 0.\, \text{if steeringGt0.2} \in S_i \text{ then leftBlinker} \in S_i\},$$

which support the intuition of Example 23 that this formulation is equivalent to the fact that for all times in the future if the vehicle is turning left then it must have its left blinker on.

3.3.3.3 Semantics of Syntactic Sugar

In Section 3.3.2.3, we described shorthands for useful pieces of formulas. In this section, we unfold the inductive definition of the semantics on these constructs.

- For the falsehood,

$$[\![\bot]\!] = [\![\neg\top]\!] = \mathbb{T} \setminus [\![\top]\!] = \mathbb{T} \setminus \mathbb{T} = \varnothing,$$

  that is, its semantics is the empty set, meaning that no traces satisfy the falsehood.

- For the disjunction,

$$[\![\varphi \vee \psi]\!] = [\![\neg(\neg\varphi \wedge \neg\psi)]\!] = \mathbb{T} \setminus ((\mathbb{T} \setminus [\![\varphi]\!]) \cap (\mathbb{T} \setminus [\![\psi]\!])) = [\![\varphi]\!] \cup [\![\psi]\!],$$

  that is, the semantics of the disjunction is the union of the semantics.

- For the implication,

$$\begin{aligned} [\![\varphi \rightarrow \psi]\!] &= [\![\psi \vee \neg\varphi]\!] = [\![\psi]\!] \cup (\mathbb{T} \setminus [\![\psi]\!]) \\ &= \{S_0, \ldots, S_n, \ldots \mid \text{if } S_0, \ldots, S_n, \ldots \in [\![\varphi]\!], \\ &\qquad \text{then } S_0, \ldots, S_n, \ldots \in [\![\psi]\!]\}. \end{aligned}$$

- For the finally,

$$\begin{aligned} [\![\Diamond\, \varphi]\!] &= [\![\top\, \mathcal{U}\, \varphi]\!] \\ &= \{S_0, \ldots, S_n, \ldots \mid \exists i \geq 0.\, S_i, S_{i+1}, \ldots \in [\![\varphi]\!] \text{ and} \\ &\qquad \forall 0 \leq k < i.\, S_k, S_{k+1}, \ldots \in [\![\top]\!]\} \\ &= \{S_0, \ldots, S_n, \ldots \mid \exists i \geq 0.\, S_i, S_{i+1}, \ldots \in [\![\varphi]\!]\}. \end{aligned}$$

- For the globally,

$$\begin{aligned} [\![\Box\, \varphi]\!] &= [\![\neg\Diamond\, \neg\varphi]\!] \\ &= \mathbb{T} \setminus \{S_0, \ldots, S_n, \ldots \mid \exists i \geq 0.\, S_i, S_{i+1}, \ldots \in \mathbb{T} \setminus [\![\varphi]\!]\} \\ &= \{S_0, \ldots, S_n, \ldots \mid \forall i \geq 0.\, S_i, S_{i+1}, \ldots \in [\![\varphi]\!]\}. \end{aligned}$$

Example 25 *In Example 23, we argued that formula* (3.5) *is a shorthand for formula* (3.4). *Unfolding the semantics of the syntactic sugar, we obtain that the semantics of the former is*

$$\{S_0, \ldots, S_n, \ldots \mid \forall i \geq 0.\, \text{if steeringGt0.2} \in S_i \text{ then leftBlinker} \in S_i\}.$$

As said in Example 24, this is an equivalent formulation of the semantics of formula (3.4). *Consequently, both formulas represent the same property.*

3.3.4 Tools

Recall from Section 3.3.3.1 that the model-checking problem is, for a given system $\mathcal{S}$ and specification φ, to decide whether $[\![\mathcal{S}]\!] \subseteq [\![\varphi]\!]$, i.e., whether all execution traces produced by $\mathcal{S}$ satisfy φ.

A strong point of LTL is that the model-checking problem is decidable for a large class of systems. Standard algorithms can be derived using automata-theoretic approaches: an LTL formula can effectively be described by an automaton (precisely, by a *Büchi automaton*), in the sense that the set of traces composing the semantics of an LTL formula can be recognised by such an automaton. Consequently, whenever a system can also be described by such an automaton, the model-checking problem is decidable by checking *trace inclusion of Büchi automata*. This decidability is used in several tools that we introduce here.

3.3.4.1 Spin[2]

Spin [13] is an open-source verification tool originally developed by the NOKIA Bell Labs. Its target is the study of multi-threaded software, described in a high level language called PROMELA. It can check deadlocks, race conditions, as well as any correctness properties expressed in LTL, and is particularly adapted to basic safety and liveness. This tool has been successfully used in domain such as telecommunications, space missions, automobile industry, and medical devices.

3.3.4.2 Spot[3]

Spot is a C++ library for the model-checking problem of LTL and automata-theoretic manipulations developed at EPITA. Its main purpose is to enrich the class of systems that can be dealt with. It also comes with Python bindings, such as Jupyter, which makes it a good candidate as an introductory tool to model-checking.

3.3.4.3 NuSMV[4]

NuSMV is an open-source symbolic model-checker jointly developed by FBK-IRST, Carnegie Mellon, Genova, and Trento Universities. This tool has been used as a cornerstone for several academic and industrial projects, including static analysers, a tool-kit for human-machine interfaces, and a language for embedded software engineering.

---

[2]https://spinroot.com/

[3]https://spot.lrde.epita.fr

[4]https://nusmv.fbk.eu

3.3.5 References

The literature on LTL is rich, as it is the base of formal verification. The textbooks [5, 11, 14] provide an exhaustive and accessible overview of it. The original work by Pnueli can be found in [19]. In the coming section, we will introduce a continuous-time extension of LTL, but the logic can be extended in many directions. To go further, one can check [7] (branching extension), [17] (extension with fixpoint operators), [12] (probabilistic extension), and [6] (parametric extension) for example.

3.4 Signal Temporal Logic

Building on the section on LTL, we define signal temporal logic (STL), study its properties, and illustrate its usefulness on examples. In Section 3.4.2 we define STL's syntax, in Section 3.4.3 its *boolean* semantics, and in Section 3.4.4 its *robust* semantics. Finally, Section 3.4.5 is dedicated to existing tools that use STL and Section 3.4.6 to references.

3.4.1 Introduction to Signal Temporal Logic

STL is inspired by LTL, but with two significant differences: it aims at modelling systems that evolve both in *continuous time* and in some *continuous space*. Properties of these systems are harder to prove, since continuous spaces are typically harder to handle than discrete ones. Notably, the *model-checking problem*, which is decidable for LTL, is undecidable for STL. On the other hand, the continuous nature of space allows us to define a quantitative (or *robust*) semantics, which proves helpful to solve other problems, such as *falsification*, explained in detail in Chapter 1.

Our first goal in this section is to familiarise the reader with specifications written in STL, so that they can understand written specifications and come up with their own. Our second goal is for the reader to understand the notion of robust semantics, which is the main addition compared to LTL, and how it can be helpful in quantising "how much" a signal satisfies a formula.

3.4.2 Syntax

We define STL's syntax and commonly used syntactic sugar constructs, mainly emphasising the differences with LTL.

3.4.2.1 Grammar

Definition 3.6 (STL syntax) The set of STL formulas is generated by the following grammar:

$$\varphi \quad ::= \quad f(x_1,\ldots,x_n) > 0 \mid \top \mid \neg\varphi \mid \varphi \wedge \varphi \mid \varphi\, \mathcal{U}_I\, \varphi, \tag{3.6}$$

where $x_1,\ldots,x_n$ are variables, and I is an interval of the form $[a,b]$ or $[a,+\infty)$ (technically, intervals of the form $[a,+\infty)$ are not in the original syntax of STL, but we include them here for their expressiveness).

Let us note the differences with LTL, on which we will elaborate later. First, there is no *next* operator, which corresponds to the fact that there is no "next" instant in continuous time. Secondly, in STL, the *until* operator comes equipped with an interval that specifies bounds on when its subformulas have to hold. This is crucial to express timed properties of real systems, since we can specify behaviours such as "the vehicle should brake before 1s", rather than "the vehicle should brake at some point". Finally, propositional variables p have been replaced by comparisons of the form $f(x_1,\ldots,x_n) > 0$. The variables $x_1,\ldots,x_n$ represent the observed values of the signals we are interested in, and $f\colon X_1 \times \ldots \times X_n \to \mathbb{R}$ is a function that evaluates "how true" the proposition is, with positive values meaning the proposition is true and negative ones meaning it is false.

Example 26 *Let us look at a few examples of atomic propositions.*

As a first example, we want to define an atomic proposition that expresses whether the ego vehicle's speed v is greater than some threshold, say $25m/s$. Obviously, we need v to be one of the variables x_i, i.e., one of the variables of interest of our system, so we will assume it is the case. In LTL, we would need to define a new atomic proposition, but here, we can simply define it as $v-25>0$. Formally, we consider the function $f_1(x_1,\ldots,x_n) = v-25$ (which can be done because $v \in \{x_1,\ldots,x_n\}$), and the corresponding atomic proposition $f_1(x_1,\ldots,x_n) > 0$. Notice that the function f_1 returns a positive value exactly when the speed is above the threshold, which is exactly what we wanted to model. We will see when we define syntactic sugar (Section 3.4.2.2) that this can simply be written $v > 25$, which is even simpler and closer to our intuition. We can do the exact same thing for discrete values, such as gears.

A powerful aspect of STL's atomic propositions is that many important properties of systems can be encoded without having to define new atomic propositions, but simply by considering other functions. For example, let us say we are interested in whether a vehicle is not too far away from its initial position $(0,0)$ (say at most $100m$). Assuming the coordinates x and y of the vehicle are among the variables of interest $\{x_1,\ldots,x_n\}$, we can use function $f_2(x_1,\ldots,x_n) = 100-\sqrt{x^2+y^2}$ and consider $f_2(x_1,\ldots,x_n) > 0$.

We can also model boolean values. Say we are interested in whether the left blinker of our vehicle is on, and we have a signal leftBlinker *that is mapped to* 1 *when the blinker is on and to* 0 *otherwise. We can use the function* $f_3(x_1,...,x_n) = 2\text{leftBlinker} - 1$ *and consider the atomic proposition* $f_3(x_1,...,x_n) > 0$. *Note that we could use the simpler function* $\tilde{f}_3(x_1,...,x_n) = \text{leftBlinker}$, *but when the blinker is off,* $\tilde{f}_3(x_1,...,x_n) = 0$, *while* $f_3(x_1,...,x_n) < 0$. *The latter property* ($f_3(x_1,...,x_n)$ *begin strictly negative) will be important when we define the robust semantics.*

Example 27 *Recall Example 3.4, in which we write an LTL formula stating that the vehicle should never turn left unless its left blinker is on. We can write a similar formula in STL. First, we need* leftBlinker *and* steering *to be among the propositions of interest to us, and let us assume that* leftBlinker *is* 0 *when the left blinker is off, and* 1 *when it is on, and* steering *is the steering angle. Then we can write*

$$\neg(\top\, \mathcal{U}_{[0,+\infty)}\, ((\neg(2\text{leftBlinker} - 1 > 0)) \wedge (\text{steering} - 0.2 > 0))). \tag{3.7}$$

Note that, contrary to LTL, we do not need a specific atomic proposition to check that steering *is above a given threshold, so the formula can be written more naturally.*

3.4.2.2 Syntactic Sugar

STL admits similar syntactic constructs as LTL. First, we can define constructs that correspond to usual (propositional) logic: $\bot$, $\varphi \vee \psi$, and $\varphi \rightarrow \psi$, which are defined exactly as in LTL. We can also define syntactic sugar for temporal operators, again as in LTL: the *eventually* operator $\Diamond_I\, \varphi$, defined as $\top\, \mathcal{U}_I\, \varphi$, and the *globally* operator $\Box_I\, \varphi$, defined as $\Diamond_I \neg\varphi$. We also use the shorthands $\varphi\, \mathcal{U}\, \psi$ for $\varphi\, \mathcal{U}_{[0,+\infty)}\, \psi$, $\Diamond\, \varphi$ for $\Diamond_{[0,+\infty)}\, \varphi$ and $\Box\, \varphi$ for $\Box_{[0,+\infty)}\, \varphi$.

We can also define the following syntactic sugar for atomic propositions: $f(x_1,...,x_n) > \lambda$ stands for $f(x_1,...,x_n) - \lambda > 0$, $f(x_1,...,x_n) < 0$ for $-f(x_1,...,x_n) > 0$, $f(x_1,...,x_n) \geq 0$ for $\neg(f(x_1,...,x_n) < 0)$, etc.

Example 28 *As in the case of LTL, we can rewrite the formula* (3.7) *as*

$$\Box\, ((\text{steering} > 0.2) \rightarrow (2\text{leftBlinker} > 1)), \tag{3.8}$$

making use both of STL's intuitive syntax and of syntactic sugar.

3.4.2.3 Differences from Linear Temporal Logic

Let us come back to the differences between STL and LTL. The absence of the *next* operator is self-explanatory, as there is no "next" instant in continuous time.

Equipping the *until* operator with an interval gives us more expressivity than LTL's *until* operator. A similar operator can be encoded in LTL by

$$\varphi\, \mathcal{U}_{[n,m]}\, \psi \equiv \bigvee_{i=n}^{m} \left(\left(\bigwedge_{j=0}^{i-1} \bigcirc^{j} \varphi \right) \wedge \bigcirc^{i} \psi \right)$$

$$\varphi\, \mathcal{U}_{[n,+\infty)}\, \psi \equiv \left(\bigwedge_{i=0}^{n-1} \bigcirc^{i} \varphi \right) \wedge \bigcirc^{n} (\varphi\, \mathcal{U}\, \psi).$$

The first one can be read as "At some instant i between n and m, ψ will hold, and φ will hold at all points until then". The second one can be read as "For the next n steps, φ will hold, and it will keep holding after that, until ψ holds". There are however two drawbacks to these encodings. First, they cannot be written in STL so they do not have a robust semantics. Moreover, discrete time only reflects continuous time in synchronous systems (in presence of a global clock), so this encoding does not correspond to a fixed amount of time in general.

In LTL, atomic propositions are either true or false in each state. In STL, they are given by functions, which should be designed so that $f(x_1,\dots,x_n) > 0$ where the proposition holds true, and $f(x_1,\dots,x_n) \leq 0$ where it does not. In general, if the zone where the proposition holds true is X, $f(x_1,\dots,x_n)$ can be defined as

- the distance to X's border if $(x_1,\dots,x_n)$ is inside X,
- minus the distance to X's border if it is outside.

Therefore, the propositions are larger (which we think of as "more true"), the further inside X they are. Note that this is how the functions f_1 and f_2 are defined in Example 27.

3.4.3 Boolean Semantics

STL can be equipped with two semantics: a *boolean* semantics that returns whether a formula is true or false, and a *robust* semantics that returns "how true" the formula is. In this section, we define the first semantics, which is very close to LTL's semantics, before we explore the robust semantics in Section 3.4.4.

3.4.3.1 Signals

Remember from Section 3.3.3.1 that the semantics of an LTL formula is a set of traces $S_0,\dots,S_n,\dots$ Similarly, the boolean semantics of an STL formula is a set of *signals*. Therefore, before defining any form of semantics, we need to define signals.

A signal represents what is observed from an execution of a system. A signal maps all execution times to specific values for the variables of interest (which may be position, velocity, but also control state, etc.).

Definition 3.7 (signal) If the variables we observe are $x_1, \ldots, x_n$, a signal is a function $\sigma\colon \{x_1, \ldots, x_n\} \times \mathbb{R}_{\geq 0} \to \mathbb{R}$ such that $\sigma(x_i, t)$ is the value of variable x_i at time t.

Note that, just like for LTL, STL formulas do not work directly on systems, but rather on what is observed of a system execution. This is a useful feature when studying complex systems, since it means that we can apply STL to black-box systems too.

One difficulty when working with signals in continuous time is to know their values at all points in time. Input signals are usually defined by the user, so they can be defined in a format that makes them finitely representable. Let us list some of the most commonly used classes of input signals:

- piecewise constant signals, which are defined by their values at a finite number of points in time $t_0, \ldots, t_n$ and is constant on $[t_i, t_{i+1})$,
- piecewise linear signals (sometimes called piecewise linear continuous signals), which are similarly defined by their values at $t_0, \ldots, t_n$ and is linear on $[t_i, t_{i+1})$, in such a way that the whole signal is continuous,
- more generally, piecewise polynomial signals (called splines), with constraints on smoothness at time points t_i.

Example signals from these three classes are illustrated in Fig. 3.1.

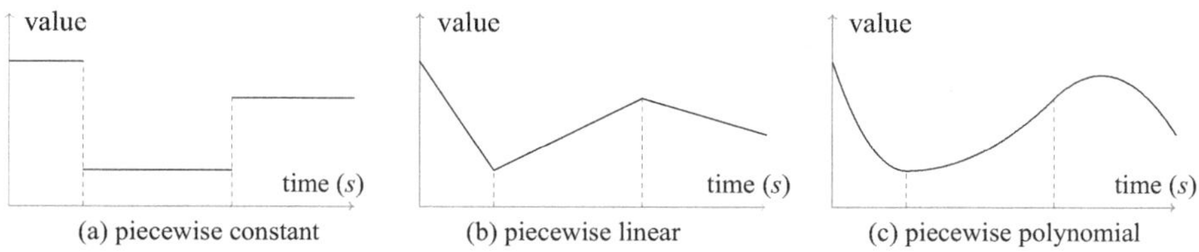

Figure 3.1: Examples of signals in different classes.

On the other hand, output signals do not belong to a class of signals fixed in advance. Still, in practice, output signals are sampled and their values interpolated as if they belonged to one of these classes (for example piecewise linear). While this is enough for most applications in practice, it is inherently limited, as it can give both false positives and false negatives.

3.4.3.2 Definition

While we could define the boolean semantics of an STL formula as a set of signals (like the semantics of LTL), we choose to define it as a function that takes a signal and returns a boolean. This is completely equivalent: given a set X, the corresponding function returns $\top$ on x if and only x is in X. Moreover, this approach will be more natural when we introduce the robust semantics later.

The *boolean semantics* of an STL formula φ is a function that takes a signal σ and answers whether σ satisfies φ.

Definition 3.8 (STL boolean semantics) STL's *boolean semantics* is a function $[\![-]\!]$ that takes as input an STL formula φ and a signal σ, and returns a boolean $[\![\varphi]\!](\sigma) \in \{\top, \bot\}$. It is defined inductively as follows:

$$
\begin{aligned}
[\![f(x_1,\ldots,x_n) > 0]\!](\sigma) &= \begin{cases} \top & \text{if } f(\sigma(x_1,0),\ldots,\sigma(x_n,0)) > 0 \\ \bot & \text{otherwise} \end{cases} \\
[\![\top]\!](\sigma) &= \top \\
[\![\neg\varphi]\!](\sigma) &= \neg[\![\varphi]\!](\sigma) \\
[\![\varphi \wedge \psi]\!](\sigma) &= [\![\varphi]\!](\sigma) \wedge [\![\psi]\!](\sigma) \\
[\![\varphi\, \mathcal{U}_I\, \psi]\!](\sigma) &= \bigvee_{t' \in I} \left(\left(\bigwedge_{t''<t'} [\![\varphi]\!](\sigma|_{t''}) \right) \wedge [\![\psi]\!](\sigma|_{t'}) \right),
\end{aligned}
$$

where $\sigma|_t$ is the signal σ shifted by t defined as $\sigma|_t\,(x_i,t') = \sigma(x_i,t+t')$.

The definition of the semantics uses operators on booleans, which is the set $\{\top,\bot\}$ ordered $\bot < \top$. The first operator is $\vee$ (the *logical or* operator) which takes the maximum of its arguments, and the second one is $\wedge$ (the *logical and* operator) which takes the minimum of its arguments. The first operator generalises to $\bigvee_{x\in X} b_x$, which takes the maximum of all b_x's (for all x in X). Similarly, the second operator generalises to $\bigwedge_{x\in X} b_x$, which takes the minimum of all b_x's (for all x in X).

This is very similar to LTL, so we only emphasise the differences. First, for atomic propositions, we use the signal values at time $t = 0$ (rather than fixed atomic propositions for each state) to compute the value of f and check whether this value is positive. For example, if $f(x_1,\ldots,x_n) = v - 50$ is supposed to represent whether the vehicle speed is above $50m/s$, then $[\![f(x_1,\ldots,x_n) > 0]\!](\sigma)$ is $\top$ if and only if $\sigma(v,0) > 50$, i.e., if the initial value of v for signal σ is greater than 50. In essence, this is the same as LTL, but using signal values instead of atomic propositions linked to states.

Secondly, the semantics for the *until* operator is different, due to the continuous nature of time, and the fact that *until* is bounded in STL. In order for the semantics $[\![\varphi\, \mathcal{U}_I\, \psi]\!]$ to hold, there must exist a time $t' \in I$ where ψ holds, and φ must hold at all time instants until t'. Again, this is in essence the same as the semantics of *until* in LTL, except for boundedness.

3.4.3.3 Boolean Semantics of Syntactic Sugar

Let us compute the boolean semantics of the different syntactic sugar constructions.

- For propositional constructs, such as $\varphi \vee \psi$ and $\varphi \rightarrow \psi$, this is computed as in the case of LTL, and we get

$$\llbracket \varphi \vee \psi \rrbracket(\sigma) = (\llbracket \varphi \rrbracket(\sigma) \vee \llbracket \psi \rrbracket(\sigma))$$
$$\llbracket \varphi \rightarrow \psi \rrbracket(\sigma) = (\llbracket \varphi \rrbracket(\sigma) \rightarrow \llbracket \psi \rrbracket(\sigma)),$$

  where, on the right-hand sides of the equations above, $\vee$ and $\rightarrow$ are operators on booleans (with $b \rightarrow b'$ defined as $\neg b \vee b'$),
- for $f(x_1,\ldots,x_n) > \lambda$, we have

$$\begin{aligned}\llbracket f(x_1,\ldots,x_n) > \lambda \rrbracket(\sigma) &= \llbracket f(x_1,\ldots,x_n) - \lambda > 0 \rrbracket(\sigma) \\ &= \begin{cases} \top & \text{if } f(\sigma(x_1,0),\ldots,\sigma(x_n,0)) - \lambda > 0 \\ \bot & \text{otherwise} \end{cases} \\ &= \begin{cases} \top & \text{if } f(\sigma(x_1,0),\ldots,\sigma(x_n,0)) > \lambda \\ \bot & \text{otherwise,} \end{cases}\end{aligned}$$

  which means that $f(x_1,\ldots,x_n) > \lambda$ holds for some signal σ if f applied to the value of σ at time $t = 0$ is greater than λ, which is indeed the intended meaning,
- for $f(x_1,\ldots,x_n) < 0$, $f(x_1,\ldots,x_n) \geq 0$, etc., we get the intended semantics by a similar computation,
- for $\Diamond_I \varphi$, we have

$$\begin{aligned}\llbracket \Diamond_I \varphi \rrbracket(\sigma) = \llbracket \top\, \mathcal{U}_I\, \varphi \rrbracket(\sigma) &= \bigvee_{t \in I} \left(\bigwedge_{t' < t} \llbracket \top \rrbracket(\sigma|_{t'}) \wedge \llbracket \varphi \rrbracket(\sigma|_t) \right) \\ &= \bigvee_{t \in I} (\llbracket \varphi \rrbracket(\sigma|_t)) \quad \text{(because } \llbracket \top \rrbracket(\sigma|_{t'}) = \top \text{ for any } t'\text{)},\end{aligned}$$

  which means that $\Diamond_I \varphi$ holds for σ at time $t = 0$ if φ holds for $\sigma|_t$ (or equivalently that φ holds for σ at time t) for some t in I, which is the intended semantics,
- for $\Box_I \varphi$, we have

$$\begin{aligned}\llbracket \Box_I \varphi \rrbracket(\sigma) = \llbracket \neg \Diamond_I \neg \varphi \rrbracket(\sigma) = \neg \llbracket \Diamond_I \neg \varphi \rrbracket(\sigma) &= \neg \bigvee_{t \in I} \llbracket \neg \varphi \rrbracket(\sigma|_t) \\ &= \neg \bigvee_{t \in I} \neg \llbracket \varphi \rrbracket(\sigma|_t) = \bigwedge_{t \in I} (\llbracket \varphi \rrbracket(\sigma|_t)),\end{aligned}$$

  which is also the intended semantics.

Example 29 *Let us compute the boolean semantics of formula* (28) *on some given signals. Let us assume that the signal for* steering *and* leftBlinker *is as*

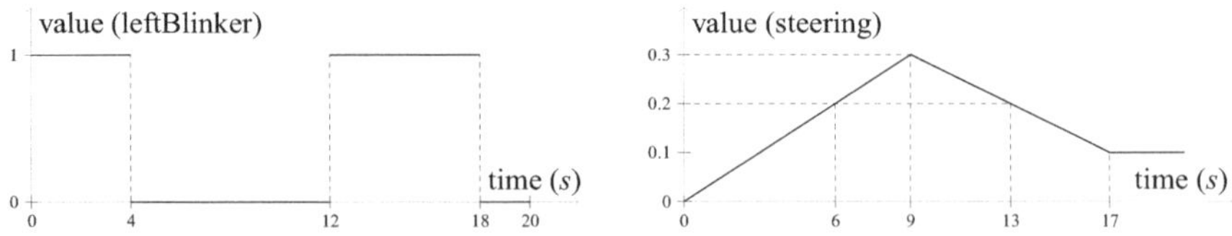

(a) LeftBlinker signal (b) Steering signal

Figure 3.2: Examples of signals.

shown in Fig. 3.2. If we call this signal σ, then for example $\sigma(\text{steering}, 6) = 0.2$ and $\sigma(\text{leftBlinker}, 6) = 0$. We adapt the formula to finite signals by changing $\square$ to $\square_{[0,20]}$ and compute its boolean semantics as follows:

$$
\begin{aligned}
&[\![\square_{[0,20]}(\text{steering} > 0.2) \rightarrow 2\text{leftBlinker} > 1]\!](\sigma) \\
&= \bigwedge_{t\in[0,20]} (\neg(\sigma|_t(\text{steering}, 0) > 0.2) \vee 2\,\sigma|_t(\text{leftBlinker}, 0) > 1) \\
&= \bigwedge_{t\in[0,20]} (\neg(\sigma(\text{steering}, t) > 0.2) \vee 2\sigma(\text{leftBlinker}, t) > 1) \\
&= \bigwedge_{t\in[4,12]\cup[18,20]} \neg(\sigma(\text{steering}, t) > 0.2) \\
&= \bot,
\end{aligned}
$$

where the first step is by inductive definition and the second one by definition of $\sigma|_t$. The last two steps are where the semantics is computed, which is done by considering each t in $[0,20]$ separately.

- *In the third step, we take the value of $\sigma(\text{leftBlinker}, t)$ into account: if t is in $[0,4]$ or $[12,18]$, then $\sigma(\text{leftBlinker}, t) = 1$, so $2\sigma(\text{leftBlinker}, t) = 2 > 1$, and the disjunct is true. For other values of t, $\sigma(\text{leftBlinker}, t) = 0$, so $2\sigma(\text{leftBlinker}, t) = 0 \not> 1$, and the disjunct is true if and only if the other branch is.*

- *In the fourth step, we take the value of $\sigma(\text{steering}, t)$ into account: if t is in $[0,6]$ or in $[13,20]$, then $\sigma(\text{steering}, t) \leq 0.2$, so $\neg(\sigma(\text{steering}, t) > 0.2)$ is true, and so is the disjunct. However this fails for other values of t. For example, if $t = 9$, then $\sigma(\text{steering}, 9) = 0.3 > 0.2$, so the negation is false. Therefore, the semantics of the whole formula is false.*

3.4.4 Robust Semantics

In this section, we first motivate the definition of a *robust* semantics of STL (Section 3.4.4.1) before defining it (Section 3.4.4.2) and mentioning some challenges in its design (Section 3.4.4.3).

3.4.4.1 Qualitative and Quantitative Semantics

The boolean semantics defined in the previous section tells us whether a specification holds or not, which is necessary for basically any application. However, we may sometimes want to know more than just whether a specification holds. For example, in falsification, in order to efficiently search for counterexamples to some specification, it is necessary to know which signals look "more promising" for the search, that is, which signals are "closer" to not satisfying the specification. STL's *robust* semantics is a *quantitative* semantics that can be used for this purpose. That is, unlike a *qualitative* semantics that answers either true or false, evaluating the robust semantics of a signal yields a number that evaluates "how true" the specification is.

Since STL's boolean semantics is on signals that have real values (rather than boolean ones, as in the case of LTL), we can use these values to give a quantitative semantics: if our specification is for a certain quantity x to be positive at time t, then a signal σ such that $\sigma(x,t) = 100$ is said to be more robust than one such that $\sigma(x,t) = 0.01$, even though they both satisfy the specification.

Formally, the robust semantics is a function $[\![-]\!]_r$ (or simply $[\![-]\!]$ if there is no ambiguity with the boolean semantics) that takes an STL formula φ and a signal σ, and returns a real number $[\![\varphi]\!]_r(\sigma)$ that can be interpreted as follows:

- if $[\![\varphi]\!]_r(\sigma) > 0$, then σ satisfies φ; the bigger $[\![\varphi]\!]_r(\sigma)$ is, the further σ is from violating φ, and the smaller it is, the closer σ is to violating φ,
- conversely, if $[\![\varphi]\!]_r(\sigma) < 0$, then σ does not satisfy φ; the smaller $[\![\varphi]\!]_r(\sigma)$ is (i.e., the bigger it is in absolute value), the further σ is from satisfying φ, and the bigger it is, the closer σ is to satisfying φ,
- finally, if $[\![\varphi]\!]_r(\sigma) = 0$, nothing can be said about whether σ satisfies φ, but it is infinitely close to both satisfying and violating it.

3.4.4.2 Definition

The *robust semantics* of STL is defined as follows:

Definition 3.9 (STL robust semantics) STL's *robust semantics* is a function $[\![-]\!]_r$ that takes as input an STL formula φ and a signal σ, and returns a real number $[\![\varphi]\!]_r(\sigma) \in \mathbb{R}$ (or an extended real, that is a value in $\mathbb{R} \cup \{+\infty, -\infty\}$). It is defined inductively as

$$
\begin{aligned}
[\![f(x_1,\ldots,x_n) > 0]\!]_r(\sigma) &= f(\sigma(x_1,0),\ldots,\sigma(x_n,0)) \\
[\![\top]\!]_r(\sigma) &= v \\
[\![\neg\varphi]\!]_r(\sigma) &= -[\![\varphi]\!]_r(\sigma) \\
[\![\varphi \wedge \psi]\!]_r(\sigma) &= \min\left([\![\varphi]\!]_r(\sigma), [\![\psi]\!]_r(\sigma)\right) \\
[\![\varphi\, \mathcal{U}_I\, \psi]\!]_r(\sigma) &= \sup_{t' \in I} \min\left(\left(\inf_{t''<t'} [\![\varphi]\!]_r(\sigma|_{t''})\right), [\![\psi]\!]_r(\sigma|_{t'})\right),
\end{aligned}
$$

where $\nu > 0$ is some arbitrary threshold decided in advance (in the rest of this section, we consider $\nu = +\infty$ in the computations).

These formulas use the notions of suprema and infima. Readers who are unfamiliar with these notions can think of them as maxima and minima respectively.

This is just a quantitative version of the boolean semantics from the previous section. This is obvious for atomic propositions, since $[\![\varphi]\!](\sigma)$ can be defined as $[\![\varphi]\!]_r(\sigma) > 0$, which respects the intuition that truth corresponds to the robust semantics being positive. The same is true for $\top$. For $\varphi \wedge \psi$ and $\varphi\, \mathcal{U}_I\, \psi$, notice that the formulas for robust semantics are the same as those for boolean semantics, where conjunction has been replaced with minimum and disjunction with maximum. This also reflects intuition, since

- x and y are both positive exactly when $\min(x, y)$ is,
- at least one of x and y is positive exactly when $\max(x, y)$ is.

Using this remark, we get a correspondence between the boolean semantics and the robust one.

Lemma 3.1

For all formulas φ and signals σ,

- *if $[\![\varphi]\!]_r(\sigma) > 0$, then $[\![\varphi]\!](\sigma) = \top$,*
- *if $[\![\varphi]\!]_r(\sigma) < 0$, then $[\![\varphi]\!](\sigma) = \bot$.*

Note that this lemma does not tell us anything if $[\![\varphi]\!]_r(\sigma) = 0$, in which case $[\![\varphi]\!](\sigma)$ may either be $\top$ or $\bot$ (but this rarely happens in practice when computing the robust semantics of physical systems).

We can also compute the robust semantics of the different syntactic sugar constructions.

- For disjunction, the semantics is computed as follows:

$$\begin{aligned}[\![\varphi \vee \psi]\!]_r(\sigma) = [\![\neg(\neg\varphi \wedge \neg\psi)]\!]_r(\sigma) &= -\min(-[\![\varphi]\!]_r(\sigma), -[\![\psi]\!]_r(\sigma)) \\ &= \max([\![\varphi]\!]_r(\sigma), [\![\psi]\!]_r(\sigma)).\end{aligned}$$

  Similarly, for implication, we get

$$[\![\varphi \to \psi]\!]_r(\sigma) = [\![\neg\varphi \vee \psi]\!]_r(\sigma) = \max(-[\![\varphi]\!]_r(\sigma), [\![\psi]\!]_r(\sigma)).$$

- For $f(x_1, \ldots, x_n) > \lambda$, we have

$$\begin{aligned}[\![f(x_1, \ldots, x_n) > \lambda]\!]_r(\sigma) &= [\![f(x_1, \ldots, x_n) - \lambda > 0]\!]_r(\sigma) \\ &= f(\sigma(x_1, 0), \ldots, \sigma(x_n, 0)) - \lambda.\end{aligned}$$

  We get similar expressions for $f(x_1, \ldots, x_n) < 0$, $f(x_1, \ldots, x_n) \geq 0$, etc.

- For $\Diamond_I \varphi$, we have

$$[\![\Diamond_I \varphi]\!]_r(\sigma) = [\![\top\ \mathcal{U}_I \varphi]\!]_r(\sigma) = \sup_{t \in I}\left(\min\left(\inf_{t'<t}[\![\top]\!]_r(\sigma|_{t'}), [\![\varphi]\!]_r(\sigma|_t)\right)\right)$$
$$= \sup_{t \in I}([\![\varphi]\!]_r(\sigma|_t)) \quad \text{(because } [\![\top]\!]_r(\sigma|_{t'}) = +\infty \text{ for any } t'\text{)}.$$

- For $\Box_I \varphi$, we have

$$[\![\Box_I \varphi]\!]_r(\sigma) = [\![\neg\Diamond_I \neg\varphi]\!]_r(\sigma) = -\sup_{t \in I} -[\![\varphi]\!]_r(\sigma|_t) = \inf_{t \in I}([\![\varphi]\!]_r(\sigma|_t)).$$

Example 30 *Let us compute the robust semantics of the formula of Example 28 on the signals of Example 29.*

$$\begin{aligned}
&[\![\Box_{[0,20]}(\text{steering} > 0.2) \to 2\text{leftBlinker} > 1]\!]_r(\sigma)\\
&= \inf_{t \in [0,20]} \max(-(\sigma|_t(\text{steering},0) - 0.2), 2\,\sigma|_t(\text{leftBlinker},0) - 1)\\
&= \inf_{t \in [0,20]} \max(0.2 - \sigma(\text{steering},t), 2\sigma(\text{leftBlinker},t) - 1).
\end{aligned}$$

This expression only depends on the values of the signals, and can be computed easily. In this particular case, we get that the infimum is reached at $t = 9$, where the robust semantics is -0.1, which implies that the boolean semantics is false by Lemma 3.1.

We invite the reader to compute the robust semantics for other signals and check that the robust semantics behaves as intended: signals that are "further away" from matching the specification (in this case, where the value of steering *is always larger, and that of* leftBlinker *always smaller than those in this example) have a smaller robust semantics, and those that are "closer" to it have a larger robust semantics, until it becomes larger than* 0*, in which case the boolean semantics becomes true.*

3.4.4.3 Challenges

Let us discuss a few challenges that arise in the definition of STL's robust semantics that the reader may encounter in practice. First, while there is a way to define equality of values in STL's syntax as $(f(x_1,\dots,x_n) \geq \lambda) \wedge (f(x_1,\dots,x_n) \leq \lambda)$, its robust semantics can never be positive:

$$\begin{aligned}
&[\![(f(x_1,\dots,x_n) \geq \lambda) \wedge (f(x_1,\dots,x_n) \leq \lambda)]\!]_r(\sigma)\\
&= \min([\![f(x_1,\dots,x_n) \geq \lambda]\!]_r(\sigma), [\![f(x_1,\dots,x_n) \leq \lambda]\!]_r(\sigma))\\
&= \min([\![f(x_1,\dots,x_n)]\!]_r - \lambda, -([\![f(x_1,\dots,x_n)]\!]_r - \lambda)) \leq 0.
\end{aligned}$$

Therefore, in practice, instead of writing equalities, one should rather specify a small error margin ε and specify the two values are not more than ε apart, as $(f(x_1,\dots,x_n) \geq \lambda - \varepsilon) \wedge (f(x_1,\dots,x_n) \leq \lambda + \varepsilon)$.

Secondly, the variables involved in a formula may have different scales, which makes some variables more influential in the computation of the robust semantics. For example, consider the formula $\varphi = (x > 1000) \vee (v < 10)$, where x (measured in meters) can grow up to thousands, while v (measured in meters by second) can grow up to a few dozens. Its robust semantics is $[\![\varphi]\!]_r = \max(x - 1000, v - 10)$, and $x - 1000$ can grow up to thousands, while $v - 10$ can only grow up to a few dozens. Therefore, $v - 10$ is nearly never the maximal term, which may cause troubles for some algorithms. A particular case is when one of the variables is discrete or $\top$ appears in the formula (usually in $\Diamond$ or $\Box$), then the robust semantics can be "blocked" at some fixed value, which is known as the *threshold problem*. The robust semantics may have two forms:

- $\min(\mathrm{v}, f(x_1, \ldots, x_n))$ for some threshold v, in which case the threshold has to be greater than other values for the robust semantics not to be constant (otherwise optimisation algorithms cannot perform well),
- $\max(\mathrm{v}, f(x_1, \ldots, x_n))$, in which case v has to be smaller than other quantities for the same reasons.

Therefore, it is advised to use threshold values for truth and discrete variables that have the same order of magnitude as other quantities used in the formula. (See Chapter 1 for other methods that deal with a similar problem, called the *scale problem*.)

3.4.5 Tools

We list some tools that make use of STL. One of the strong points of having a robust semantics (rather than a boolean one) is that it is a function into the reals, which can thus be optimised. That is the principle of *falsification* (elaborated upon in Chapter 1), whose goal is to find a counter-example to some property φ by finding a control signal σ that minimises $[\![\varphi]\!]_r(\sigma)$: if $[\![\varphi]\!]_r(\sigma) < 0$, then σ is a counter-example. Some of the tools listed here are falsification tools.

3.4.5.1 STLInspector[5]

STLInspector is a tool to check STL specifications. The user specifies an informal specification in natural language and an STL formula that they think reflects the specification. The tool then automatically generates signals using mutation testing and asks whether they satisfy the intended specification. If at least one signal is wrongly classified by the user, then they can rework their STL specification. If all signals are classified correctly, then the user can have some confidence that their specification is correct (although it does not guarantee that the formal specification corresponds to the informal one).

[5]https://pypi.org/project/STLInspector/

3.4.5.2 S-TaLiRo[6]

S-TaLiRo is a Matlab toolbox for falsification. The user provides a description of the system, either as a Simulink / Stateflow diagram, or as a user-defined function. They also provide a specification, in the form of an metric temporal logic (MTL) formula. MTL is a simpler version of STL, where atomic propositions return a boolean, rather than a real. The tool then tries to find system inputs that minimise the robust semantics of the specification using stochastic optimisation algorithms, such as Monte-Carlo methods or ant colony optimisation. See Chapter 1 for further description of the tool.

3.4.5.3 python-monitors[7]

This is a python package that monitors STL specifications, as well as properties expressed in other formalisms (for example regular expressions). It works on discrete sequences (like LTL), rather than in continuous time (like STL). Given a signal σ and a specification φ, monitoring is the problem of knowing for which times t, $[\![\varphi]\!](\sigma|_t)$ is true (this problem is discussed in more details in Chapter 2). The user provides a discrete sequence of atomic propositions (for example given by some system execution) and a specification to monitor, and the tool returns the set of times at which the specification holds.

Several monitoring algorithms are discussed in Chapter 2. The form of specifications these algorithms accept are *timed automata*, which STL formulas translate to.

3.4.5.4 Breach[8]

Breach is a Matlab toolbox that can serve multiple purposes: writing informal requirements as formal specifications, monitoring temporal properties, and generating tests. For the latter, if the system is given as a Simulink model, then the tool can use falsification to generate inputs with low robust semantics, in order to find counter-examples to the specification. See Chapter 1 for further description of the tool.

3.4.6 Conclusion

Logics of real-time systems have been studied for decades, see [3] for a survey. Apart from LTL, STL was inspired by a few logics, notably MTL [16] and metric interval temporal logic (MITL) [2]. STL itself was introduced in [18]. The idea of robust semantics was first introduced in [10] for MITL, and later adapted to STL. STL is a relatively specific topic, and there are no general textbooks or

[6] https://sites.google.com/a/asu.edu/s-taliro/s-taliro/
[7] https://github.com/doganulus/python-monitors/
[8] https://github.com/decyphir/breach/

surveys about it. However, [15] is a general survey of methods used in cyber-physical system (CPS) development, and it contains a part on STL. The interested reader can also look at extensions of STL, such as a parametric extension [4], or an extension with frequency analysis [9]. Moreover, the robust semantics we introduced in this section is the most standard, but not necessarily the best-suited for every purpose, for example [8] defines a robust semantics that takes time into account, while [1] extends the expressivity of STL to give a robust semantics that takes both space and time into account.

References

[1] Akazaki, T. and Hasuo, I. 2015. Time robustness in MTL and expressivity in hybrid system falsification. *In*: International Conference on Computer Aided Verification, pages 356–374. Springer.

[2] Alur, R., Feder, T. and Henzinger, T.A. 1996. The benefits of relaxing punctuality. Journal of the ACM (JACM), 43(1): 116–146.

[3] Alur, R. and Henzinger, T.A. 1991. Logics and models of real time: A survey. *In*: J.W. de Bakker, Cornelis Huizing, Willem P. de Roever and Grzegorz Rozenberg, (eds). Real-Time: Theory in Practice, REX Workshop, Mook, The Netherlands, June 3–7, 1991, Proceeding, volume 600 of Lecture Notes in Computer Science, pages 74–106. Springer.

[4] Asarin, E., Donzé, A., Maler, O. and Nickovic. D. 2011. Parametric identification of temporal properties. *In*: Sarfraz Khurshid and Koushik Sen, (eds). Runtime Verification—Second International Conference, RV 2011, San Francisco, CA, USA, September 27–30, 2011, Revised Selected Papers, volume 7186 of Lecture Notes in Computer Science, pages 147–160. Springer.

[5] Baier, C. and Katoen, J-P. 2008. Principles of Model Checking. MIT Press.

[6] Chakraborty, S. and Katoen, J-P. 2014. Parametric LTL on Markov Chains. *In*: Josep Díaz, Ivan Lanese and Davide Sangiorgi, (eds). Theoretical Computer Science—8th IFIP TC 1/WG 2.2 International Conference, TCS 2014, Rome, Italy, September 1–3, 2014. Proceeding, volume 8705 of Lecture Notes in Computer Science, pages 207–221. Springer.

[7] Clarke, E.M. and Allen Emerson, E. 1981. Design and synthesis of synchronization skeletons using branching time temporal logic. *In*: Dexter Kozen, (ed). Logics of Programs, Workshop, Yorktown Heights, New York, USA, May 1981, volume 131 of Lecture Notes in Computer Science pages 52–71. Springer.

[8] Donzé, A. and Maler, O. 2010. Robust satisfaction of temporal logic over real-valued signals. *In*: Krishnendu Chatterjee and Thomas A. Henzinger, (eds). Formal Modeling and Analysis of Timed Systems—8th International Conference, FORMATS 2010, Klosterneuburg, Austria, September 8–10, 2010. Proceeding, volume 6246 of Lecture Notes in Computer Science, pages 92–106. Springer.

[9] Donzé, A., Maler, O., Bartocci, E., Nickovic, D., Grosu, R. and Smolka, S. 2012. On temporal logic and signal processing. *In*: Supratik Chakraborty and Madhavan Mukund, (eds). Automated Technology for Verification and Analysis—10th International Symposium, ATVA 2012,

Thiruvananthapuram, India, October 3–6, 2012. Proceeding, volume 7561 of Lecture Notes in Computer Science, pages 92–106. Springer.

[10] Fainekos, G.E. and Pappas, G.F. 2009. Robustness of temporal logic specifications for continuous-time signals. Theoretical Computer Science, 410(42): 4262–4291.

[11] Gabbay, D.M., Hodkinson, I. and Reynolds, M. 1994. Temporal Logic: Mathematical Foundations and Computational Aspects, volume 28 of Oxford Logic Guides. Oxford University Press.

[12] Hansson, H. and Jonsson, B. 1994. A logic for reasoning about time and reliability. Formal Aspects of Computing, 6: 512–535.

[13] Holzmann. G.J. 2004. The SPIN Model Checker—primer and reference manual. Addison-Wesley.

[14] Huth, M. and Ryan, M. 2004. Logic in Computer Science: Modelling and Reasoning about Systems (2nd ed.). Cambridge University Press.

[15] Kapinski, J., Deshmukh, J.V., Jin, X., Ito, H. and Butts, K. 2016. Simulation-based approaches for verification of embedded control systems: An overview of traditional and advanced modeling, testing and verification techniques. IEEE Control Systems Magazine, 36(6): 45–64.

[16] Koymans, R. 1990. Specifying real-time properties with metric temporal logic. Real-time Systems, 2(4): 255–299.

[17] Kozen. D. 1983. Results on the Propositional μ-Calculus. Theory of Computer Science, 27: 333–354.

[18] Maler, O. and Nickovic, D. 2004. Monitoring temporal properties of continuous signals. *In*: Yassine Lakhnech and Sergio Yovine, (eds). Formal Techniques, Modelling and Analysis of Timed and Fault-Tolerant Systems, Joint International Conferences on Formal Modelling and Analysis of Timed Systems, FORMATS 2004 and Formal Techniques in Real-Time and Fault-Tolerant Systems, FTRTFT 2004, Grenoble, France, September 22–24, 2004, Proceeding, volume 3253 of Lecture Notes in Computer Science, pages 152–166. Springer.

[19] Pnueli, A. 1997. The temporal logic of programs. *In*: Proceeding of the 18th Annual Symposium on Foundations of Computer Science (FOCS'77), pages 46–57.

[20] Sakarovitch, J. 2009. Elements of Automata Theory. Cambridge University Press.

[21] Winskel, G. 1993. The formal semantics of programming languages: An introduction. Foundations of Computing Series. MIT Press.

Chapter 4

Testing for Machine Learning-Based Systems

Xiao-Yi Zhang
University of Science and Technology Beijing, Beijing, China

Fuyuki Ishikawa
National Institute of Informatics, The University of Electro-Communications and SOKENDAI, Tokyo, Japan

Paolo Arcaini
National Institute of Informatics, Tokyo, Japan

4.1 Introduction

Testing is an "activity in which a system or component is executed under specified conditions, the results are observed or recorded, and an evaluation is made of some aspect of the system or component" (defined in ISO/IEC/IEEE 26153). Testing has been one of the key activities for quality assurance in software-intensive systems.

The recent advance of machine learning (ML) techniques, especially deep learning, has led to active investigation towards their industrial applications. Beforehand, ML techniques were target of laboratory research and the primary concern was on the prediction performance. As the prediction performance has evolved rapidly, increasing demand is put more on the dependability and quality of the ML applications.

With the aforementioned trend, it is natural to see increasing attention on testing for ML components and systems [13, 20]. We here refer to ML components/systems as those that were constructed with ML techniques to provide prediction functions. In the ML terminology, such components are called models that reflect mathematical relations between inputs and outputs. Examples of such functions include detection of pedestrian in images obtained by car cameras. On the other hand, ML components have different characteristics from traditional software components (programs) as their behaviors are inductively derived from training data, not by direct human design. This point is especially critical when we use deep neural networks that consist of millions or more parameters and the behavior is too complex to interpret.

This chapter describes the key principles in testing of ML components and ML-based systems. We first focus on difference from testing traditional programs. We then introduce how automated testing techniques have evolved starting from techniques for traditional programs.

4.2 Preliminary: Machine Learning and Deep Neural Networks

Machine learning (ML) refers to techniques to build functionality of prediction by experience or data. The current trend from late 2010s for applications of ML techniques primarily focuses on supervised learning. In supervised learning, training data is given that contains input-output pairs so that the relationship between them can be learnt via training. Practical application of supervised learning has been actively investigated as its role is often clear: supervised learning allows to build software components by deriving their behaviors from training data even if we cannot rigorously specify the behaviors.

Other types of ML techniques include unsupervised learning and reinforcement learning. Unsupervised learning allows for extracting mathematical relations from given data. An example of unsupervised learning techniques is clustering, which divides the given data into multiple clusters by some similarity metric. Unsupervised learning is considered as data analysis activities rather than software system development. Reinforcement learning allows us to build agents that take proper actions in an evolving environment. Here what is learnt is the value of each action in each state or what action should be taken in each state. Although there have been active efforts on reinforcement learning, it is still in an early stage before discussing practical testing activities or quality assurance in industrial use.

In this chapter, we use the term ML to refer to supervised learning and focus on testing activities of components (models) constructed by supervised learning and systems including such components.

We illustrate the use of ML techniques, i.e., supervised learning, in Fig. 4.1. The mathematical structures to capture the prediction functionality are called *models*. In a simple case, models can have the form of $y = ax + b$ to predict y from the input value of x. Values of the parameters a, b are derived through the training process: typically to minimize the mean prediction error to the given points in the training data. We sometimes distinguish the models whose parameter values are determined as *trained models*. The model is represented as a program object and integrated into the whole system.

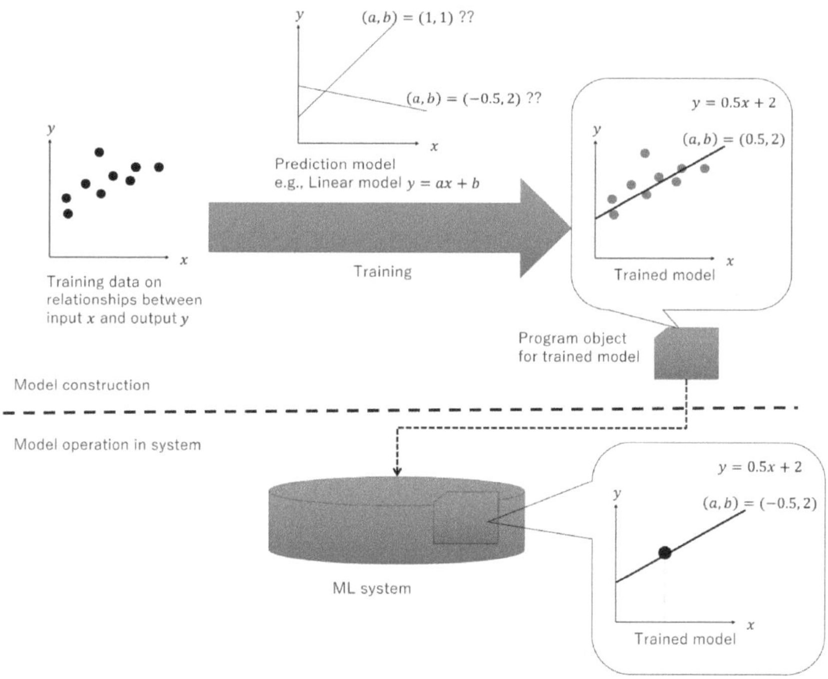

Figure 4.1: Machine learning.

This use of the term *model* is different from that for software engineering where we consider human-designed models to capture the essence of the target system at different abstraction levels. In this chapter, we explicitly use the word *ML components* as the same meaning as *trained models* in the ML context.

There are many different models, together with the learning mechanisms. The previous example of $y = ax + b$, called linear regression, was simplistic. More sophisticated model structures are often used such as Support Vector Machine and Random Forest to capture more complex relationships.

The trend on ML was driven by technical advance in Deep Learning. Deep Learning uses the model structure called Deep Neural Networks (DNNs). Fig. 4.2 illustrates the structure. A DNN contains multiple layers and each layer consists of multiple neurons as shown in the left side of the figure. Each neuron collects output values from the neurons $x_{i,1}, x_{i,2}, \cdots$ in the previous i-th layer. Then, the output values are aggregated by using weight parameters w_i and the result is passed to an *activation* function[1] to determine the final output of the neuron passed to the neurons in the $i+1$-th layer (right side of the figure). The weight parameters w_i, often of millions or more, are the target of optimization via training in the case of DNNs. DNNs are especially effective to deal with complex high-dimension data such as images and natural-language documents. We don't describe all the detail on DNNs. The structure in Fig. 4.2 should be noted as we may consider this structure upon testing as we consider control structure when we test traditional program code.

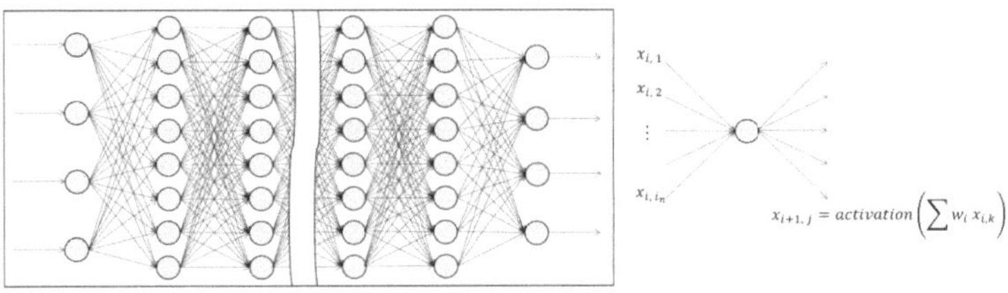

Figure 4.2: Deep neural networks.

4.3 Testing of ML Components: Viewpoint from Prediction Performance and Data

4.3.1 Foundation for Quality Evaluation of ML Components

The baseline for quality evaluation of ML components is to measure the prediction performance by using metrics such as accuracy. Specifically, we prepare a test dataset, which must be independent of training data so that we can evaluate the generalization ability. The test dataset contains input-output pairs and we compare the output from the ML component under test for each input with the expected output.

Let us consider the case of anomaly detection for products in a factory by using images from cameras. The input-output pair consists of a camera image

[1]We don't get into detail of activation functions. There are several functions known as effective such as ReLU and Sigmoid functions, often non-linear to express complex functions.

and a binary label that tells us whether the product has anomaly or not (positive or negative). This example thus is a binary classification task.

Suppose we had 500 of test input-output pairs and the ML component could make correct outputs for 465 of them. The det ailed counts were shown below.

ML Output	Positive (total 45)	Negative (total 455)
Positive in ground truth (total 20)	15	5
Negative in ground truth (total 480)	30	450

Then, we say the *accuracy* is $(15+450)/500 = 93\%$ by counting the correct outputs, i.e., true-positive and true-negative cases. We usually want to investigate more detail of the performance by distinguishing wrong detection (false-positive, 30 cases) and oversight (false-negative, 5 cases). Specifically, first we count the number of outputs correctly said as positive (anomaly) among all the outputs of positive. This metric is called *precision*: $15/45 = 33\%$. Second, we count the number of outputs correctly said as positive (anomaly) among all the really positive products. This metric is called *recall*: $15/20 = 75\%$. It depends on the stakeholders which metrics are significant while there is trade-off between precision and recall. A typical way of thinking is to put higher priority in recall so that possibly anomaly products are reported with little oversight and then human experts can check false-positives or non-anomaly products.

We will not introduce many other essential concepts in evaluation of ML components such as overfitting and ROC curve, or evaluation criteria for other types of tasks, e.g., regression tasks to predict continuous output values. Foundation of machine learning can be found in books such as [10].

What should be noted here is that in this way the performance evaluation is relative to the test dataset. It is essential to explicitly argue what are included or not in the 500 images of the test dataset. For example, do we cover different anomaly patterns expected to appear in the operation? Are the photographic conditions unified or diverse as expected in the operation such as camera position and lighting? The validity and impact of evaluation thus depends on the data quality.

4.3.2 Quality Guidelines

As practical application of ML systems is intensively investigated, more focus has been put on quality or dependability of such systems. In Japan, two guidelines called QA4AI [6][2] and AIQM [11][3] have been published and being

[2]http://www.qa4ai.jp/

[3]https://www.cpsec.aist.go.jp/achievements/aiqm/index_en.html

actively updated since 2019 and 2020, respectively.[4] QA4AI is driven by practitioners, especially, QA (quality assurance) engineers. QA4AI thus includes more detailed practices through concrete examples. AIQM provides more normative description that is generic and serves as the standard with clear definition of relevant concepts.

Both of the guidelines clarify the targets of evaluation. QA4AI defines five targets of evaluation as data, model, system, process, and customer expectation. Data and model are the core deliverables in ML development as described in Section 4.3.1. Quality of the whole system should be argued including other components, possibly those that complement incompleteness of the ML components such as filters of improper outputs from the ML components. Process quality is significant specifically in terms of agility as ML imposes various types of uncertainty, e.g., it is impossible to precisely estimate beforehand what level of accuracy will be achieved, inputs in the operation may have an unknown different distribution from the one in training and test datasets. It is therefore significant to have agility to continuously iterate experiments and improvements. Finally, as quality investigated by the developer side should match with the customer expectation, we should explicitly evaluate it, e.g., how much the customers can accept unavoidable failures or how much the customers require assertive explanations.

AIQM defines three external quality characteristics, observable from user, as AI performance, safety, and fairness. AI performance refers to the traditional ML quality aspect shown in Section 4.3.1. Although average performance of prediction over the dataset is high as AI performance, occurrences of highly severe hazards are not accepted in some applications, e.g., automated driving. Safety deals with this aspect by evaluating the risks. Fairness refers to a lack of undesirable bias over sensitive features such as gender and race.

AIQM also defines eight internal quality characteristics handled by engineers to achieve the three external characteristics. These internal characteristics start with requirements coverage to explicitly consider case decomposition of originally fuzzy target of prediction. This is exactly the same point discussed in Section 4.3.1. Then, we can think of data design, data amount, and model performance.[5] AIQM also mentions the soundness of the programs as well as maintaining quality through the operation time.

In both of the guidelines, the essence lies in intensive analysis of requirements in the form of case decomposition so that we can argue the quality of data and model not only on the whole dataset but with more fine-grained viewpoints.

---

[4]QA4AI stands for Quality Assurance for AI and AIQM stands for AI Quality Management

[5]These aspects are classified into several quality characteristics but we omit the detail here

4.4 Focus of This Chapter

As demonstrated in QA4AI and AIQM, evaluation approaches for data and model performance have been discussed well and established to some extent. In this chapter, we further discuss application of traditional testing techniques for program code. We assume data collection based on requirements has been done to some extent but cannot be perfect due to the fuzzy nature of the target as well as the heavy cost of data collection and preprocessing such as labelling. Traditional testing techniques can enhance quality investigation by aiming at effective tests in terms of diversity or capability to discover potential problems.

4.5 Automated Testing for Traditional Programs

The ultimate goal of testing is to detect bugs (defects) effectively and efficiently and ensure quality of the target product or service. On the other hand, it is impossible to test all the possible execution scenarios or to certainly expose unknown bugs. Various testing techniques have been investigated under these limitations to efficiently reach bug-exposing scenarios or realize a certain level of completeness.

Coverage criteria in white-box testing are one representative of testing techniques. Coverage criteria allow for evaluating a test suite *TS* (a set of test cases) by measuring how much of the "elements" in the target program are covered by executing the test suite. For example, suppose there is a conditional branch such as `if(A or B) then ... else ...`. In this case, we can think of coverage criteria such as:

Decision coverage (Branch coverage) We should prepare test cases in the test suite *TS* that cover both of the two branches as true and false, i.e., $\exists t \in TS\colon (t \models \texttt{(A or B)})$, and $\exists t \in TS\colon (t \models \texttt{not(A or B)})$.

Condition coverage We should prepare test cases in the test suite *TS* that cover different values for each of the two conditions `A` and `B`, i.e., $\exists t \in TS\colon (t \models \texttt{A})$, $\exists t \in TS\colon (t \models \texttt{not A})$, $\exists t \in TS\colon (t \models \texttt{B})$, and $\exists t \in TS\colon (t \models \texttt{not B})$.

Modified Condition/Decision Coverage (MC/DC) A condition value is said to be *covered* for MC/DC only when it makes "impact" on the decision, i.e., flipping its value also flips the value of the decision (it turns it from *true* to *false*, or from *false* to *true*); note that a condition is required to impact the decision in both directions. For example, test $t = (\mathit{true}, \mathit{true})$ is not useful for MC/DC, as flipping any of the two conditions does not affect the overall decision (which remains *true*). The test $t = (\mathit{true}, \mathit{false})$, instead, covers `A` in one direction, i.e., flipping its value makes the decision *false*.

By designing tests with high coverage values, we expect to detect hidden bugs by executing relevant part of the code. We also have a certain level of confidence when we don't find any bugs after executing high-coverage tests.

We can say coverage criteria are examples of **adequacy criteria** that serve as the test objectives. We aim at obtaining a test suite that meets the adequacy criteria as much as possible by human design or automated test generation. In this chapter, we are specifically interested in automated **input generation methods** as an advanced approach to testing. By applying such methods, we generate test inputs that achieve high scores of the adequacy criteria.

The above two aspects are used to derive inputs of test cases. We also need to define the way to judge success or failure of test execution, called test **oracle**.

For traditional programs an enormous amount of work has been conducted for automated test generation. One line of such work is Search-based Testing described in Chapter 10. For example, EvoSuite is a tool for Java unit testing with the following characteristics [5].

- Adequacy criteria: common coverage criteria such as statement coverage and branch coverage
- Oracle: optional application of mutation-based search to identify assertions[6]
- Input generation method: metaheuristic optimization that searches for best suites by iterative improvement

Another direction of automated test generation is Concolic testing [15], which combines the *concrete* approach such as random input values with the *symbolic* approach such as use of solvers to find input values that meet a certain set of conditions.

In this chapter, we discuss testing techniques for ML components and systems with the same scheme of the three characteristics.

4.6 Challenges and Approaches in Testing of ML Components and Systems

We need to be aware of how different characteristics of ML components affect the existing principles and techniques for testing of traditional software programs. Although there have been many discussions on this point, e.g., [20], we focus on the three aspects discussed in the previous section for automated test generation.

Adequacy criteria for ML components and systems have been investigated

[6] Roughly speaking, we can search for assertions if we assume the current code is correct, basically only when we generate a test suite for regression testing.

actively as traditional code coverage criteria do not work effectively. For example in deep neural networks, there is no essential conditional branch and computation just goes straightforwardly. Instead, the results of complex, interconnected numerical computation decide the output from the component.

The oracle problem is often critical in testing of ML components and systems as in many cases they are non-testable [19]. That is, it is too costly or even impossible to prepare the expected output for every possible input value. For example, for testing image recognition model, we need to give a label to each input image, which is costly and cannot be done for many images. There is no absolutely right answer for some applications such as salary prediction. Sometimes we apply ML because we humans cannot find answers from the enormous data, e.g., book recommendations.

Finally, the input generation method may be changed according to characteristics of the target, the components or systems under test and the adequacy criteria.

We review the key studies for testing of ML components and systems though this area is still evolving at the time of writing.

4.7 Automated Testing Methods for ML Components

There have been intensive studies on automated testing for ML components, specifically, deep neural networks (DNNs). Below we review initial key studies in the evolving area.

4.7.1 *Initial Studies: DeepXplore and DeepTest*

DeepXplore is a key study that has invoked active investigation of testing methods for DNNs [12]. It was applied for testing of image classification models and its approach is characterized as below.

Adequacy criteria DeepXplore adopts a new concept called "neuron coverage". This criterion measures how many of the neurons have been activated[7] by a test suite. By maximizing the neuron coverage, the obtained input images have more diversity, causing different behaviors of the target DNN.

Oracle DeepXplore adopts N-version testing, that is, use of other implementations of the same functionality. This is pseudo-oracle as different DNNs cannot always have exactly same outputs. However, use of pseudo-oracle helps identify potential problems of the target DNN.

Input generation method DeepXplore adopts a gradient-based search method.

[7] Judged by a threshold value for output from each neuron.

Input images are generated by adding perturbations to the original ones. The generation is guided by gradients to increase both of the neuron coverage and distances of outputs from the pseudo-oracles. Constraints are defined by domain knowledge to limit the perturbations so that they are not too destructive and unrealistic.

DeepXplore is said as the first whitebox-testing method for DNN for its use of neuron coverage.

DeepTest is the successor from the same research group [17]. The essential idea is similar but DeepTest adopts so-called “metamorphic testing” for the oracle. DeepTest was applied to a DNN that outputs a steering control value from an input traffic image. The output is considered "problematic" for a perturbed input image if the output is too different from the output for the original input image. We thus do not need different implementations in the case of DeepTest, or even the expected right answer for each input image. This is because we compare two outputs for two inputs rather than checking one output for one input. We will discuss this idea of metamorphic testing later in this chapter.

In the following sections, we will review other key studies that followed DeepXplore and DeepTest. Although each study has different design choices on the three aspects (and others) but we will pick up only specific part of the individual studies.

4.7.2 Input Generation Methods

We showed that DeepXplore adopts a gradient-based search method to generate input images by adding perturbations to the original ones. Below are examples of other notable approaches.

4.7.2.1 DeepRoad

Perturbed images do not essentially extend the space of input images used for testing as they only apply limit types of small changes such as scaling to the original images. The work called DeepRoad used Generative Adversarial Networks (GANs) to convert input images to have features of specific weather [21]. GANs learn distributions of natural images and can generate new images from the distributions. The state-of-the-art GANs also allow for specifying or transferring “styles” of images and image features for different weather conditions are transferred in the case of DeepRoad.

4.7.2.2 DeepConcolic

In testing of traditional software programs, Concolic testing has been one of key approaches for input generation [15]. Concolic testing combines the *concrete* approach such as random input values with the *symbolic* approach such as use of solvers to find input values that meet a certain set of conditions. Concolic testing thus has advantages of the two approaches: efficient computation for easy cases and targeted generation of rare or corner cases that cannot be easily reached.

The work called DeepConcolic applies the same way of thinking to input generation for DNNs [16]. Here, the symbolic approach is Linear Programming to solve optimization problems defined by different types of adequacy criteria. We will review different criteria used by DeepConcolic in the following discussion on the adequacy criteria.

4.7.3 Test Adequacy Criteria

DeepXplore defined the new concept of a neuron coverage that evaluates whether each neuron has been activated by the test suite or not. The following studies investigated such criteria in depth.

4.7.3.1 DeepGauge

DeepGauge [9] investigated more fine-grained coverage criteria for deep neural networks. The original neuron coverage only checked whether each neuron has been activated or not by the test suite execution. This is simplistic as continuous values of neuron outputs are captured in a binary way and relationships between neurons are not considered. DeepGauge defined various coverage criteria such as one about how diverse values were covered for each neuron output, e.g., how many of four output areas of [0, 0.25), [0.25, 0.5), [0.5, 0.75), [0.75, 1] were covered. Another example is which different combinations of neurons were activated in each layer (a unit including many neurons), e.g., neurons [0, 4, 10] were activated together, but [0, 1, 3] were not.

4.7.3.2 DeepConcolic

DeepConcolic [16] we described for input generation methods used different types of adequacy criteria, including the neuron coverage. Another adequacy criterion adopted by DeepConcolic is based on Lipschitz Continuity. Lipschitz Continuity intuitively says that input changes do not lead to radical output changes, which is judged by a threshold c. An adequacy criterion was defined to count how many test instances were found that prove Lipschitz Continuity does not hold for a certain value of c.

4.7.3.3 Surprise Adequacy

The key aspect of test inputs is how they differ from the distribution of training data. It is expected that the trained models (DNNs) cannot make proper inference if the inputs are far from the training data. Surprise adequacy is a adequacy criterion that tries to capture this aspect [8]. Not simply looking at the distances from training data in the input space, it tries to capture differences of neuron activations of the target DNN. It thus aims to measure differences of how the input data are perceived by the target DNN.

4.7.4 Oracle

DeepXplore used N-version testing or other implementations of the same functionality. DeepTest used metamorphic testing to check output changes are within a certain range given variations of inputs. The latter approach has been used also by many other studies as metamorphic testing only requires the definition of the checked properties (not other implementations). Below we discuss different directions to define the test oracle, that is, what are problems we want to detect.

4.7.4.1 Metamorphic Testing

As discussed in Section 4.6, it is often difficult or impossible to prepare the expected right answer for each individual test input. Metamorphic testing is a technique used for such situations [14]. Instead of checking individual outputs, we check certain relationships, called metamorphic relations (MRs) hold between two test cases. Specifically, we consider a relationship in the form "if we change the input this way, the output should change in that way".

The simple case is shown in Fig. 4.3. Suppose we obtained output y for a certain input x with the function under test f. We apply a certain transformation M_{in} to the input x and obtain another input x_{follow} and then the output $y_{follow} = f(x_{follow})$ as the follow-up test case. We define M_{in} so that we can expect a relation or constraint holds between these values as a metamorphic relation and use this relation as a criterion to check the function f.

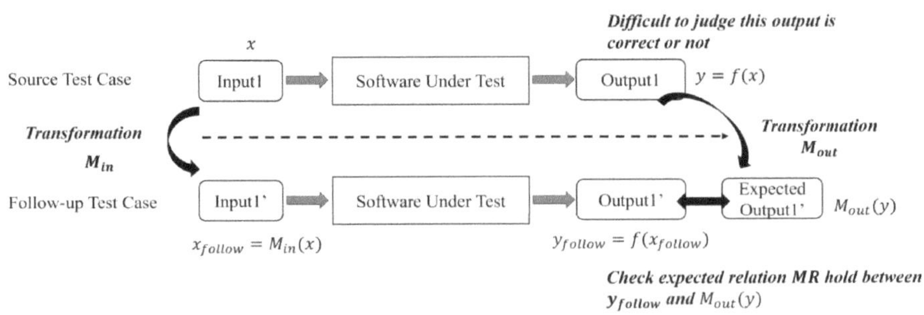

Figure 4.3: Metamorphic testing.

A simple example is shown for the `sin` function:

$$y = sin(x) \Rightarrow y_{follow} = f(x_{follow}) = y \texttt{ where } x_{follow} = \pi - x$$

We can also have:

$$y = sin(x) \Rightarrow y_{follow} = f(x_{follow}) = -y \texttt{ where } x_{follow} = -x$$

If such conditions do not hold, we can expect there is some problem, bug or sometimes other types of reasons such as misunderstanding of the specification. We can execute this pair of test cases even if we don't have the way to calculate the right `sin` value for any individual input.[8] The second test case (the right side of the formula) is called follow-up test cases. We execute any test case and at this point we cannot check whether the output is right or not. We then generate and execute the follow-up test case according to a specific MR, then the output of the follow-up test case can be judged to be OK or not (PASS or FAIL).

4.7.4.2 Fairness

There have been many other studies that focus on the oracle, or requirements specific to ML. One example of such oracle with increasing attention is fairness. As ML sometimes has large impact on organizational or social decision making, specific types of bias are not acceptable, e.g., bias based on gender or race. There are several ways to define as concrete, measurable criteria for fairness [2]. Sometimes fairness is argued by comparing individuals. For example, in salary prediction or loan approval decision, for a woman with very similar attributes (career etc.) to a certain man, the system should make similar outputs. Otherwise, the output for the woman is considered as undesirable. For such a definition of individual fairness, testing is suitable to detect undesirable cases as studied in [18].

4.8 Methods for Different Testing Targets

We have discussed active studies for testing ML components, especially DNNs. Below we discuss testing of other targets such as the whole system including ML components as well as the training pipeline to build ML components.

4.8.1 Testing of Whole System including ML Components

Some systems include many functional components other than those built with ML techniques. For example, automated driving systems may use ML only for part of the perception functionality, specifically, for image recognition on obstacle objects, signals, and lanes. In such systems, failures of ML components do not always lead to failures of the whole system. Other sensors may detect objects even if image recognition with cameras fails. Failures in recognizing

[8]This example is just for illustration. Nowadays, we have many libraries to calculate the `sin` values and we can use them to check the output values directly.

objects far away from the ego-vehicle may not lead to any problem. It is therefore significant to test ML components from the viewpoint of the system-level requirements or hazards. This direction is significant also because it is an endless activity to improve ML components that cannot have 100% accuracy.

There have been studies in this direction primarily targeting automated driving systems. Below we describe two key studies in this direction.

The work in [1] considered the testing of a braking system with an ML-based perception component. Various environmental conditions are considered as the input space such as levels of snow, rain, or fog as well as road topology. Conditions that led to critical situations were discovered by the search-based approach, i.e., use of an evolutionary algorithm. Specifically, the work builds a decision tree to model how much critical situations tend to occur in what conditions.

Another work used uses formal analysis to identify positions of the preceding car such that recognition failure of the car leads to the failure of driving requirements (crash or near-crash). Then, physical conditions can be mapped to constraints on the images, e.g., the position of the preceding car, especially if we use computer-graphic images rather than natural images [3].

4.8.2 Metamorphic Testing for Various Testing Targets

Metamorphic testing described in Section 4.7.4.1 is a generic approach not limited to testing of ML components. It has been used for many applications such as mathematical computations. We can use it to test the training pipeline used for building ML components, including the training data and the code for running the training task.

The work in [4] investigated the effectiveness of metamorphic testing for training pipeline implementations. An example of the metamorphic relations is given as follows:

1. Prepare a certain training dataset. This is the input to the training pipeline, and then a ML component is built. Finally, prediction results are also made by using the component. Suppose for example the object in *image*0 is judged as a dog.
2. Then, apply a certain conversion on each image in the training dataset, such as shuffling RGB channels or rotating orthogonally. Then, this modified input is given to the training pipeline to built another ML component. The *image*0 is modified to *image*0′ and passed to the second ML component. Then, it is expected *image*0′ (color-shuffled or rotated object) should be judged as a dog.

We do not need to assume the object in *image*0 is really a dog, as we are only checking the metamorphic relation. It was reported [4] that these kinds of tests could reveal many implementation bugs experimentally embedded in the training pipeline code.

4.9 Concluding Remarks

In this chapter, we have discussed emerging techniques of automated test generation for ML components and systems. We have focused on automated input generation methods primarily for image classification tasks and discussed how the concepts for testing of programs have been evolved for ML components and systems, especially, DNNs.

On the other hand, practical effectiveness of these testing techniques is still under investigation. For example, there are several studies that checked effectiveness of the neuron coverage or similar structural coverage criteria [7].

Although more empirical studies are required, investigation on ML testing is further accelerating. We hope the content of this chapter provides readers the foundation for tackling the challenges in ML testing. We envision effective disciplines and techniques will be established soon thorough academia-industry collaboration.

References

[1] Abdessalem, R.B., Nejati, S., Briand, L.C. and Stifter, T. 2018. Testing vision-based control systems using learnable evolutionary algorithms. *In*: The 40th International Conference on Software Engineering (ICSE 2018), pages 1016–1026, May 2018.

[2] Corbett-Davies, S. and Goel, S. 2018. The measure and mismeasure of fairness: A critical review of fair machine learning. https://arxiv.org/abs/1808.00023, August 2018.

[3] Dreossi, T., Donz, A. and Seshia, S.A. 2017. Compositional falsification of cyber-physical systems with machine learning components. *In*: The 9th NASA Formal Methods Symposium (NFM 2017), pages 357–372, May 2017.

[4] Dwarakanath, A., Ahuja, M., Sikand, S., Rao, R.M., Jagadeesh Chandra Bose, R.P., Dubash, N., et al. 2018. Identifying implementation bugs in machine learning based image classifiers using metamorphic testing. *In*: The 27th ACM SIGSOFT International Symposium on Software Testing and Analysis (ISSTA 2018), pages 118–120, July 2018.

[5] Fraser, G. and Arcuri, A. 2013. Whole test suite generation. IEEE Transactions on Software Engineering, 39(2): 276–291, February 2013.

[6] Fujii, G., Hamada, K., Ishikawa, F., Masuda, S., Matsuya, M., Myojin, T., et al. 2020. Guidelines for quality assurance of machine learning-based artificial intelligence. *In*: International Journal of Software Engineering and Knowledge Engineering, Volume 30, pages 1589–1606.

[7] Harel-Canada, F., Wang, L., Gulzar, M.A., Gu, Q. and Kim, M. 2020. Is neuron coverage a meaningful measure for testing deep neural networks? In The 28th ACM Joint Meeting on European Software Engineering Conference and Symposium on the Foundations of Software Engineering (ESEC/FSE 2020).

[8] Kim, J., Feldt, R. and Yoo, S. 2019. Guiding deep learning system testing using surprise adequacy. *In*: The 41st International Conference on Software Engineering (ICSE 2019), pages 132–142, May 2019.

[9] Ma, L., Juefei-Xu, F., Zhang, F., Sun, J., Xue, M., Li, B., et al. 2018. DeepGauge: Multi-granularity testing criteria for deep learning systems. *In*: The 33rd ACM/EEE International Conference on Automated Software Engineering (ASE 2018).

[10] Mueller, A.C. and Guido, S. 2016. Introduction to Machine Learning with Python: A Guide for Data Scientists. O'Reilly Media, 2016.

[11] National Institute of Advanced Industrial Science and Technology. 2021. Machine Learning Quality Management Guideline, 1st English edition. Technical Report, CPSEC-TR-2020002, February 2021.

[12] Pei, K., Cao, Y., Yang, J. and Jana, S. 2017. Deepxplore: Automated whitebox testing of deep learning systems. *In*: The 26th Symposium on Operating Systems Principles (SOSP 2017), pages 1–18, October 2017.

[13] Riccio, V., Jahangirova, G., Stocco, A., Humbatova, N., Weiss, M. and Tonella, P. 2020. Testing machine learning based systems: a systematic mapping. Empirical Software Engineering, 25(6): 5193–5254.

[14] Segura, S., Fraser, G., Sanchez, A.B. and Ruiz-Corts. A. 2016. A survey on metamorphic testing. IEEE Transactions on Software Engineering, 42(9): 805–824, September 2016.

[15] Sen, K. 2007. Concolic testing. *In*: The 22nd IEEE/ACM International Conference on Automated Software Engineering (ASE 2007), pages 571–572, November 2007.

[16] Sun, Y., Wu, M., Ruan, W., Huang, X., Kwiatkowska, M. and Kroening. D. 2018. Concolic testing for deep neural networks. *In:* The 33rd ACM/EEE International Conference on Automated Software Engineering (ASE 2018).

[17] Tian, Y., Pei, K., Jana, S. and Ray, B. 2018. DeepTest: automated testing of deep-neural-network-driven autonomous cars. *In*: The 40th International Conference on Software Engineering (ICSE 2018), pages 303–314, May 2018.

[18] Udeshi, S., Arora, P. and Chattopadhyay, S. 2018. Automated directed fairness testing. *In*: Proceeding of the 33rd ACM/IEEE International Conference on Automated Software Engineering, ASE 2018, pages 98–108, New York, NY, USA, 2018. Association for Computing Machinery.

[19] Weyuker, E.J. 1982. On testing non-testable programs. The Computer Journal, 25(4): 465–470, November 1982.

[20] Zhang, J.M., Harman, M., Ma, L. and Liu, 2022. Y. Machine learning testing: Survey, landscapes and horizons. IEEE Transactions on Software Engineering, 48(1): 1– 36, January 2022.

[21] Zhang, M., Zhang, Y., Zhang, L., Liu, C. and Khurshid, S. 2019. DeepRoad: GAN-based metamorphic testing and input validation framework for autonomous driving systems. *In*: The 33rd ACM/IEEE International Conference on Automated Software Engineering (ASE 2018), pages 132–142.

Chapter 5

Safety Standards and Safety Assurance Framework for ADS

Kenji Taguchi
Underwriters Laboratories, Tokyo, Japan

Fuyuki Ishikawa
National Institute of Informatics, The University of Electro-Communications and SOKENDAI, Tokyo, Japan

Tsutomu Kobayashi
Japan Aerospace Exploration Agency (JAXA), Tsukuba, Japan

Ichiro Hasuo
National Institute of Informatics and SOKENDAI, Tokyo, Japan

Artificial Intelligence (AI)/Machine Learning (ML) systems greatly impact the current safety regulatory framework, including domestic/international standards and guidelines on safety critical systems such as automobiles, railways and aeroplanes. In this chapter, we will briefly explore the current safety standards and guidelines around AI/ML-related systems, especially for automated driving systems (ADS).

The chapter is organised as follows. Current safety standards and guidelines for ADS are reviewed in Section 5.1, and safety assurance for ADS based on safety cases is discussed in Section 5.2. In particular, in Section 5.2.2, we compare the safety case framework from Uber ATG [9] and the ANSI/UL 4600 [4], based on our results in [25].

5.1 Safety Standards and Guidelines for ADS

Traditional safety standards for safety critical systems focus on the functional safety of the target systems. Functional safety standards such as ISO 26262 [6] for automotive electric and electronic (E/E) systems have dealt with systematic failures and random hardware failures found in E/E systems. They also mandate the system life-cycle to reduce risks involved in those failures. However, AI/ML systems introduce unexpected behaviours that threaten those systems' safety. In this section, we will explain some standards by ISO, ANSI/UL and an industrial consortium and discusses how they address safety issues in ADS.

5.1.1 ANSI/UL 4600

ANSI/UL 4600 [4] is a safety standard for autonomous products standardised by Underwriters Laboratories (UL). This standard has several unique characteristics compared with other safety standards. One of the main characteristics is to rely on the *safety case* as the primary safety assurance framework, and its use is carefully and meticulously specified in the standard. We will explain the history of the notion of safety case and its applications to ADS in later sections including Section 5.2.1. The contents of a safety assurance requirement (which is a clause in the standard) are classified into the following categories:

- **General:** General description of the requirement.
- **Mandatory:** More detailed description of the requirement. Addressed by the safety case. No deviation is permitted.
- **Required:** Addressed by the safety case. The deviation is only permitted in some particular cases.
- **Highly Recommended:** Best practices, which should be followed. May be omitted in case of low-risk items.
- **Recommended:** Optional. Good practices and/or suggestions for helpful techniques.
- **Conformance** Conformance measure for the requirement via self-audit and independent assessment following Chapter 17 in the standard.

The category choice for an assurance requirement depends on the evaluated level of risks. For instance, a Mandatory requirement should be faithfully followed and addressed in a safety case. On the other hand, a Recommended requirement can be omitted if the level of risk is low.

Assurance requirements stated in ANSI/UL 4600 range from Safety Case (in Chapter 5) to (Risk) Assessment (in Chapter 17).

5. Safety Case and Arguments
6. Risk Assessment
7. Interaction with Humans and Road Users
8. Autonomy Functions and Support
9. Software and System Engineering Processes
10. Dependability
11. Data and Networking
12. Verification, Validation, and Test
13. Tool Qualification, COTS, and Legacy Components
14. Lifecycle Concerns
15. Maintenance
16. Metrics and Safety Performance Indicators (SPIs)
17. Assessment

We briefly explain Chapters 5 and 8 of ANSI/UL 4600. Chapter 5 discusses a safety case format and a style, such as the use of Goal Structuring Notation (GSN) [7], avoidance of argument defects by using a checklist of logical fallacies and the use of epistemic defeaters. It also deals with risk acceptance, such as an acceptable level of risk by acceptance criteria, and more general issues on safety management, such as safety culture throughout the development activities. Chapter 8 deals with hazards related to the pipeline (sensing, perception, planning, etc.) of autonomy functions and safety constraints, e.g., coverage of relevant environment information on the Operational Design Domain (ODD), acceptable robustness on machine learning and AI techniques. Even though this standard does not aim solely at ADS, it discusses how to map assurance requirements in the standard to ISO 26262.

5.1.2 SOTIF: ISO 21448

ISO 21448 [5] is a safety standard referred to as the Safety of Intended Functionality (SOTIF), which deals with *the absence of unreasonable risk due to hazards resulting from functional insufficiencies of the intended functionality or by reasonably foreseeable misuse by persons*. On the other hand, functional safety deals with the malfunctioning behaviour of the electric/electronic systems.

The basic approach of the SOTIF is to argue the system's safety based on the scenarios which take interactions between the system and its surrounding environment into account. The scenarios are classified into the four areas in Fig. 5.1. These four areas are combinations of Known/Unknown and Hazardous/Not Hazardous in which *Not hazardous* scenarios represented by Areas 1 and 4 can be excluded since they do not impose any risk. The SOTIF aims to reduce the probability of scenarios in Areas 2 and 3, which would confine hazardous behaviour to an acceptable level of risk.

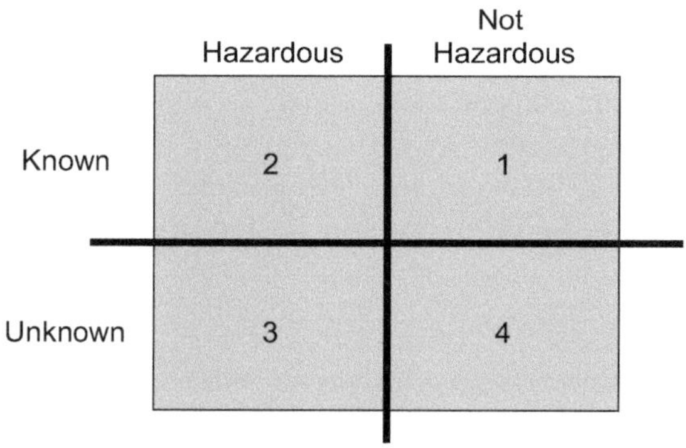

Figure 5.1: SOTIF four areas [5].

5.1.3 SaFAD: Safety First for Autonomous Driving

The Safety First for Autonomous Driving (SaFAD) [2] is an industrial consortium-based guideline for safety principles, design methods, validation, and verification of ADS. The design focuses on Fail-Safe (FS) capabilities (e.g., FS-1: Determine the location, which uses perception information from the environment, map data and motion data of an ego vehicle to ensure that the automated driving systems operate within ODD) and Fail-Degraded (FD) capabilities (e.g., FD-3: Detect when degraded performance is not available, which continuously check the availability of the degraded mode), and lists the following twelve principles for the development for ADS.

- Safe Operation (Degradation and fail-operation)
- Operational Design Domain (ODD determination)
- Vehicle Operator-Initiated Handover (Interaction for engaging and disengaging)
- Security (Protection from security threats)
- User Responsibility (Clarification of driver's responsibility)
- Vehicle-Initiated Handover (Minimal risk condition, takeover requests)

- Interdependency between the Vehicle Operator and the ADS (Fail-operation)
- Safety Assessment (V&V to ensure that the safety goals are met)
- Data Recording (Recording the relevant data and compliance with data privacy laws)
- Passive Safety (Crash scenarios)
- Behavior in Traffic (Manners on the road)
- Safe Layer (Recognition of system limits)

The essence of the safety design is represented by how Fail-Safe and Fail-Degrade capabilities are mapped to the logical architecture of ADS. A case study demonstrates how this architecture works.

SaFAD proposes a lifecycle and an assurance pipeline to collect safety (assurance) artefacts to build the safety case for DNN (Deep Neural Net). See Fig. 5.2.

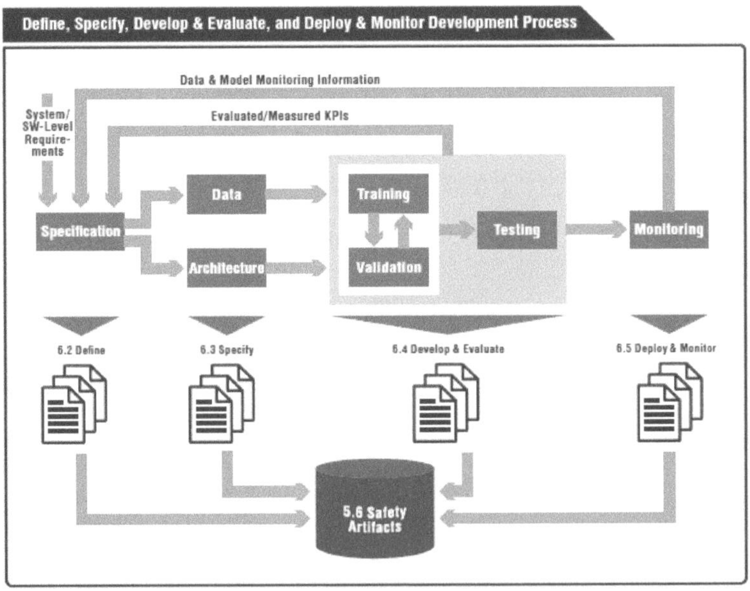

Figure 5.2: DNN Process and Safety Case (a figure from [2]).

SaFAD was issued as ISO TR 4804 [20] and is further developed as a new ISO standard ISO/AWI TS 5083 [19].

5.2 Safety Assurance for ADS

How safe is safe enough? is the perennial question we ask about the safety of ADS. We are now witnessing a new technology trend where safety cases are being adopted as the primary safety assurance framework for ADS. We will explain its historical background, related research projects, and an experimental conformance evaluation of a safety case for ADS by Uber ATG (currently Aurora Innovation).

5.2.1 Safety Case for ADS

The safety case concept was introduced after the Piper Alpha (an offshore oil platform in the North Sea) incident occurred in 1988. It killed 167 people and cost approximately 1.8 billion pounds in financial losses. The incident report by Lord Cullen [13] emphasised the importance of the goal-based regulatory approach as opposed to the traditional prescriptive approach to ensuring safety. In recent years, the goal-based regulatory approach (the so-called safety case regime) has been introduced in various industrial sectors such as automotive (ISO 26262 and ISO 21448), railway (IEC 62425), avionics (ICAO guidance for a safety case [18]) and medical devices (FDA infusion pump safety case). This has not been welcomed without any criticism. The most notable criticism of the safety case-based approach comes from N. Leveson [21], who pointed out that a safety argument can be misguided by psychological biases (e.g., confirmation bias) and logical fallacies, which would jeopardise the legitimacy of the safety case as a safety assurance framework. As discussed in the previous section, some new safety case-based standards, such as ANSI/UL 4600, take this criticism seriously and provide strict guidelines for properly using safety cases.

A safety case can take several forms, but most often, it is a structured document that can be supported by a graphical notation such as Goal Structuring Notation (GSN) [7]. Using a notation such as GSN helps present a safety case in a concise and structured manner that is easy for reviewers and assessors to understand and review.

The use of safety cases is also adopted in various safety assurance frameworks for ADS, such as Pegasus and Uber ATG's Safety Case Framework [9].

To construct a safety argument for ADS, we need to address the following structural issues when designing a safety argument structure:

- Macro argument structure
    - the top-level argument structure, which reflects the fundamental principles for safety assurance

- Micro argument structures
    - Sub-argument structures that address specific issues, e.g., safety assurance of ML components, systems engineering process specifics, etc.
- Connection argument structure
    - Connecting micro-argument structures that address different safety issues based on some consideration of harmonising cross-cutting concerns

The Safety Case Framework by Uber ATG in Section 5.2.1.2 presents an example of the macro argument structure. A macro argument structure should be carefully designed to cover the entire safety assurance framework consistently and comprehensively. An example of a micro argument is given in the SOTIF argument structure, which should be integrated into the macro argument structure later in a complete safety case. To connect each micro-argument structure, it is necessary to harmonise cross-cutting concerns that may be addressed independently in multiple micro-argument structures. For example, suppose we consider functional safety and cybersecurity in separate micro argument structures. In that case, some interference between hazards and threats should be captured in some way to connect multiple argument structures. This type of interference and the resulting argument structure is discussed in [26].

5.2.1.1 Pegasus

The Pegasus (Project for the Establishment of Generally Accepted quality criteria, tools and methods as well as Scenarios and Situations for the release of HAD functions) [1] is a research project for the safety assurance of Highly Automated Driving (HAD) functions for ADS. It has developed the scenario description language OpenScenario [8] and a safety assurance framework, among other research results. The safety assurance framework for HAD proposed by Pegasus is the Pegasus Safety Argumentation, which is a conceptual framework to support tight integration of its process and associated pipeline for assurance artefacts production, representing how they support the safety argumentation. The Pegasus safety argumentation is a layered approach which consists of four layers (layer 1: ADS acceptance model (technology (social) acceptance of the HDS functions), layer 2: Logical structure of the safety argument (in GSN), layer 3: Methods and tools (e.g., V&V) and layer 4: Evidence (results from the application of layer 3). Fig. 5.3 shows the process where evidence to support safety argumentation is generated [22].

5.2.1.2 The Safety Case Framework (Uber ATG)

The Safety Case Framework (SCF) is a safety case for ADS developed by then Uber ATG (currently Aurora Innovation). It was a detailed yet publicly

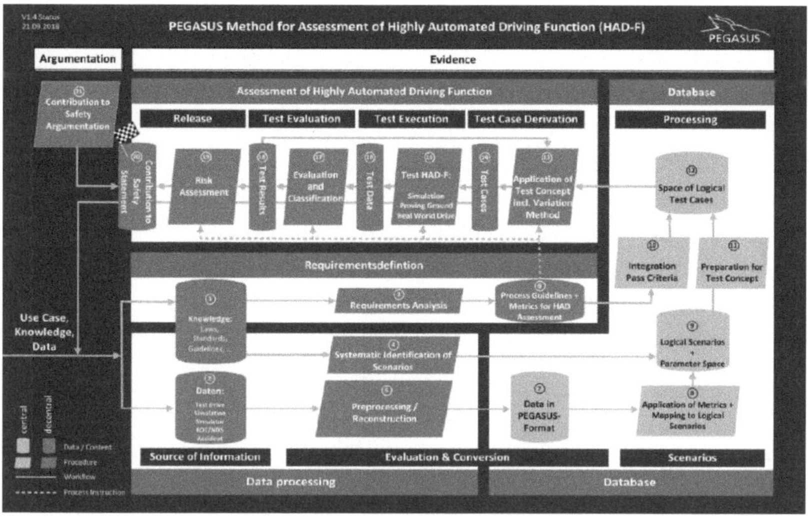

Figure 5.3: Pegasus Method for Assessment of HAD-F (a figure from [22]).

available safety case modelled in GSN. Unfortunately, the SCF we used for our study on conformance evaluation against ANSI/UL 4600 in [25] and explained in Section 5.2.2 is no longer available on the internet, and the current version by Aurora Innovation [11] does not provide the whole argument structure including evidence as solutions. In what follows, we will refer to the SCF by Uber ATG, not by Aurora Innovation.

Figure 5.4: The SCF top-level argument structure (a figure taken from [9]).

Figure 5.4 shows the top-level argument structure of the SCF. Its ultimate goal is to establish that "our self-driving vehicles are acceptably safe to operate on public roads." This goal is further broken down into the following sub-goals G1 to G5:

- **G1: Proficient** The Self-Driving Vehicle is acceptably safe during nominal operation,
- **G2: Fail-Safe** The Self-Driving Vehicle is acceptable safe in presence of faults and failures,
- **G3: Continuously improving** Any anomaly that could affect the safety of the Self-Driving Vehicle is identified, evaluated, and resolved with appropriate corrective and preventative actions,
- **G4: Resilient** The Self-Driving Vehicle is acceptably safe in case of reasonable foreseeable misuse and unavoidable events,
- **G5: Trustworthy** The Self-Driving Enterprise is trustworthy.

The argument structure under G1 is devoted to systems engineering and SOTIF, including autonomy functions. G2 is for functional safety, i.e., handling safety functions under faults and failures. G3 deals with any risks involved in the operational process and their mitigation measures, G4 is for the system's dependability, and G5 is for satisfying regulatory requirements.

This macro argument structure is very reasonably presented because it covers all the necessary safety principles required to assure the entire safety of an ADS. An attractive characteristic of this structure is that it faithfully follows the safety principles in Uber ATG's safety report [10], which is submitted to NHTSA (National Highway Traffic Safety Administration) as a Voluntary Safety Self-Assessment Report (VSSA) for ADS [23]. Maintaining consistency between a safety case and a safety report is beneficial for an automotive manufacturer since these two documents/artefacts are essential to claim its ADS's safety without reworking.

5.2.1.3 SOTIF Argument Structure

Annex A (General Guidance on SOTIF) of ISO 21448 [5] provides two examples of SOTIF-related safety arguments in GSN. They are very generic and avoid system-specific features, which helps one build their safety argument. We will explain the first example in Fig. 5.5.

The top goal is *SOTIF is achieved and upheld for the AD (Automated Driving)/DA (Driver Assistant) functionality*, which is placed in the context of *Context C1-G-SOTIF* and *Context C2-G-SOTIF*. The former context refers to the function and system specification produced as WP (Work Product) 5.5 and WP 8.5, and the latter to ODD and associated WP 5.5 and WP 8.5. Referring to some specific work products helps to understand which part of the standard is related to the argument.

The top goal is decomposed into two subsequent sub-goals. One is G-Dev; it claims that the development activities carried out faithfully follow the SOTIF-related assurance requirements. The other is G-field, which claims that if any safety issues arise in the field operation (e.g., trial runs), suitable measures are taken to keep the AD/DA functionality free from unreasonable risk. The latter sub-goal has an assumption A1-G-Field, which states that new risks referred to in the G-Field sub-goal are either scaling effects or changes in the ODD. Each goal has further decompositions referred to as Figs. A.2 and A.10 in the subsequent pages; we do not discuss them here.

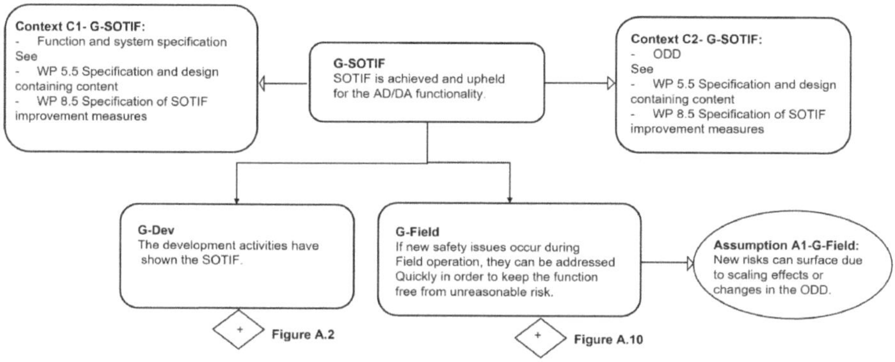

Figure 5.5: Figure A.1 in [5].

SOTIF micro argument structure needs to be built in the macro argument structure, which covers the whole safety argument for ADS without causing any inconsistencies in an argument.

5.2.2 Conformance Evaluation on the SCF against ANSI/UL 4600

We review our results in [25] that experimentally assess whether the SCF conforms to ANSI/UL 4600. The main reasons for this exercise are that we would like to develop and establish a methodology for assessing a safety case to comply with a certain standard and an analysis method on the structural characteristics of that safety case. To our knowledge, there are no well-established criteria to assess a given safety case compliant with a certain standard or analysis methods to support it.

We must emphasise that ANSI/UL 4600 we used for this study is the first edition published in 2020 [3] (currently, the third edition [4] is available).

We analysed assurance requirements specified as General (the most general assurance requirement) and Mandatory (more general assurance condition than the General and no deviation is permitted) from Chapter 5 to Chapter 17 in ANSI/UL 4600. We compared them with goals and solutions in the SCF. As was stated previously, the choice of categories of assurance requirements depends on the risk assessment result of the item. Still, it is unavailable from the SCF, so we chose the most mandatory part of the assurance requirements.

5.2.2.1 Evaluation

The SCF was not developed in conformance with ANSI/UL 4600 per se. We must emphasise that our preliminary aim of this exercise is not for a judgement of its conformance to 4600 but to get insights into the construction method of a safety case for the ADS as a whole and to establish an assessment method for a safety case in compliance with a certain safety standard and a guideline.

Some general difficulties we encountered while analysing the SCF are as follows:

1. Interpretation of a standard is inherently ambiguous. Generally, a newly issued standard such as ANSI/UL 4600 needs several years to establish a common understanding among engineers and assessors. We had difficulties understanding some safety requirements in the standard.
2. The SCF represented in GSN only provides an argument structure without any evidence, such as V&V results and with some reference to substantial evidence, such as test results, etc., so that we can only judge whether statements expressed as goals conform to some assurance requirements in the standard (we may call our approach the black-box assessment).

We used the format shown in Table 5.1 for assessment, where the content of the first column is an assurance requirement in ANSI/UL 4600, the second column for identified goal(s) and solution(s) from the SCF, which we judged being complied with the assurance requirement, and the third column for our verdict (Conformed or Nonconformed).

Table 5.1: Assessment format

ANSI/UL 4600	Safety Case Framework	Verdict
Assurance requirement in ANSI/UL 4600	Identified goal(s) and solution(s), if any	Conformed or Nonconformed (with some remarks)

We present some cases to show how we analysed and judged the conformance.

Table 5.2 shows the straightforward conformance relation between the assurance requirement 7.3.3 in ANSI/UL 4600 with a goal G2.1.3.1.1.6 in the SCF, even though slightly different terms *human-settable item parameters* and *human adjustable parameters* are used. This type of mismatch in terminology occurs in several places, but we did not take a stringent policy to interpret the terms.

Table 5.3 shows the lack of supporting argument relevant to the assurance requirement 8.3.3 for sensor fusion and redundancy management techniques with

Table 5.2: Example of assessment result #1

ANSI/UL 4600	Safety Case Framework	Verdict
7.3.3 Hazards which can be contributed to by human-settable item parameters shall be acceptably mitigated.	G2.1.3.1.1.6 Hazards and risks due to human adjustable parameters have been mitigated.	Conformed

relation to the ODD. Our judgement, shown in the third column, is that even though a goal G1.2.2.1.2.12 and a solution Sn1.2.2.1.2.12 are relevant to the assurance requirement 8.3.3, they are not sufficiently argued to comply with it.

Table 5.3: Example of assessment result #2

ANSI/UL 4600	Safety Case Framework	Result
8.3.3 Sensor fusion and redundancy management techniques shall be used as necessary to result in acceptable sensor performance for the defined ODD.		Unconfirmed (No clearly defined goals for sensor fusion, but the relevant argument we found is a goal *G1.2.2.1.2.12 Perception system has sufficient availability and coverage to detect objects in the ODD appropriately and a solution Sn1.2.2.1.2.12.5 Perception fusion architecture.)*

Table 5.4 shows that the standard has a wider scope than ADS, which might be useful for autonomous products in general but not necessarily compliant with the SCF. A Minimum Equipment List (MEL) is a list for the operation of aircraft and neither this notion nor associated assurance requirements are found in safety standards for automotive systems.

Table 5.4: Example of assessment result #3

ANSI/UL 4600	Safety Case Framework	Result
10.3.5 A Minimum Equipment List (MEL) shall be defined for each autonomous operational mode.		Unconfirmed (No argument on MEL)

ANSI/UL 4600 introduces new safety metrics called the Safety Performance Indicator (SPI), which can be defined for various properties/components to quantify safety performance. SPIs play an important role in validating a safety argument, and its definition is elaborated in edition 3. Table 5.5 shows that our judgement is done with a goal and its associated solutions (evidence).

Table 5.5: Example of assessment result #4

ANSI/UL 4600	Safety Case Framework	Result
16.3.1 Data for each defined SPI shall be collected.	*G3.2.5.2 Safety performance indicator data measurements are valid, Sn3.2.5.2.3* SPI *data collection tools, Sn3.2.5.2.4 SPI data collection justification of robustness*	Confirmed (Data collection related to SPIs can be judged as conforming indirectly from a combination of a goal and associated solutions)

As can be seen from the sample analysis results, the verdict for conformance is straightforward in some cases and difficult to judge in others.

5.2.2.2 Analysis Results

The first analysis result shown in Table 5.6 summaries the conformance ratio for each chapter in 4600, where **C** stands for conformed and **N** for Nonconformed.

Table 5.6: Conformance ratio for each chapter in ANSI/UL 4600

Chapter of ANSI/UL 4600	C	N
5. Safety Case and Arguments	8	8
6. Risk Assessment	9	11
7. Interaction with Humans and Road Users	3	6
8. Autonomy Functions and Support	6	4
9. Software and System Engineering Processes	8	1
10. Dependability	19	6
11. Data and Networking	5	5
12. Verification, Validation, and Test	11	9
13. Tool Qualification, COTS, and Legacy Components	6	0
14. Lifecycle Concerns	6	8
15. Maintenance	0	6
16. Metrics and Safety Performance Indicators (SPIs)	5	4
17. Assessment	2	10

As Table 5.6 shows, the conformance ratio varies significantly from one chapter to another. The highest score in conformance is 8:1, which is marked by Chapter 9 of Software and System Engineering Process. This analysis result is reflected by very detailed process-related arguments in the SCF. The lowest score is 0:6 by Chapter 15 of Maintenance. This is mainly because many arguments only partially conform with assurance requirements in ANSI/UL 4600. For instance, an assurance requirement 15.1.1 in ANSI/UL 4600 for mitigation of hazards and risks related to maintenance and inspection is only argued by maintenance in the SCF; no argument for inspection is found in the SCF.

We mapped the previous analysis results onto the top-level argument structure of the SCF, which comprises five sub-goals; **G1: Proficient**, **G2: Fail-Safe**, **G3: Continuously Improving**, **G4: Resilient** and **G5: Trustworthy** shown in Fig. 5.4. The result is displayed in Table 5.7.

This table can be interpreted horizontally and vertically. The horizontal interpretation shows which chapter in ANSI/UL 4600 deals with which sub-goal of the SCF. The vertical interpretation shows which sub-argument structure addresses issues specified in corresponding chapters in ANSI/UL 4600.

Some observations from the horizontal analyses are listed below:

- H-1 Chapter 5. Safety case: Goals are spread across almost all sub-argument structures since the fundamental mechanism for safety assurance is based on the safety case.
- H-2 Chapter 6. Risk Assessment: It is covered by G2: Fail-Safe and G3: Continuously Improvement. This is plausible since risk assessment is mainly carried out by Fail-Safe operations and operational processes addressed in G2 and G3.
- H-3 Chapter 8. Autonomy Functions and Support: It is mainly covered in G1: Proficient, which is plausible for its intention.
- H-4 Chapter 9. Software and System Engineering Process: It is mostly covered by G1: Proficient. We cannot judge its plausibility, but there is no reason against it.
- H-5 Chapter 10. Dependability: It is either safety-related (G2: Fail-Safe) or resilient (G4: Resilient), which results from the very nature of dependability.
- H-6 Chapter 16. Metrics and SPIs: It is entirely covered by G3: Continuously Improving. It is plausible since SPIs are used for risk assessment in the operation phase.
- H-7 Chapter 17. Assessment: It is only covered by G5: Trustworthy due to its role in assessment.

According to this analysis, we can conclude that the distribution of intended use of assurance requirements is very plausible.

Likewise, some observations on the vertical analyses are listed below:

- V-1 G1: Proficient: It covers almost all Chapters 8, 9 and 12 since it intends to cover systems engineering development as a whole and autonomy,

Table 5.7: Mapping analysis results to the top-level structure of the SCF

	G1: Proficient	G2: Fail-Safe	G3: Continuously Improving	G4: Resilient	G5: Trustworthy
5. Safety Case	G1.1.1.8.1	G2.1.1, G2.1.1.1.2	G3.2.1.6		G5.1.1.1.1
6. Risk Assessment		G2.1.3, G2.1.1.2, G2.1.1.3, G2.1.1.4	G3.3.1.1.4, G3.3.1.1.3		
7. Interaction with Humans and Road Users		G2.1.3.1.1.6, G2.1.1.4.12		G4.2.1.1,	
8. Autonomy Functions and Support	G1.3.1.6.2, G1.2.2.1.1, G1.2.2.6.3	G.2.1.1.2.2			
9. Software and System Engineerin	G1.1.1.6.4, G1.1.1.1, G1.1.1.5.4, G1.1.1.5.4, G1.1.1.6, G1.3.1.4, G1.1.1.6		G3.2.1.4		
10. Dependability	G1.2.2.7,	G2.1.3.1.5, G1.2.2.5.1.9, G2.1.3.3, G2.1.3.3.1, G2.1.3.4.3.3		G4, G4.1.4, G4.1.1.5.1, G4.1.3.1, G4.3	
11. Data and Networking		G2.1.3.1.1.10, G2.1.3.1.11, G2.1.1.14			
12. Verification, Validation, and Test	G1.3.1.4.1.3.2, G1.3.1.4.1.2.1, G1.2.2.1.2.12, G1.3.1.4.1.6.4, G1.3.1.4.1.6.4, G1.1.	G2.1.3.5.1.2.2.5	G3.4.1.6	G4.3.2.2,	
13. Tool Qualification, COTS, and Legacy Components	G1.1.1.5.4.6, Sn1.1.1.9.4.7.1	G2.1.1.3.5, G2.1.1.4.7, G2.1.3.1.17			
14. Lifecycle Concerns		G2.1.3.1.4, G2.1.3.1.18, G2.1.3.1.19, G2.1.3.1.21, G2.1.3.1.22, G2.1.3.1.23.2			
15. Maintenance					
16. Metrics and Safety Performance Indicators (SPIs)			G3.2.1.2, G3.2.3, G3.2.5.2, Sn3.2.5.2.3, G3.3.2, Sn3.3.2.1		
17. Assessment					G5.1.3.2.2, G5.1.3.2.2, Sn5.1.3.2.2.1

V-2 G2: Fail-Safe: It covers almost all chapters since safety analysis is carried out in almost all activities,

V-3 G3: Continuously Improving: It covers a part of Chapter 6 and all of Chapter 16, which matches its intention,

V-4 G4: Resilient: It mostly covers Chapter 10, which can be predictable from its main role,

V-5 G5: Trustworthy: It covers Chapters 5 and 17. Particularly why G5 covers Chap 17 is based on the same reason as H-7.

These vertical analyses show that each sub-goal and its argument structure fulfils its intention.

5.2.2.3 Related Work

A few research works address the issues we pursued in our evaluation in Section 5.2.2. The closest is that by Graydon et. al [17], which discusses issues on conformance to software assurance standards and presents a conformance arguments structure in which how a part of an argument is linked to objectives in software safety standard in avionics DO-178B [24]. This approach appears useful since assessing compliance with a given safety argument is time-consuming for assessors. This approach would help them understand the intent of the safety argument and its relation with assurance requirements/ objectives in the standard.

Submission of a safety case is mandated in several safety standards, as was stated previously. Many research works propose their methods of constructing a safety case for a certain standard. For instance, Dargar et. al [14, 15] focuses on a safety case for ISO 26262, but it does not deal with conformance criteria for the standard. This kind of research result assumes that conformance to a standard is built into its approach and does not provide any explicit criteria for how the approach assures conformance with a standard.

The safety case approach to assure the safety of AI/ML systems is carried out by several authors. For instance, Gauerhof [16] focused on the inherent nature of functional inefficiencies which could be caused by AI/ML systems and provided a safety argument structure that covers their root causes. As was explained in a previous chapter, the Pegasus is a European research project on autonomous driving and developed the safety argumentation framework for Highly Automated Driving Functions. Its presentation material [22] includes reference to standards, e.g., ISO 26262, ISO 21448 and NHTSA guidelines. Still, it does not detail how the framework can conform to those standards and guidelines.

5.2.2.4 Final Remarks

To obtain more objective judgement on our evaluation results in Section 5.2.2, we consulted various industry experts (OEMs, suppliers and tool vendors) in functional safety, SOTIF, and ADS on our evaluation results by providing them with the check sheet (Table 5.1). Their overall response was positive due to our detailed analyses on the conformance relation between ANSI/UL 4600 and the SCF, which they think would help to understand ANSI/UL 4600 and the role of a safety case for documents for future certification of their ADS. They found difficulty assessing the results due to terminology mismatches between the two and the lack of deep understanding of ANSI/UL 4600. Still, they provided helpful comments to improve our results.

Here are our observations from the survey on related work and our experience with this experimental evaluation on the SCF in Section 5.2.2.

- Safety principles affecting the top-level argument structure may vary from one company to another. There remains a question of whether there would be a generic and one-fit-all top-level argument structure.
- Micro-argument patterns adopted in the SCF are very traditional, as seen in many existing works, e.g., papers and guidelines. There are several proposals for micro-argument structures for machine learning and SOTIF so that these results could be used as micro-argument patterns.
- Safety case approaches in the Pegasus project and SaFAD could be useful to connect the assurance artefact production pipeline (simulation results, etc) with the argument structure.
- Interrelations between functional safety, SOTIF, and cybersecurity are not presented in the SCF. Security-informed safety case [12] type of approach or argument patterns for safety and security [26] could be used as micro-argument patterns.

References

[1] Pegasus Project, https://www.pegasusprojekt.de/en/home.

[2] Safety First for Autonomous Driving, 2019. https://group.mercedesbenz.com/documents/innovation/other/safety-first-for-automateddriving. pdf.

[3] ANSI/UL 4600: 2020. Standard for Safety for Evaluation of Autonomous Products. ed 1.

[4] ANSI/UL 4600: 2023. Standard for Safety for Evaluation of Autonomous Products. ed 3.

[5] ISO 21448:2022. Road Vehicles—Safety of the intended functionality.

[6] ISO 26262:2018. Road Vehicles—Functional Safety.

[7] ACWG. 2021. Goal Structuring Notation Community Standard (ver. 3).

[8] ASAM. 2021. OpenScenario (ver. 1.1.0).

[9] Uber ATG. 2020. Safety Case Framework.

[10] Uber ATG. 2020. Safety Report—A principled approach to safety.

[11] Aurora. 2022. Safety Case Framework, https://safetycaseframework.aurora.tech/gsn.

[12] Bloomfield, R.E., Netkachova, K., and Stroud, R.J. 2013. Security-informed safety: If it's not secure, it's not safe. *In*: Software Engineering for Resilient Systems, 5th International Workshop, SERENE 2013, LNCS 8166, pages 17–32. Springer.

[13] Hon Lord Cullen. 1990. The Public Inquiry into the Piper Alpha Disaster. Dept. of Energy, 1 and 2.

[14] Dardar, R. 2013. Building a Safety Case in Compliance with ISO 26262 for Fuel Level Estimation and Display System. Master Thesis, Mälardalen University.

[15] Dardar, R., Gallina, B., Johnsen, A., Lundqvist, K., and Nyberg, M. 2012. Industrial Experiences of Building a Safety Case in Compliance with ISO26262. *In*: 23rd IEEE International Symposium on Software Reliability Engineering Workshops, ISSRE Workshops, pages 349–354.

[16] Gauerhof, L., Munk, P., and Burton, S. 2018. Structuring Validation Targets of a Machine Learning Function Applied to Automated Driving. *In*: Safecomp 2018, LNCS 11093, pages 45–58. Springer.

[17] Graydon, P., Habli, I., Hawkins, R., Kelly, T., and Knight, J. 2012. Arguing Conformance. *In*: IEEE Software, number 3, pages 50–57. IEEE .

[18] ICAO. 2011. Guidance Material on Building a Safety Case for Delivery of an ADS-B Separation Service. version 1.0.

[19] ISO. ISO/AWI TS 5083:Road vehicles—Safety for automated driving systems—Design, verification and validation.

[20] ISO. 2019. TR 4804:2020: Road vehicles—Safety and cybersecurity for automated driving systems—Design, verification and validation.

[21] Leveson, N. 2012. White Paper on the Use of Safety Cases in Certification and Regulation. http://sunnyday.mit.edu/SafetyCases.pdf.

[22] Maus, A. 2019. PEGASUS Safety Argumentation. 2019.

[23] NHTSA. 2017. Automated Driving Systems: A Vision for Safety 2.0, https://www.nhtsa.gov/sites/nhtsa.gov/files/documents/13069aads2.0_090617_v9a_tag.pdf.

[24] RTCA. 1992. DO-178B: Software Consideration in Airborne System and Equipment Certification.

[25] Taguchi, K. and Ishikawa. F. 2021. Experimental Conformance Evaluation on UBER ATG Safety Case Framework with ANSI/UL 4600. *In*: Safecomp Workshop, WAISE 2021, LNCS 12853, pages 272–283. Springer.

[26] Taguchi, K., Souma, D., and Nishihara, H. 2015. Safe & Sec Case Patterns. *In*: Safecomp Workshop, ASSURE 2015, LNCS 9338, pages 27–37. Springer.

Chapter 6

Uncertainty-wise Testing

Man Zhang
Kristiania University College, Oslo, Norway

Shaukat Ali
Simula Research Laboratory, Oslo, Norway

Tao Yue
Simula Research Laboratory, Oslo, Norway and Nanjing University of Aeronautics and Astronautics, Nanjing, China

Fuyuki Ishikawa
National Institute of Informatics, The University of Electro-Communications and SOKENDAI, Tokyo, Japan

6.1 Introduction

There have been intensive discussion on uncertainty in software engineering as more and more systems operate deeply in the real world [1]. In the context of software engineering, uncertainty refers to aspects that make it impossible to fix everything at development time. There has been various classifications of uncertainty, 26 types of uncertainty were discussed in [5] in the context of dynamically adaptive systems. One example of uncertainty type is unpredictable environment where events and conditions relevant with the state transition cannot be fully predicted and fixed beforehand. This chapter presents an approach to explicitly handle the uncertain aspects through the development process from requirement models to executable tests.

Uncertainty is fairly common in Cyber-Physical Systems (CPSs) due to their nature. Thus, various types of uncertainties in the CPS development must be identified, understood, and properly handled to eliminate (if possible) any safety and security issues during their operation. Figure 6.1 shows a typical CPS together with various places at which different uncertainties may occur in the CPS. A CPS consists of a set of *physical units* that communicate with each other, e.g., via information networks (e.g., 4G or directly wired connections). Each *physical unit* itself may consist of dedicated hardware with deployed software or could be a simple sensor reading data from the environment.

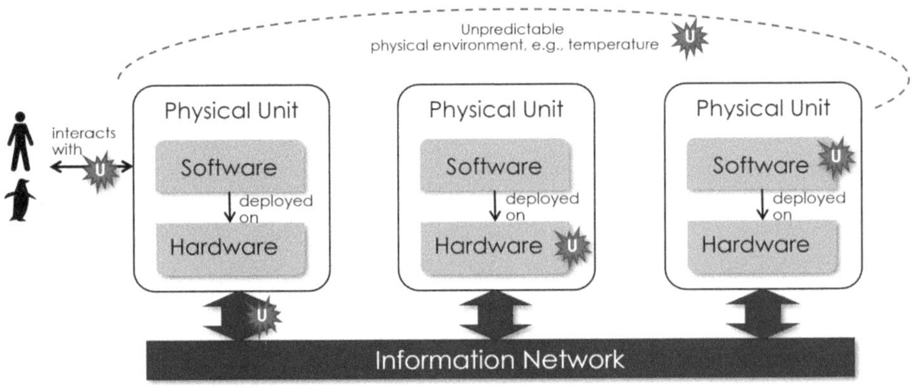

Figure 6.1: Cyber-physical system and uncertainties.

Uncertainties may appear at various situations in a CPS. First, uncertainties may be introduced by agents, e.g., humans, animals, or other systems that interact with a CPS. A typical example is humans interacting with the CPS in an unexpected manner. Second, uncertainties may occur in physical units such as in software, or hardware, or both. Examples include software and hardware errors leading to uncertain behaviors of a CPS. Third, uncertainties may occur in the information network used by the physical units to communicate with each other, such as due to packet loss or slow internet connection. Fourth, uncertainties may happen in the physical environment of a CPS, such as unexpected high temperatures. Fifth, uncertainties can occur due to interactions of more than one of the uncertainties described in the first four points.

6.2 Running Example

To explain our approach from the beginning to the end in a unified manner, we will use a running example of *Safe Home* which is inspired by a case study from Pressman [4]. This case study is about smart features implemented in a sensor-based smart home. In particular, we will focus on functionalities related to ensuring the safety and security of the system. Examples of such functionalities include detecting intrusion and other disasters such as fire or water leaks based on specialized sensors. For the sake of simplicity, we only focus on one such functionality, i.e., related to *intrusion detection*. This functionality aims at detecting the intrusion at home with continuous monitoring via dedicated motion sensors installed on windows and doors of the home. If a motion is detected, the home owner is notified and an alarm is enabled. If it is a false alarm, the home owner can simply enter her/his secret code and stop the alarm. If the secret code is entered wrong more than three times, a notification is sent to the police automatically. If the home owner is away and the alarm keeps ringing for more than the configured number of minutes (e.g., 5 minutes), a notification is also sent to the police automatically.

A use case diagram corresponding to this functionality is shown in Fig. 6.2. It has four actors, i.e., *Home Owner*, *Control Panel*, *Alarm*, and *Sensor*. In addition, it has two use cases *Monitor Windows and Doors* and *Configure Monitoring Area*. *Home Owner* can configure the monitoring area through a dedicated control panel via the *Configure Monitoring Area* use case. This use case is included by the *Monitor Windows and Doors* use case, which in turn detects an intrusion via dedicated sensors, consequently activates the alarm and notifies the home owner or the police, in case an intrusion is detected. Additionally, *Home Owner* can enable/disable the detection via *Control Panel*.

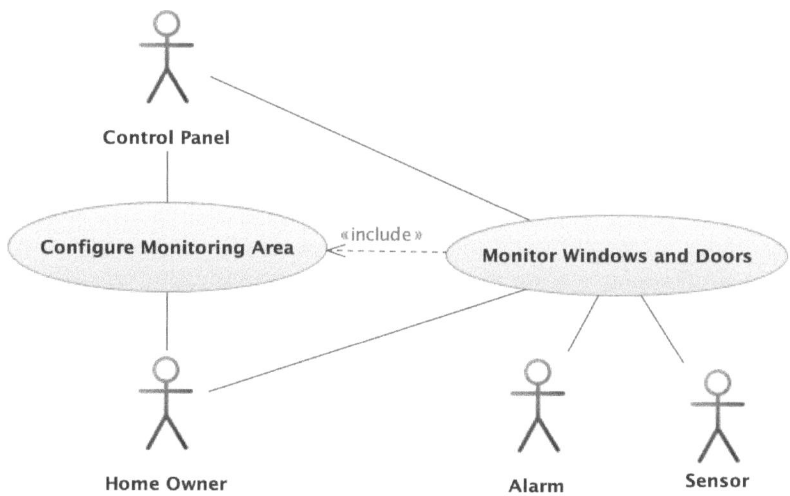

Figure 6.2: Use case diagram of safe home.

6.3 Overview of Uncertainty-wise Testing Approaches

This chapter focuses on testing CPSs under uncertainties, which is composed of a series of methodologies from specifying and modeling uncertainties, and all the way down to test generation and test execution. An overview of these methodologies and their relationships are shown in Fig. 6.3. The figure shows that the overall process has four main methodologies, which were proposed based on an uncertainty conceptual model (see Section 6.4 for more details).

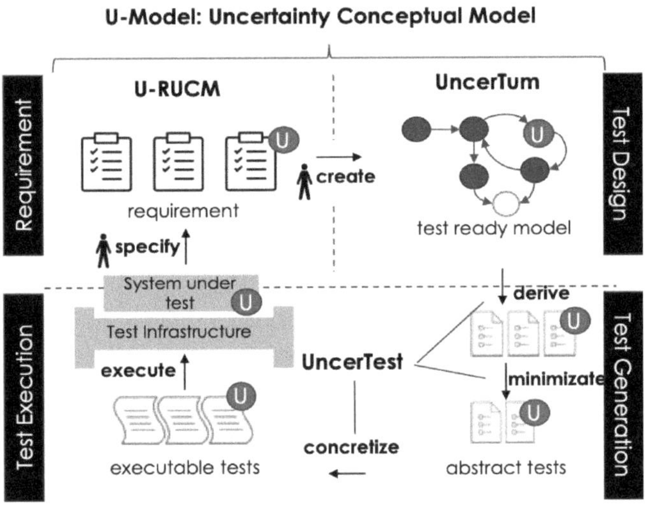

Figure 6.3: Overview of uncertainty-wise approaches.

The uncertainty conceptual model is called *U-Model* and it conceptualizes uncertainties in the context of developing and testing CPSs [14]. U-Model was specifically designed to identify various uncertainties in CPSs at a higher level of abstraction for the purpose facilitating common understanding of various uncertainty related concepts. Later on, *U-Model* was used as the basis to support the development of specialized uncertainty identification, definition, and classification mechanisms tailored for developing and testing CPSs of different domains and for achieving various purposes such as model-based testing (MBT) of CPSs under uncertainty and uncertainty-wise requirements verification and validation. In this chapter, we will demonstrate its application to support MBT of CPSs under uncertainty. Concretely, first, we will discuss a methodology for specifying uncertainty-related testing requirements in Section 6.5. Second, we will present an uncertainty-wise test design to develop test ready models using specified uncertainty testing requirements in Section 6.6. Finally, test generation (including test optimization) based on defined test ready models, and test execution will be discussed together in Section 6.7.

6.4 Conceptual Understanding of Uncertainty

This section describes conceptual understanding of uncertainty in the context of CPSs. Given the complexity of the topic, it is important to conceptually understand various concepts related to uncertainty at a higher-level of abstraction. We present detail of *U-Model* in Section 6.4.1 to support conceptual modeling and understanding.

6.4.1 U-Model—The Conceptual Model for Uncertainty

U-Model was proposed with the aim to take the subjective perspective of understanding uncertainties; thus, it was built around the concept of *Belief*. Thus, at a high-level, U-Model consists of *BeliefModel*, which is imported by *MeasureModel* and *UncertaintyModel*, as shown in Fig. 6.4. *BeliefModel* describes necessary concepts and their relationships related to belief. *UncertaintyModel* focuses on concepts and their relationships concerning uncertainty. Finally, *MeasureModel* formulates concepts related to measurements that can be associated with beliefs and uncertainties. In the following subsections, we will describe each of these models and illustrate them with the running example (Section 6.2).

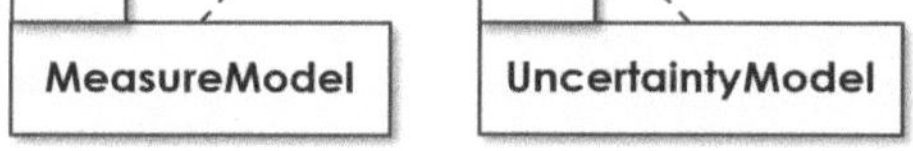

Figure 6.4: Architecture of *U-model*.

6.4.1.1 Belief Model

The various concepts and relationships among them from *BeliefModel* are shown in Fig. 6.6, whereas their definitions are provided in Table 6.1. Note that these definitions are reused from the original paper [14]. Since *U-Model* takes the subjective perspective when defining uncertainties in the context of the development and testing of CPSs, *U-Model* defines *BeliefAgent* (see B2 in Table 6.1). For example, in our context, a belief agent can be a tester who wants to test *SafeHome*. The tester wants to test her/his *Belief* (see B1 in Table 6.1) about *SafeHome*, e.g., whether motion sensor on windows can detect intrusion. As shown in the figure, *Belief* is an abstract concept and it has to be concertized as *BeliefStatement* (see B3 in Table 6.1). A belief statement could be a simple statement in English or a complex model (e.g., a UML state machine). An example of such a statement is: Tester believes that "*the system can detect the intrusion with 99% probability.*"(see Fig. 6.5) and tester can state that she/he is highly confidence (i.e., a measurement of the belief) about her/his statement. The belief may be supported by *Evidence* (see B4 in Table 6.1). For example, the evidence for the statement could be logs of previous versions of the product. The belief statement has a source, because of which a belief agent has *Uncertainty* (see U1 in Table 6.2) in her/his statement, i.e., *IndeterminacySource* (see B5 in Table 6.1). In this example, the source of the uncertainty might be related to the sensors or the communication between the sensors and the system. Both *IndeterminacySource* and *Uncertainty* can have associated *Measurement* (see B6 in Table 6.1), of a particular *Measure* (see Fig. 6.8). With this statement, the measurement of the uncertainty (i.e., "the system can detect intrusion") is 99%.

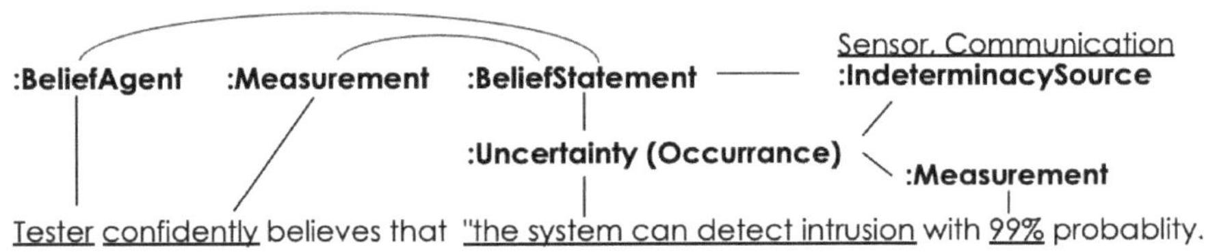

Figure 6.5: An example of identifying and defining uncertainties with *U-model*.

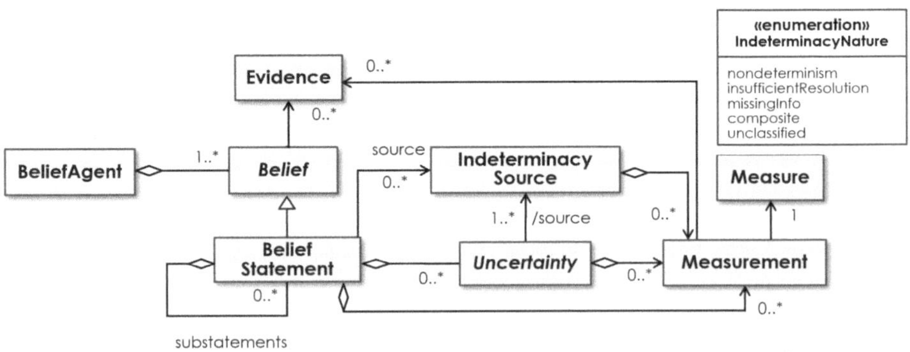

Figure 6.6: Belief model.

6.4.1.2 Uncertainty Model

UncertaintyModel in Fig. 6.7 shows various concepts and relationships among them. An uncertainty can be of any of the following types: *Occurrence* (see U2

Table 6.1: Definition of the belief model.

Concept#	Name	Definition
B1	Belief	"A belief is an implicit subjective explanation or description of some phenomena or notions that is held by a BeliefAgent"
B2	Belief Agent	"BeliefAgent represents an individual, a community of individuals sharing the same set of beliefs, or a technology, such as a software system, with built-in beliefs."
B3	Belief Statement	"A BeliefStatement is an explicit specification of some Belief about a possible phenomenon or notions belonging to a given subject area."
B4	Belief Evidence	"Evidence is either the observation of or record of a real-world event occurrence or, alternatively, the conclusion of some formalized chain of logical inference, which provides information that may contribute to determining the validity (i.e., truthfulness) of a BeliefStatement."
B5	Indeterminacy Source	"IndeterminacySource represents a situation whereby the information required to ascertain the validity of a BeliefStatement is indeterminate in some way, resulting in uncertainty being associated with that statement."
B6	Measurement	"Measurement represents the optional quantification (or qualification) that specifies the degree of Belief/ Uncertainty/IndeterminacySource."

in Table 6.1), *Content* (see U3 in Table 6.1), *Time* (see U4 in Table 6.1), *Environment* (see U5 in Table 6.1), and *Geographical Location* (see B2 in Table 6.1). In the belief statement, the tester holds that the system has uncertainty about the detection of intrusion; thus, the uncertainty type is *Occurrence*. In addition, an uncertainty can have several other properties, i.e., *Pattern*, *Lifetime*, *Locality*, *Risk*, and *Effect* (see definitions U7-U11 in Table 6.2). For this uncertainty (i.e., detect the intrusion successfully), its *Effect* is none and the probability of the occurrence uncertainty is 99%, thus the *Risk* would be *Low*. But for its alternative (i.e., fail to detect the intrusion), its *Effect* is extreme and the probability is low (1%), thus the *Risk* would be *High*.

6.4.1.3 Measure Model

MeasureModel of *U-Model* is shown in Fig. 6.8. *Measure* can be of many different types; however, at a high level, these can be classified into *Probability*, *Ambiguity*, and *Vagueness*. These measures are standard textbook concepts and can be consulted here [2, 3, 6, 12]. In this example, the measure is *Probability* and the measurement is 99%.

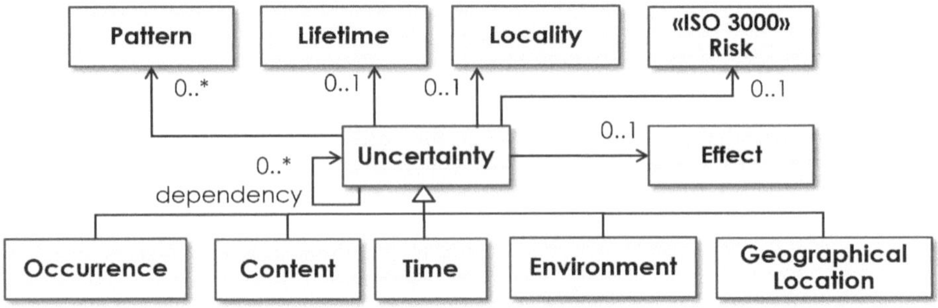

Figure 6.7: Uncertainty model.

Table 6.2: Definition of the uncertainty model

Concept#	Name	Definition
U1	Uncertainty	"Uncertainty (lack of confidence) represents a state of affairs whereby a BeliefAgent does not have full confidence in a Belief that it holds."
U2	Occurrence	"Occurrence represents a situation whereby a BeliefAgent lacks confidence in occurrence existing in a BeliefStatement."
U3	Content	"Content represents a situation whereby a BeliefAgent lacks confidence in content existing in a BeliefStatement."
U4	Time	"Time represents a situation whereby a BeliefAgent lacks confidence in time existing in a BeliefStatement."
U5	Environment	"Environment represents a situation whereby a BeliefAgent lacks confidence in environment existing in a BeliefStatement."
U6	Geographical Location	"GeographicalLocation represents a situation whereby a BeliefAgent lacks confidence in geographical location existing in a BeliefStatement."
U7	Pattern	"Pattern represents an intelligible way in which an Uncertainty appears."
U8	Lifetime	"Lifetime represents the duration of time for which an Uncertainty remains active."
U9	Locality	"A particular place or a position where Uncertainty occurs in the BeliefStatement."
U10	Risk	"Risk measures the risk associated with Uncertainty."
U11	Effect	"Effect represents the result of Uncertainty in the BeliefStatement."

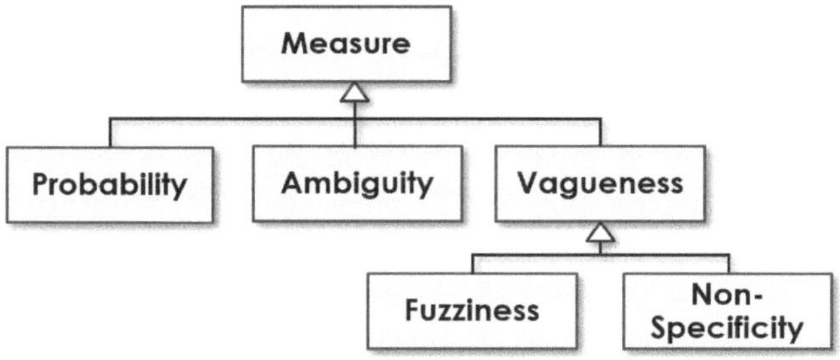

Figure 6.8: Measure model.

6.5 Uncertainty Requirements

U-Model defines concepts at a high level for the purpose of facilitating the understanding of uncertainty from the subjective perspective. However, we need to instantiate the conceptual model such that more detailed and concrete uncertainty specifications can be achieved. To this end, Uncertainty-wise Restricted Use Case Modeling (*U-RUCM*) was developed [15]. *U-RUCM* is an extension of (*RUCM*) [10], which is a restricted language-based use case specification language, aiming to reduce ambiguity of natural language and facilitate automated generation of other software development artifacts such as UML analysis models [11] and test cases [9]. RUCM has been demonstrated to be successful in many areas. Interested readers may check [7–9].

The architecture of *U-RUCM* is shown in Fig. 6.9. *U-RUCM* is based on UML's use case notation. Key features of *U-RUCM* are formalized as *Belief-UCMeta*—a metamodel for formalizing *U-RUCM*'s concepts and their relationships. *Belief-UCMeta* imports *UCMeta* [10] and *U-Model* (Section 6.4.1). *UCMeta* is the underlying metamodel that formalizes *RUCM*. *UCMeta* consists of four main packages: (1) *UCTemplate* implementing the RUCM template with various fields, e.g., use case name and description (see an example in Fig. 6.11); (2) *SentenceStructure* defining sentence constructs from the English linguistic perspective, e.g., noun phrases and subjects of sentences; (3) *SentenceSemantics* focusing on providing semantics of sentences in terms of their roles in flows of events, e.g., the sentence describing a *system validation* with given data; and (4) *SentencePattern* distinguishing sentences in terms of their structures such as sentences with the structure of subject-verb-direct object.

U-RUCM extends *RUCM* by providing two additional templates to specify *Belief Use Case Specifications* (BUCS) and uncertainties by implementing concepts from *U-Model*. Examples include indeterminacy sources, risk, and belief degree. A BUCS is the result of an instantiation of *Belief-UCMeta*.

At a high-level, *U-RUCM* consists of a use case diagram, for instance, the one shown in Fig. 6.2 for *Safe Home*. Corresponding to each use case in the use case diagram, e.g., *Monitor Windows and Doors*, one can define a detailed use case

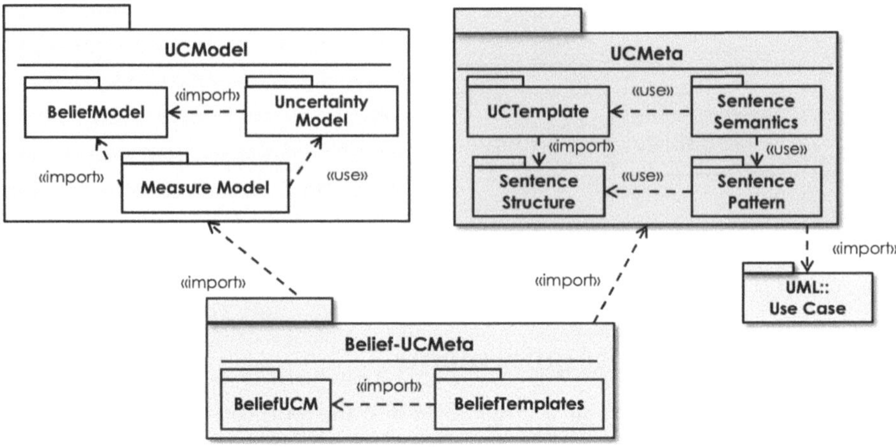

Figure 6.9: Architecture of *U-RUCM*.

specification based on a template defined in [10] and extended in [15] to specify uncertainty requirements. The template includes a well-defined structure, a set of keywords, and restrictions such that various automation can be supported, such as the automated generation of UML diagrams and tests.

The templates for specifying use case specifications and uncertainties are shown in Fig. 6.10, which are composed of the basic fields such as use case name, description, actors, and relationships. In addition, the template for use case specifications describes the key fields for specifying uncertainty related information such as belief agent, time points and duration, and indeterminacy source, i.e., the implementation of the concepts defined in *U-Model*. The template for uncertainty also shows specific structures that are related to the specification of uncertainty (see the bottom of the figure). Note that Fig. 6.10 describes the purpose of each field.

For our running example *Safe Home*, we show the uncertainty requirements specification corresponding to the "Monitor Windows and Doors" in Fig. 6.11, Fig. 6.12, and Fig. 6.14.

Figure 6.11 describes the basic fields of the use case and uncertainty related information. For example, it describes that this use case includes *Configure Monitoring Area* use case, as also shown in Fig. 6.2. It also shows the uncertainty related information such as a tester being the belief agent and various indeterminacy sources such as *Broken Control Panel (Button or Screen)*, due to which uncertainty may occur when monitoring windows and doors.

In addition, Fig. 6.11 shows a *Belief Basic Flow* named as *BB*, that is, a set of "happy day" scenarios. In total, it has 13 steps and a postcondition. A set of keywords are highlighted in different colors such as VALIDATES THAT, which checks a condition performed by the system or the testing framework. For example, in Step 6, the system checks whether the system is monitoring the

The template for specifying a BUCS

Use Case Name	The name of the use case. It usually starts with a verb.	
Brief Description	Summarizes the use case in a short paragraph.	
Primary Actor	The actor who initiates the use case.	
Secondary Actor(s)	Other actors the system relies on to accomplish the services of the use case.	
Dependency	Include and extend relationships to other use cases.	
Generalization	Generalization relationships to other use cases.	
Belief Agent(s)	One or more agents who hold belief about this BUCS.	
Time Point and Duration	The time point when the BUCS is specified and the duration in which the belief agent(s)'s belief on the BUCS holds.	
Belief Degree	The degree to which the belief agent(s) believe the BUCS.	
Indeterminacy Source(s)	The set of indeterminacy sources related to the BUCS (REF is used).	
Evidence	Evidence to support the BUCS, and its contained belief and uncertainty elements (REF is used).	
Belief Precondition	Belief agent(s)' belief on the precondition of the BUCS, which describes what should be true before the use case is executed.	
Belief Basic Flow (*Belief degree*)	Specifies the main successful path, also called "happy path".	
	Steps (numbered)	A set of ordered **belief sentences**.
	Belief Postcondition	Belief agent(s)' belief on what should be true after the basic flow executes.
Belief Specific Alternative Flow (*Belief degree*)	Applies to one specific step of the reference flow.	
	URFS	The reference flow step where the belief agent(s) believe there are uncertainties.
	Alternative Step	An alternative to the reference flow step.
	Steps (numbered)	A set of ordered belief sentences.
	Belief Postcondition	Belief agent(s)' belief on what should be true after the specific alternative flow executes.
Belief Bounded Alternative Flow (*Belief degree*)	Applies to more than one step of the reference flow, but not all of them.	
	URFS	A list of reference flow steps where the belief agent(s) believe there are uncertainties.
	Alternative Steps	A set of alternatives to the reference flow steps.
	Steps (numbered)	A set of ordered **belief sentences**.
	Belief Postcondition	Belief agent(s)' belief on what should be true after the bounded alternative flow executes.
Belief Global Alternative Flow (*Belief degree*)	Applies to all the steps of the reference flow.	
	Belief Branching Condition	Belief agent(s)' belief on the condition, which describes what should be true when branching from any of the steps of the reference flow.
	Steps (numbered)	The set of ordered **beliefs sentences.**
	Belief Postcondition	Belief agent(s)' belief on what should be true after the global flow executes.

The template of specifying a belief sentence (BS) with one and more Uncertainties

Time Point and Duration	The time point when the BS is specified and the duration in which the belief agent(s)'s belief on the BS holds.	
Belief Degree	The degree to which the belief agent(s) believe the BS.	
Indeterminacy Source(s)	The set of indeterminacy sources related to the BS (REF is used).	
Evidence	Evidence to support the BS, and its contained belief and uncertainty elements (REF is used).	
Uncertainty Details	Specifies the details of the Uncertainty in the BS.	
	Type	The type of this uncertainty (Occurrence/Content/Time/Environment/GeographicalLocation)
	Indeterminacy Source(s)	The set of indeterminacy sources related to this Uncertainty (REF is used).
	Measure Value	The measurement of this uncertainty.
	Risk	The possible risk led by this uncertainty.
	Pattern	The pattern of the occurrence of this uncertainty

Figure 6.10: *U-RUCM* templates.

home. If the check fails, then the flow is transferred to *SAlt1 Specific Alternate Flow* shown in Fig. 6.12. Similarly, there is another check at Step 10, if it fails then the flow branches to *SAlt1 Specific Alternate Flow* shown in Fig. 6.12.

In *RUCM*, alternative flows should be branched with specific conditions. For instance, Fig. 6.12 shows a *Specific Alternative Flow* named SAlt1 which branches from a condition check (i.e., VALIDATE THAT) referring to Step 6 of *BB* using keyword RFS (i.e., RFS BB 6). In this example, Step 6 is to check

Belief Use Case Specification

Use Case Name	Monitor Windows and Doors
Brief Description	The system monitors the status of windows and doors when Home Owner enables the monitoring function.
Primary Actor	Home Owner
Secondary Actors	Sensor, Control Panel, Alarm
Dependency	INCLUDE USE CASE Configure Monitoring Area
Generalization	None
Belief Agent(s)	Believer
Timepoint and Duration	16-07-2021, After
Belief Degree	UModel.Measure.Fuzziness::High Confidence
Indeterminacy Source(s)	**REF** Broken Control Panel (Button or Screen), **REF** Improper Operation of Home Owner, **REF** Improper Implementation, **REF** Unavailable Communication, **REF** Broken Sensor
Evidence	**REF** Custom Feedback of Pervious Version
Belief Precondition	The system is idle. The system is displaying "The system is idle." message to Home Owner.

Belief Basic Flow "BB" ▼	Steps	
	1	Home Owner closes all windows and doors properly.
	2	INCLUDE USE CASE Configure Monitoring Area
	3	Home Owner presses the "Enable Monitoring" button on the Control Panel properly.
	4	The system displays "Please wait. Enabling Monitoring." message to Home Owner.
	5	The system enables the monitoring function.
	6	The system VALIDATES THAT the status of the system is Monitoring.
	7	The system displays the "The home is under monitoring." message to Home Owner.
	8	DO
	9	WinDoorSensor detects the existence of intrusion.
	10	IF the windows and doors are closed THEN
	11	The system VALIDATES THAT the status of Sensor is Closed.
	12	ENDIF
	13	UNTIL the status of the system is Running.
	Postcondition	The system is Running. No intrusion is detected.

Figure 6.11: Safe home use case specification–part 1.

whether the status of the system is updated to *Monitoring* after the system invokes a method to enable the monitoring function (see Step 5 in *BB*). If the condition check is evaluated as false (i.e., the status is not updated improperly), the steps in SAlt1 will be executed. Considering that the result of the condition check is uncertain (but it is determined), an occurrence of executing SAlt1 could be regarded as an uncertainty that relies on the referred steps (see *rfsSentence* in Fig. 6.12). As shown in Fig. 6.13, the uncertainty might be due to improper implementation of the system (i.e., *IndeterminacySource*) and the probability of its occurrence could be 5%.

Figure 6.14 shows more constructs of *U-RUCM*. First, it shows another example of *Belief Specific Alternative Flow*, i.e., BSAlt1 capturing alternative steps which might replace the referred step(s), i.e., SAlt2 Step 2 in Fig. 6.12. As one can see from SAlt2, Step 1 describes a condition: the windows and doors are open, which can be referred as an intrusion. Step 2 is to a system check on whether the Sensor detects the intrusion. However, due to some unknown reasons, the Sensor might fail to detect the intrusion. Thus, the Step 2 in SAlt2 can be replaced by the Steps A1-A2 in BSAlt1 in Fig. 6.14, then the flow will

Flow		
Belief Bounded Alternative Flow (URFS) "BBAlt1" ▼	URFS BB 5-7	
	A1	The system does not enable the monitoring function.
	A2	The system VALIDATES THAT the status of the system is Running.
	A3	The system does not display any message to Home Owner.
	1	Home Owner contacts the company to fix the system.
	2	ABORT.
	Postcondition	The system is out of order.
Specific Alternative Flow (RFS) "SAlt1" ▼	RFS BB 6	
	1	The system VALIDATES THAT the status of the system is Running.
	2	The system displays the "Error: fail to enable monitoring, please contact the company." to Home Owner.
	3	ABORT.
	Postcondition	The system is out of order.
Specific Alternative Flow (RFS) "SAlt2" ▼	RFS BB 10	
	1	ELSEIF the windows and doors are open THEN
	2	The system VALIDATES THAT the status of Sensor is Open.
	3	The system activates the Alarm.
	4	The system VALIDATES THAT the Alarm is ringing.
	5	The system sends the intrusion notification to Home Owner.
	6	The system displays "Intrusion is detected." message.
	7	DO
	8	The system VALIDATES THAT the status of Alarm is Started.
	9	UNTIL the status of Alarm is Stopped and the status of the system is Monitoring.
	10	ENDIF
	11	RESUME STEP BB 8
	Postcondition	The intrusion is detected and released.

Figure 6.12: Safe home use case specification–part 2.

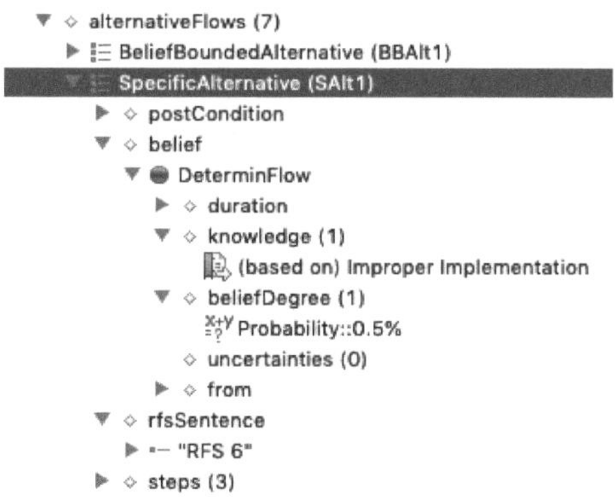

Figure 6.13: Uncertainty flow.

resume to Step 8 of BB. Note that, in the view of the system, there is no any error since the Sensor gives the incorrect signal to the system. To specify this situation, *U-RUCM* defines URFS, and it can be used to specify alternative steps referring to behaviors which are scope of the system (e.g., Control Panel (BBAlt1 in Fig. 6.12), Sensor (BSAlt1 in Fig. 6.14)). Regarding RFS, it typically refers to the check performed by the system, e.g., the system Status should be Monitoring once the monitoring function is invoked (see SAlt1 in Fig. 6.12).

Belief Specific Alternative Flow (URFS) "BSAlt1" ▼	URFS SAlt2 2	
	A1	The system VALIDATES THAT the status of Sensor is Closed.
	A2	The system displays "The home is under monitoring."message.
	1	RESUME STEP BB 8
	Postcondition	The system failed to detect the intrusion. The system is displaying "The home is under monitoring." message.
Belief Global Alternative Flow "BGAlt1" ▼	The warning has been triggered for more than 5 minutes or the system has received a wrong password 3 times.	
	1	The system calls police.
	2	The system displays "Calling police." message.
	3	ABORT.
	Postcondition	The system is displaying "Calling police." message.
Belief Global Alternative Flow "BGAlt2" ▼	Home Owner cancels the Alarm.	
	1	The system displays "Enter your password." message to Home Owner.
	2	Home Owner types the password to cancel the Alarm
	3	The system VALIDATES THAT the password is valid.
	4	The system deactives the Alarm.
	5	The system VALIDATES THAT the status of Alarm is Stopped.
	6	The system displays the "The Intrusion is released. The Alarm is canced." message.
	7	ABORT.
	Postcondition	The Alarm is stopped.
Specific Alternative Flow (RFS) "SAlt3" ▼	RFS BGAlt2 3	
	1	The system displays "Please enter valid password." message to Home Owner.
	2	RESUME STEP BGAlt2 1
	Postcondition	The password is not valid.

Figure 6.14: Safe home use case specification–part 3.

In *U-RUCM*, each step in the flows can have associated uncertainty which can be characterized with the specific template shown in Fig. 6.10. For example, the associated specification with Step A1 is shown in Fig. 6.15. The specification describes a possible uncertainty in Step A1 in *BSAlt1*, which is related to the inability of the system to identify the intrusion. The details of the uncertainty include: (1) information about associated indeterminacy sources, i.e., *Broken Sensor*, (2) belief degree, e.g., 1% in this example, and (3) risk, e.g., *High*. Figure 6.14 also shows two Belief Global Alternative Flows (i.e., BGAlt1 and BGAlt2). These can be invoked from any step of the belief basic flow presented in Fig. 6.11.

6.6 Uncertainty Modeling

Uncertainty-wise Test Modeling (*UncerTum*) is a model-based testing approach to support testing CPSs under uncertainty. *UncerTum* supports creating test ready models in UML that explicitly specify uncertainty. *UncerTum* can be applied as a follow-up of the uncertainty specification in *U-RUCM* or one can start creating test ready models directly with *UncerTum*.

UncerTum implements a UML profile for uncertainty specification, called the UML Uncertainty Profile (*UUP*). Just like *U-RUCM*, *UUP* implements *U-Model* described in Section 6.4.1. As shown in Fig. 6.16, *UUP* has four UML

Belief Details	
Sentence	The system VALIDATES THAT the status of Sensor is Closed.
Brief Description	The system fails to detect the intrusion.
Timepoint and Duration	16-07-2021, After
Belief Degree	UModel.Measure.Probability::1%
IndeterminacySource	REF Broken Sensor
Evidence	None

Uncertainty Details		
(Untitled) ▼	Type	UModel.Uncertainty.Occurrence::the status of Sensor is Closed
	Decription	The sensor fails to identify the real status of the window or door.
	Measured Value	UModel.Measure.Probability::1%
	Derived Source(s)	REF Broken Sensor
	Risk	UModel.Risk.High
	Pattern	UModel.Pattern.Random

Figure 6.15: Uncertainty specification.

profiles corresponding to *Belief*, *Measurement*, *Uncertainty*, and *Indeterminacy* based on *U-Model*. In addition, *UncerTum* includes a set of model libraries to allow test modelers to quantify uncertainties with various uncertainty measures such as ambiguity, probability, and vagueness from *U-Model*. However, in *U-Model*, these concepts are abstract, but *UncerTum* implements them with concrete measures from the literature, or reuse some from the UML profile for Modeling and Analysis of Real-Time and Embedded Systems (*MARTE*).[1] To keep the approach based on standards, *UncerTum* also uses the UML Testing Profile (*UTP*) V.2. Finally, *UncerTum* also defines a set of guidelines, i.e., a step-wise process to create test ready models such that our accompanying model-based testing tool can be used to generate test cases from these models.

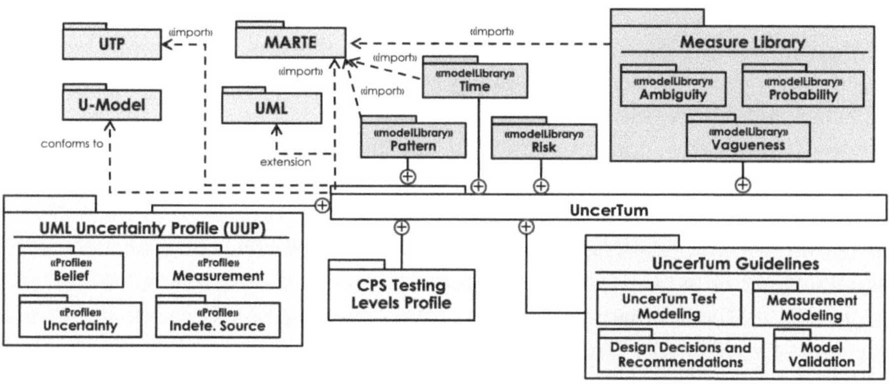

Figure 6.16: UML-based uncertainty modeling solution–uncertum.

[1] https://www.omg.org/omgmarte/

To explain how to create test ready models with *UncerTum*, we will use the running example presented in Section 6.2. To use *UncerTum*, one needs to create a set of UML class diagrams and state machines with stereotypes from *UUP*. Figure 6.17 and Fig. 6.19 show the example class diagram and state machine respectively, which correspond to the intrusion detection feature belonging to the *Security System* of Safe Home.

To implement intrusion detection, the security system needs sensors and an alarm, shown as the composition relationships between *SecuritySystem* and *Sensor*, and *SecuritySystem* and *Alarm*, respectively. Both *Sensor* and *Alarm* are stereotyped as «InfrastructureElement» for being hardware components, whereas *SecuritySystem* is stereotyped as «IntegrationElement» since it integrates elements from different levels. *SecuritySystem* is also composed of a *ControlPanel* with a dedicated interface *IControlPanel* that can be accessed by home owner to configure the security system. *IControlPanel* is stereotyped as «ApplicationElement» since it is related to the application use of the system. The class diagram also shows two indeterminacy sources associated with *ControlPanel* and *Sensor* annotated with the «IndeterminacySource» stereotype. The indeterminacy source related to *ControlPanel* is about whether the buttons on the control panel are broken or not. The indeterminacy source related to *Sensor* is about whether or not the sensor can detect intrusion correctly. To evaluate the occurrence of such indeterminacy sources, we define constraints specified with OCL representing specifications of the occurrences annotated with the «IndeterminacySpecification» stereotype. For instance, *Sensor* could have two possible statuses defined with *The sensor is normal* constraint and *The sensor is out of the order* constraint, respectively. Here, *Sensor* has two operation stereotyped with «IndeterminacySourceInput» (i.e., *deactivateDetection()* and *activateDectection()*) represents methods that can introduce indeterminacy sources when testing CPSs. The methods could be implemented in the testing environment for testing CPS under various environmental configurations. Figure 6.18 shows the specification of *The sensor is out of the order* in detail. As shown in the figure, the specification is not an expected status (see the value for the *Is Expected* property in the Properties window); the status could occur with 1% probability (see the value for the *Measurement* property); the status can be manipulated by *deactivateDetection()* for triggering the occurrence of the indeterminacy source (see value for the *Triggered By* property) and *activateDectection()* for releasing its occurrence (see the value of the *Released By* property).

To specify the behavior, an example UML state machine is shown in Fig. 6.19 with stereotypes from *UUP*. The state machine starts from the *S1: Idle* state, from where it leaves from two transitions. The first transition is related to the inability of the system to enable monitoring and leading the system to the *S3: Failed to enable monitoring* state. The second transition is related to the monitoring being enabled successfully. In this case, the system transits to the *S2: Monitoring* state.

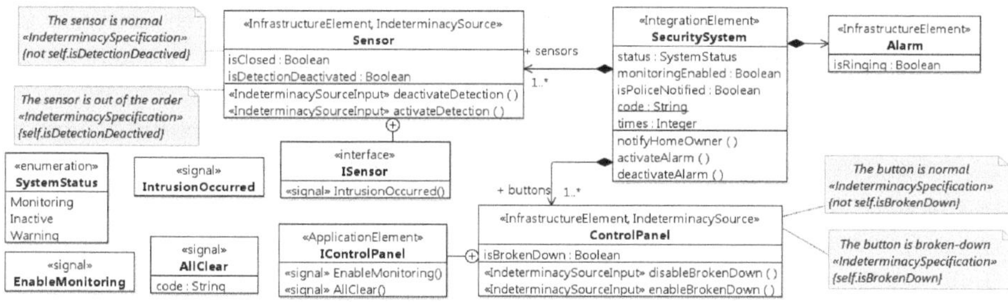

Figure 6.17: Uncertum–example class diagram of safe home.

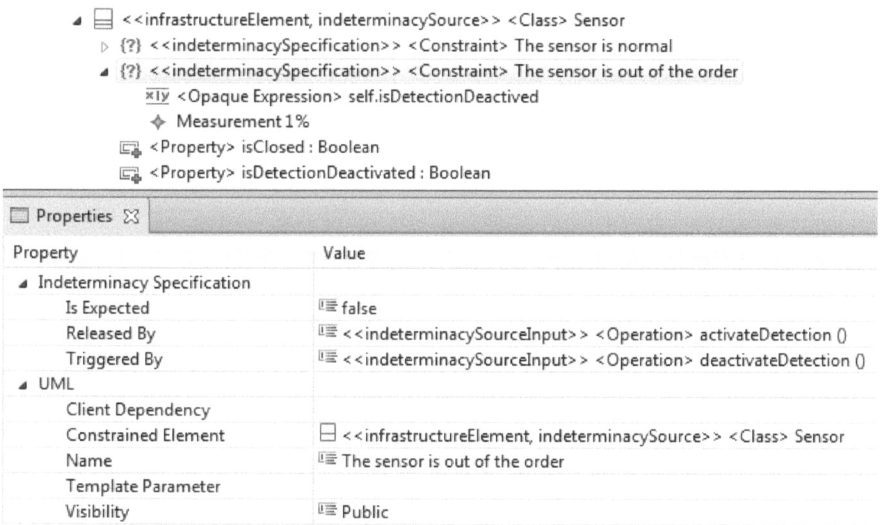

Figure 6.18: U ncertum–example of specifying indeterminacy source.

From *S2*, when an intrusion is triggered via the *T3: IntrusionOccurred()* signal, there are two possible transitions. First, due to a system failure, the warning is not issued and the system transits to the *S7: Not Warning* state. Otherwise, the system activates the alarm (the *activateAlarm()* effect on *T3*), notifies the home owner with the *notifyHomeOwner()* effect, and transits to the *S5: Waiting for resolving warning* state. If the home owner resolves the warning via the secret code, the system transits via the *exit1* point to the *S2* state, and continues the monitoring. If the warning is not cleared by the home owner, then the police is notified after 5 minutes and the system transits to the *S6: Police notified* state. Moreover, if the secret code entered is wrong for more than three times, the police is notified.

Note that various states and transitions are stereotyped «BeliefElement» from *UUP*. Corresponding to each belief element, we can specify various information such as belief degree and indeterminacy source. Figure 6.20 shows detailed information of state *S4: Intrusion Warning and Alarm Ringing* stereotyped with «BeliefElement». The element contains a

Measurement and an *Uncertainty*. As properties of the *Uncertainty* shown in the figure, the uncertainty is specified with the *Low* risk and of type *Occurrence*. In addition, the uncertainty is caused by transition *T3: IntrusionOccurred()* (see *Referred Cause*, led by the indeterminacy source referring to *ControlPanel* (see *Referred Indeterminacy Source*). The corresponding specification of the source is *The sensor is normal* (see *Related Ind Specs*). *Enable Pattern*, *Find Position* and *Select Specification* are used to specify how to introduce the indeterminacy source in order to enable the occurrence of the uncertainty. As shown in the figure, the *Specified* indeterminacy source is *Always* introduced at a position where is *Just_Previous* of this uncertainty, i.e., *T3*.

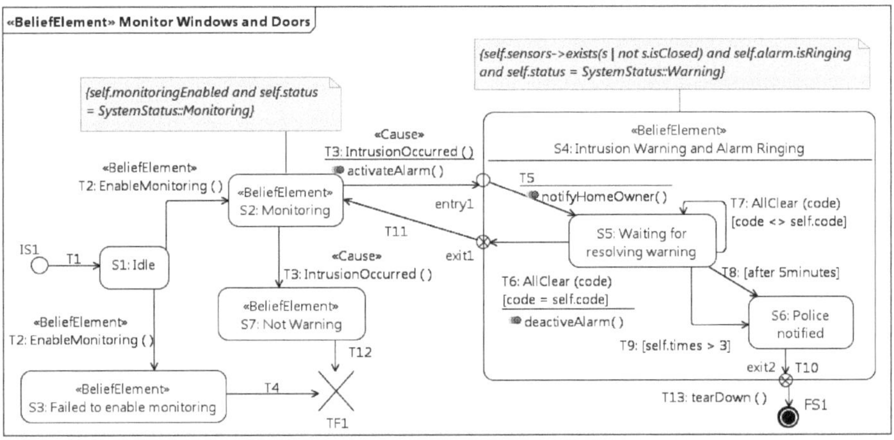

Figure 6.19: UncerTum–Example state diagram of safe home.

6.7 Uncertainty Testing

The UML class diagram and state machines developed as test ready models with *UncerTum* shall be used to generate test cases with suitable testing strategies. To this end, *UncerTest* was proposed in [13], which consists of uncertainty-wise test case generation and test case minimization approaches as shown in Fig. 6.21. *UncerTest* has four steps. As shown in the figure, the creation of test ready models is step 1 of *UncerTest*. A test ready model is an abstract representation of the CPS under test, with a particular focus on modeling the expected behavior of the CPS under various uncertainties. Step 2 generates abstract test cases, whereas Step 3 converts the generated abstract test cases into executable test cases. Finally Step 4 executes test cases on CPS being tested. Step 1 was discussed in Section 6.6, whereas Step 3 is about simply

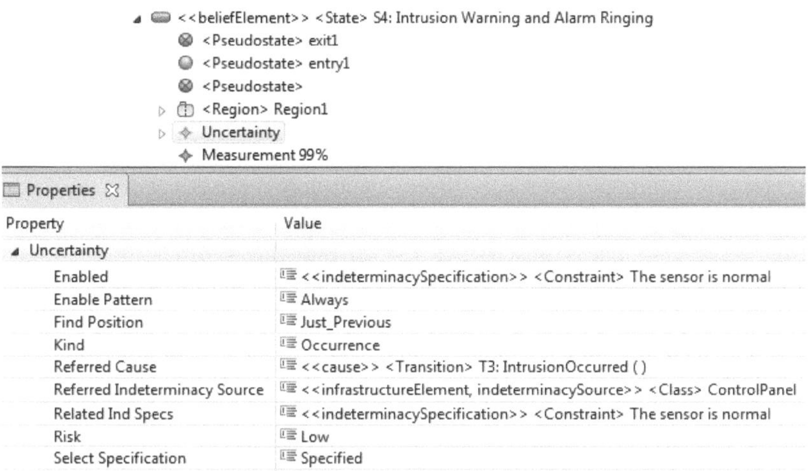

Figure 6.20: UncerTum–example of specifying uncertainty.

converting abstract test cases into executable test cases with a software. The rest of the subsections explain Uncertainty-wise Test Case Generation and Uncertainty-wise Test Case Minimization from Step 2 and test case execution details of Step 4.

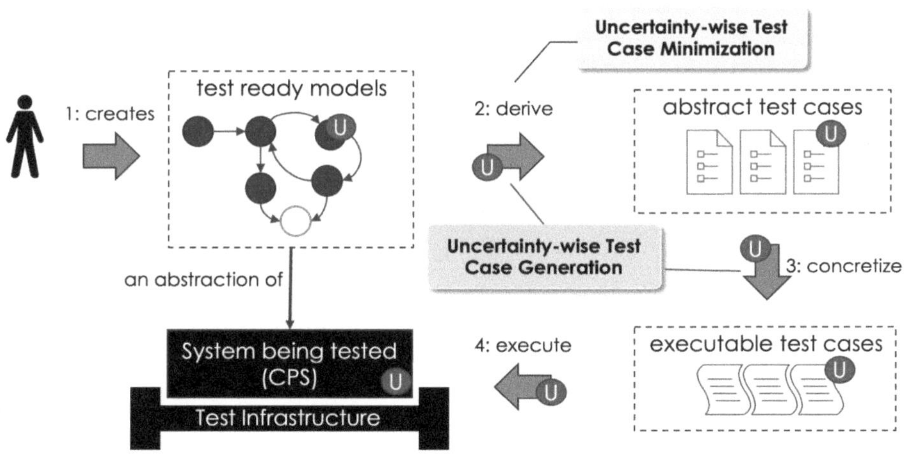

Figure 6.21: Uncertainty-wise testing.

6.7.1 *Uncertainty-wise Test Case Generation*

The second step of *UncerTest* generates a set of abstract test cases with uncertainty-wise test case generation consisting of two coverage criteria.

Abstract test cases are those that do not have concrete data and are independent of implementation details. For the abstract test case generation, we defined two coverage criteria on UML state machines (i.e., Belief State Machines (BSMs)) developed with *UncerTum*: (1) the *All Simple Belief Paths* coverage (ASiBP), which generates all simple paths in a BSM containing unique states and transitions. By simple, we mean there is no loop. An example is shown in Fig. 6.22; and (2) the *All Specified Length Belief Paths* coverage (ASlBP), which generates all paths in a BSM. Given that the number of all paths in a BSM can be very large, one need to specify the maximum length. For both of the coverage, a path is one abstract test case.

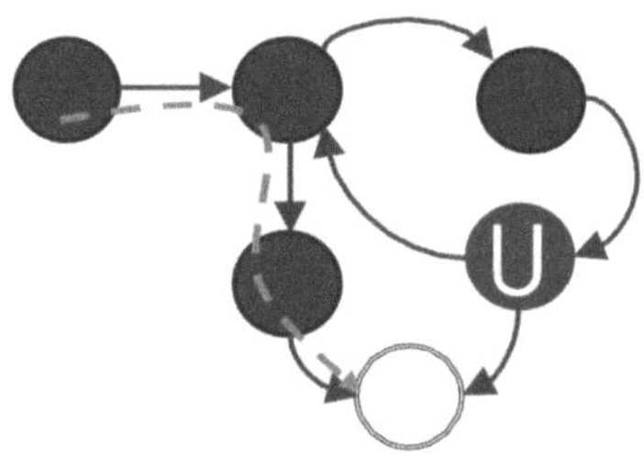

Figure 6.22: All simple belief path coverage (ASiBP).

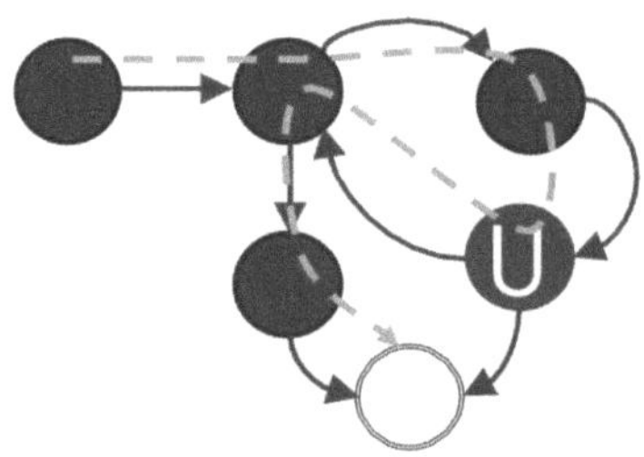

Figure 6.23: All specified length belief path coverage (ASlBP).

Figure 6.24 represents an example of a test (generated by *UncerTest*) based on the state machine shown in Fig. 6.19. The left represents a path of the state machine which is regarded as a test. It starts from an initial state *IS1* and ends with the final state *FS1*. The right side represents a snippet of the executable test case corresponding to the sub-path, i.e., *S2* (lines 1–3) $\xrightarrow{T3(lines7-12)}$ *S4* (lines 13–41). Lines 4–6 introduce the specified indeterminacy source with *activateDetection()* (see Fig. 6.20). To evaluate the occurrence of the uncertainty specified in *S4*, lines 14–22 first report whether the specified uncertainty is occurred with the related indeterminacy source specification.

Since the uncertainty might not always occur, it is necessary to report the occurrence of its alternative. In this example, an alternative of this uncertainty is $S2 \xrightarrow{T3} S3$, and its occurrence is reported at lines 23–32. Next, lines 33–34 record the current state of the SUT. Then, lines 36–41 are to check whether the uncertainty is occurred as expected.

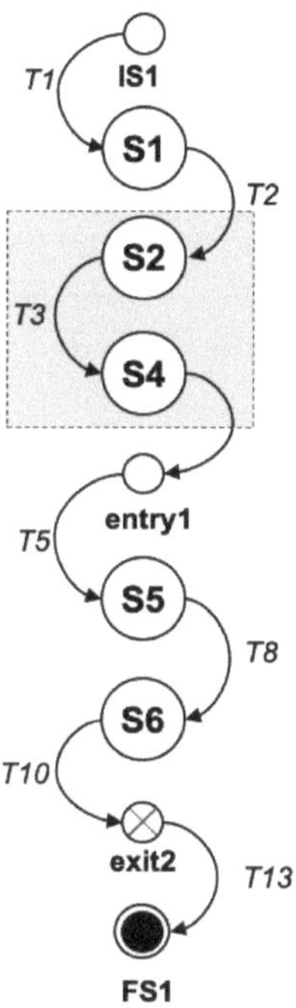


```
1  //evaluate state invariant: S2: Monitoring
2  assertTrue("S2: Monitoring", MOCLUtil.getInstance().evaluateOCL(system.eClass(),
3          "self.monitoringEnabled and self.status = SystemStatus::Monitoring", system));
4  //enable indeterminacy input: activateDetection
5  UncerTestReporter.reportTestStep("execute activateDetection");
6  SafeHomeExecutionUtil.getInstance().opTestAPI("activateDetection");
7  //execute tranision: T3 IntrusionOccurred
8  UncerTestReporter.reportTestStep("execute IntrusionOccurred");
9  SafeHomeExecutionUtil.getInstance().opTestAPI("IntrusionOccurred");
10 //execute effect: activateAlarm
11 UncerTestReporter.reportTestStep("execute activateAlarm");
12 SafeHomeExecutionUtil.getInstance().opTestAPI("activateAlarm");
13 //uncertainty occurrence and its source evaluation
14 UncerTestReporter.setUncerInfo(
15         "<S2: Monitoring, T3: IntrusionOccurred, S4: Intrusion Warning and Alarm Ringing>",
16         MOCLUtil.getInstance().evaluateOCL(system.eClass(),
17                 "self.sensors->exists(s | not s.isClosed) and
18                         self.alarm.isRinging and self.status = SystemStatus::Warning", system),
19         Arrays.asList("<The sensor is normal, not self.isDetectionDeactivated, sensor>")
20                 .toArray(new String[1]),
21         MOCLUtil.getInstance.toboolenArray(Arrays.asList(MOCLUtil.getInstance().evaluateOCL(
22                 sensor.eClass(), "not self.isDetectionDeactivated", sensor))));
23 //alterative uncertainty occurrence and its source evaluation
24 UncerTestReporter.setAlterUncerInfo(
25         "<S2: Monitoring, T3: IntrusionOccurred, S4: Intrusion Warning and Alarm Ringing>",
26         "<S2: Monitoring, T3: IntrusionOccurred, S7: Not Warning>",
27         MOCLUtil.getInstance().evaluateOCL(system.eClass(),
28                 "(not self.alarm.isRinging) and self.status = SystemStatus::Monitoring", system
29         Arrays.asList("<The sensor is out of the order, self.isDetectionDeactivated, sensor>")
30                 .toArray(new String[1]),
31         MOCLUtil.getInstance.toboolenArray(Arrays.asList(MOCLUtil.getInstance().evaluateOCL(
32                 sensor.eClass(), "not self.isDetectionDeactivated", sensor))));
33 //report current status of sut
34 UncerTestReporter.reportUncerInfo(monPars, MOCLUtil.getInstance().getCurrentSUT(monEObjs),
35         "system", MOCLUtil.getInstance().getCurrentSUT(system));
36 //evaluate state invariant: S4: Intrusion Warning and Alarm Ringing
37 collector.checkThat("S4: Intrusion Warning and Alarm Ringing",
38         MOCLUtil.getInstance().evaluateOCL(system.eClass(),
39                 "self.sensors->exists(s | not s.isClosed) and
40                         self.alarm.isRinging and self.status = SystemStatus::Warning", system),
41         isEqual.equalTo(true));
```

Figure 6.24: *UncerTest*–example of a test with snippet of an executable test case generated with *UncerTest*.

6.7.2 Uncertainty-wise Test Case Minimization

To minimize the number of abstract test cases to be concertized and executed, we devised multi-objective search-based approaches. In particular, the four approaches were developed. All the four approaches shared two objectives, i.e., minimizing the number of test cases to be executed and maximizing the transition coverage. However, the approaches differ in terms of the four uncertainty-related objectives to be maximized as follows.

Average number of uncertainties Cover more uncertainties by each test case.

Uncertainty space coverage Cover more combinations by the whole test suite.

Uncertainty measurement Increase confidence on normal execution by more likely test runs.

Coverage of unique uncertainties Cover more unique uncertainties by the whole test suite.

The NSGA-II search algorithm was employed for test case minimization. Depending on the preferences of a user, one or more of these minimization techniques can be employed.

6.7.3 Test Case Execution

When test cases are executed, one needs to determine the pass and fail of the test cases and determine whether uncertainties are discovered. To test CPS under uncertainty, we introduce indeterminacy sources (see Fig. 6.24) and determine whether expected uncertainties occurred due to the introduced indeterminacy sources. *UncerTest* focuses on evaluating known uncertainties and identifying unknown uncertainties and possible indeterminacy sources.

In particular, we defined the following types of uncertainty verdicts:

- Known uncertainty occurred with known indeterminacy source introduced;
- Known uncertainty occurred without introduction of known indeterminacy source;
- Known uncertainty not occurred with the introduction of known indeterminacy source;
- Known uncertainty not occurred without the introduction of indeterminacy source;
- Known uncertainty occurred with unknown indeterminacy source;
- Known uncertainty not occurred with unknown indeterminacy source; and
- Unknown uncertainty occurred.

Figure 6.25 represents results of the occurrence of an uncertainty (in the test case shown in Fig. 6.21) specified with such a verdict. As shown in the figure, the uncertainty and related indeterminacy source occurred as expected (see lines 3 and 19), and none of the alternative uncertainties occurred. Thus, for the uncertainty, its results is "KnOccurred-With-In" (i.e., known uncertainty occurred with known indeterminacy source introduced).

In addition, we implemented the following test case verdicts by further considering occurrences of the uncertainties in test cases: *Known uncertainty occurred*, *Unknown uncertainty occurred*, *Not occurred*, *Passed*, *Failed*, *Error*, *Inconclusive*, and *None*.

```
{
  "id": "<S2: Monitoring, T3: IntrusionOccurred, S4: Intrusion Warning and Alarm Ringing>",
  "occurred": true,
  "alterUns": [
    {
      "id": "<S2: Monitoring, T3: IntrusionOccurred, S7: Not Warning>",
      "occurred": false,
      "relatedIndSpecs": [
        {
          "id": "<The sensor is out of the order, self.isDetectionDeactivated, sensor>",
          "occurred": false
        }
      ]
    }
  ],
  "relatedIndSpecs": [
    {
      "id": "<The sensor is normal, not self.isDetectionDeactivated, sensor>",
      "occurred": true
    }
  ],
  "pars": [
    "system",
    "sensor",
    "alarm"
  ],
  "eobjs": [...],
  "curPar": "system",
  "uncertaintyResult": "KnOccurred-With-InS"
}
```


Figure 6.25: *UncerTest*–Example of a result of the occurrence of an uncertainty

6.8 Conclusions

This chapter presented uncertainty-wise testing approaches that were designed for automated and systematic testing of Cyber-Physical Systems (CPSs) under various uncertainties. The complete solution starts with specifying uncertainty test requirements in a restricted natural language-based approach, called *U-RUCM*. Then, followed by test requirement specifications, a tester can design test ready models with a dedicated uncertainty-wise test modeling solution, called UncerTum, using the requirements specified with *U-RUCM*. Once we have test ready models, the accompanying model-based testing tool can generate test cases from test ready models and further minimize them, if required. Finally, the test cases (with or without optimization) are concertized to be executed on CPSs for testing.

References

[1] Baresi, L. and Carlo Ghezzi, C. 2010. The disappearing boundary between development-time and run-time. *In*: Proceeding of the FSE/SDP Workshop on Future of Software Engineering Research, FoSER '10, page 17–22, New York, NY, USA, 2010. Association for Computing Machinery.

[2] Dempster, A.P. 2008. Upper and lower probabilities induced by a multivalued mapping. *In*: Classic works of the Dempster-Shafer theory of belief functions, pages 57–72. Springer.

[3] Klir, G.J. 2013. Facets of systems science, volume 7. Springer Science & Business Media.

[4] Pressman, R.S. 2010. Software Engineering: A Practitioner's Approach. McGraw-Hill, European edition, 2010. Adapted by Darrel Ince.

[5] Ramirez, A.J., Jensen, A.C. and Cheng. B.H.C. 2012. A taxonomy of uncertainty for dynamically adaptive systems. *In*: 2012 7th International Symposium on Software Engineering for Adaptive and Self-Managing Systems (SEAMS), pages 99–108.

[6] Taylor, B.N. and Kuyatt, C.E. 1994. Guidelines for evaluating and expressing the uncertainty of nist measurement results.

[7] Wang, C., Pastore, F., Goknil, A., Briand, L. and Iqbal, Z. 2015. Automatic generation of system test cases from use case specifications. *In*: Proceeding of the 2015 International Symposium on Software Testing and Analysis, ISSTA 2015, page 385–396, New York, NY, USA. Association for Computing Machinery.

[8] Yue, T. and Ali, S. 2012. Bridging the gap between requirements and aspect state machines to support non-functional testing: Industrial case studies. *In*: Antonio Vallecillo, Juha-Pekka Tolvanen, Ekkart Kindler, Harald Störrle and Dimitris Kolovos, (eds). Modelling Foundations and Applications, pages 133–145, Berlin, Heidelberg. Springer Berlin Heidelberg.

[9] Yue, T., Ali, S. and Zhang, M. 2015. Rtcm: A natural language based, automated, and practical test case generation framework. *In*: Proceeding of the 2015 International Symposium on Software Testing and Analysis, ISSTA 2015, page 397–408, New York, NY, USA. Association for Computing Machinery.

[10] Yue, T., Briand, L.C. and Labiche, Y. 2013. Facilitating the transition from use case models to analysis models: Approach and experiments. ACM ACM Transactions on Software Engineering and Methodology, 22(1), March 2013.

[11] Yue, T., Briand, L.C. and Labiche, Y. 2015. Atoucan: An automated framework to derive uml analysis models from use case models. ACM

ACM Transactions on Software Engineering and Methodology, 24(3), May 2015.

[12] Zadeh. L.A. 1996. Fuzzy sets. *In*: Fuzzy sets, fuzzy logic, and fuzzy systems: selected papers by Lotfi A Zadeh, pages 394–432. World Scientific.

[10] Zhang, M., Ali, S., Yue, T. and Norgre, R. 2017. Uncertainty-wise evolution of test ready models. Information and Software Technology, 87: 140–159.

[13] Zhang, M., Ali, S., Yue, T., Okariz, O. and Norgren, R. 2017. Uncertainty-wise evolution of test ready models. Information and Software Technology, 87: 140–159, 2017.

[14] Zhang, M., Selic, B., Ali, S., Yue, T., Okariz, O. and Roland Norgren, R. 2016. Understanding uncertainty in cyber-physical systems: A conceptual model. *In*: Andrzej Wa̧sowski and Henrik L.nn, (eds). Modelling Foundations and Applications, pages 247–264, Cham, 2016. Springer International Publishing.

[15] Zhang, M., Yue, T., Ali, S., Selic, B., Okariz, O., Norgre, R. and Intxausti. K. 2018. Specifying uncertainty in use case models. Journal of Systems and Software, 144:573–603.

Chapter 7

Decision Making for Automated Driving

Sasinee Pruekprasert
National Institute of Advanced Industrial Science and Technology (AIST), Tokyo, Japan

Ahmet Cetinkaya
Shibaura Institute of Technology, Tokyo, Japan

Ichiro Hasuo
National Institute of Informatics and SOKENDAI, Tokyo, Japan

7.1 Introduction

In the last few decades, we have witnessed the increasing popularity and rapid development of automated driving systems (ADSs). New sensing, perception, planning, and control techniques are being introduced, and as a result, vehicles are becoming increasingly more capable to achieve automated driving tasks.

In this chapter, we discuss decision making components in ADS, which provide guidance on the driving behavior of vehicles. In our discussion, we focus our attention to safety, as it is one of the major concerns for decision making. This issue is emphasised in standards such as ANSI/UL 4600 [67], which addresses safety requirements during the operation of ADSs, and

ISO/DIS 21448 [35], which identifies measures to be taken to achieve the safety of the intended functionality (SOTIF) in ADSs.

We start by introducing the decision-making hierarchy of a contemporary ADS. Then, we focus on the motion planning component, which is responsible for computing feasible trajectories for the vehicle. Designing motion planners is challenging because the computed trajectory must be dynamically feasible for the vehicle and comfortable for the passenger. In addition, trajectories must avoid collisions with obstacles detected by sensors, which may be subject to uncertainties. As a result, motion planning for ADSs is a large research area with a variety of approaches. In this chapter, we discuss optimisation-based, model predictive, game theory-based, and machine learning-based motion planning approaches.

Next, we discuss formal verification techniques that can be combined with decision-making components to ensure the safety of the ADS. We discuss two reachability analysis techniques, a set-based one and a probabilistic one, that can be used to compute road occupancy of all road participants subject to uncertainties such as sensor noise and environmental disturbances. Based on this information, we can verify the safety of the vehicle for some future time interval if the vehicle's road occupancy area does not intersect with that of other road participants.

Finally, we discuss a recent rule-based approach to the safety of ADSs, namely the responsibility-sensitive safety (RSS) framework. RSS introduces a set of mathematically defined safety rules, each of which is adapted to an individual driving scenario. In general, there is a big gap between driving safely and arguing that vehicles can do so—a gap that is currently hindering the public acceptance of automated driving. Rule-based safety arguments, such as those enabled by RSS, bridge the gap and thus play a crucial role in the deployment of automated driving in the real world.

7.2 Hierarchy of Decision Making

As depicted in Fig. 7.1, the decision making of a ADS mainly consists of four hierarchical components: a route planner, a behavioral planner, a motion planner, and a local feedback controller.

7.2.1 Route Planner

A route planner is at the highest level of the decision-making hierarchy. The role of this component is to plan a static route through the road network toward a given target destination, which is represented as a directed graph with weighted edges, where a weight can represent the travel time or the cost of taking a particular road. Therefore, a route planning problem can be formulated as finding a path with minimum cost in a directed graph. Since a road network

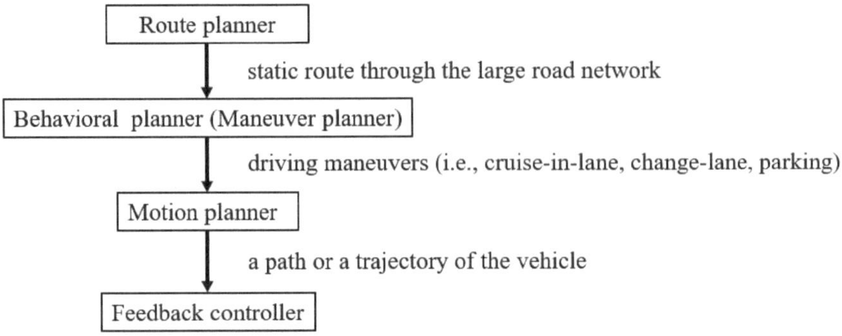

Figure 7.1: The decision-making hierarchy of an autonomous vehicle.

graph may contain millions of nodes and edges, the application of route-planner is an important driving force behind the research on speedup techniques for shortest path algorithms [8]. Efficient route planning algorithms are up to three million times faster than Dijkstra's algorithm [18], and are capable of computing an optimal route on continent-scale benchmark instances in milliseconds [8]. We invite an interested reader to see [8, 18] for comprehensive surveys of practical route-planning algorithms.

7.2.2 Behavioral Planner (Maneuver Planner)

Given a sequence of road segments of the planned static route, the task of a behavioral planner is to plan appropriate driving maneuvers (e.g., cruise-in-lane, change-lane, parking) based on perceived information of other traffic participants, obstacles, road conditions, and signals from infrastructure. For instance, at an intersection, the behavioral planner of the ego vehicle needs to decide between "passing" and "yielding" maneuvers based on its perceived behaviors and predicted intentions of other traffic participants [34]. Several techniques for intention prediction and future trajectories estimation of other road participants have been studied, e.g., learning-based techniques [29, 41] and probabilistic planning formalisms [10, 34].

7.2.3 Motion Planner

After the behavioral layer decides the driving maneuver of the ego vehicle, a motion planner is responsible for computing a path or a trajectory, where a trajectory is a path with a prescribed velocity profile along the path. The computed path or trajectory must be dynamically feasible for the vehicle model, comfortable for the passenger, and avoid collisions with obstacles and other traffic participants. The motion planning problems are generally formulated as optimisation problems, whose exact solutions are mostly computationally intractable. Therefore, numerical approximation and heuristic methods are typically used in practice. We will discuss the motion planning layer in more detail in Section 7.3.

7.2.4 Feedback Controller

A feedback controller's task is to select appropriate actuator inputs that execute the reference path or trajectory computed by the motion planner. The controller needs to execute the planned motion while dealing with tracking errors due to the inaccuracies of the vehicle model used by the motion planner and other road participants' trajectories prediction errors. A survey of effective feedback controllers is discussed in [53].

7.3 Motion Planning

In this chapter, we focus on the motion planning layer, in which various challenges of an ADS are the most manifest. There are many motion planning techniques designed for various driving scenarios such as highway driving, parking, and urban driving in cities with intersections and roundabouts (see [27] and the references therein). In what follows, we start with an overview of some optimisation-based and model-predictive motion planning techniques. Then we discuss game theory-based and other planning approaches.

7.3.1 Optimisation-Based Approaches

We provide an overview of optimisation-based trajectory planning approaches for ADSs. Specifically, we explain some techniques introduced by the earlier works [71] and [48], and provide a discussion on the more recent literature that employ related techniques. In the literature, planned trajectories are commonly described by using Frenet frames. Moreover, Frenet frames are also used for characterising lattice-based discretisations of the road. For this reason, we start our exposition with an explanation of Frenet frames defined for smooth planar curves [42].

7.3.1.1 Planar Curves and Frenet Frames

Many trajectory planning algorithms use smooth planar curves to represent possible paths on a road. Such curves can also be used for characterising the roads themselves (see, e.g., [12]). Let $p\colon [0,L] \to \mathbb{R}^2$ be a smooth planar curve of length $L > 0$. We use $s \in [0,L]$ to denote the arc length along this curve and

$e_1\colon [0,L] \to \mathbb{R}^2$ to denote the *tangential vector* that satisfies

$$e_1^{\mathsf{T}}(s)e_1(s) = 1, \tag{7.1}$$

$$\frac{\mathrm{d}p(s)}{\mathrm{d}s} = e_1(s), \tag{7.2}$$

where $(\cdot)^{\mathsf{T}}$ denotes transpose. Now, let $e_2\colon [0,L] \to \mathbb{R}^2$ be given by

$$e_2(s) = \begin{bmatrix} 0 & -1 \\ 1 & 0 \end{bmatrix} e_1(s). \tag{7.3}$$

The vector $e_2(s)$ is called the *normal vector* at arc length s and the pair $(e_1(s), e_2(s))$ is referred as the *Frenet frame* [42]. An illustration of the Frenet frame is given in Fig. 7.2.

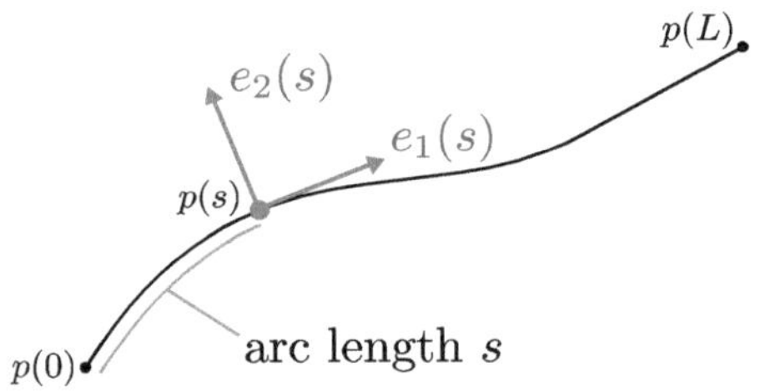

Figure 7.2: A planar curve $p\colon [0,L] \to \mathbb{R}^2$ and the Frenet frame $(e_1(s), e_2(s))$ at arc length s.

It follows respectively from (7.1) and (7.3) that we have $e_1^{\mathsf{T}}(s)\frac{\mathrm{d}e_1(s)}{\mathrm{d}s} = 0$ and $e_1(s)e_2(s) = 0$. As a consequence, for each arc length s, there exists a unique scalar $\kappa(s) \in \mathbb{R}$ such that

$$\frac{\mathrm{d}e_1(s)}{\mathrm{d}s} = \kappa(s)e_2(s).$$

Here, $\kappa\colon [0,L] \to \mathbb{R}$ is called the *curvature*. Using the curvature, we can obtain the orientation $\theta(s) \in \mathbb{R}$ of the curve at arc length s as $\theta(s) = \theta_0 + \int_0^s \kappa(\tau)\mathrm{d}\tau$, where θ_0 is the initial orientation.

In the following, we first look at a trajectory generation approach where the goal is to find the optimal movements along the axes of the Frenet frame.

7.3.1.2 Optimal Trajectory Generation Using Frenet Frames

The work [71] uses planar curves and Frenet frames for generating optimal trajectories for a vehicle on a highway. In particular, a planar curve p is used as an ideal path along a given road. This ideal path is also referred as the *center line*. Frenet frames along the curve p are then used for representing locations of points on generated trajectories.

Consider a trajectory $x\colon\, [0,T] \to \mathbb{R}^2$ parameterised with time $t \in [0,T]$. The location of a point on this trajectory at any given time t can be represented using the Frenet frames that move along p. Specifically, $x(t)$ can be given by

$$x(t) = p(s(t)) + d(t)e_2(s(t)),$$

where $s(t)$ denotes the arc length at time t, and $d(t)$ is the lateral offset from p at that time, as shown in Fig. 7.3. Notice that given the ideal path p, the point $x(t) \in \mathbb{R}^2$ can be represented equivalently with the pair $(d(t), s(t))$.

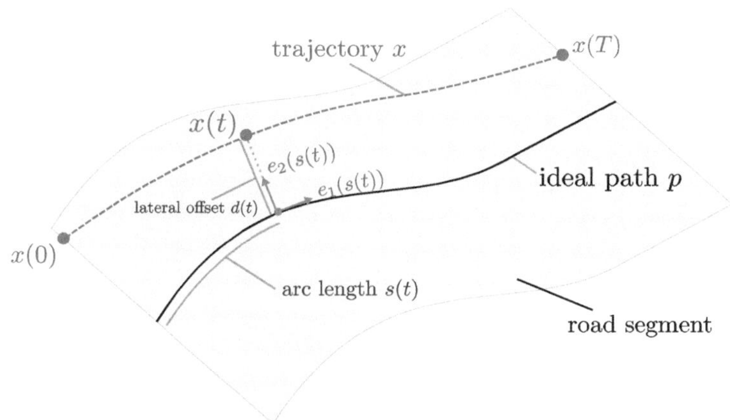

Figure 7.3: Representation of a point $x(t)$ on a trajectory $x\colon\, [0,T] \to \mathbb{R}^2$ through arc length $s(t)$ and lateral offset $d(t)$ with the help of Frenet frame $(e_1(s(t)), e_2(s(t)))$ on a given ideal path p.

The key idea in [71] is to separately design the lateral offset d and the arc length s as functions of time t. The design is based on an optimisation framework with different parameters for different on-road tasks such as velocity keeping, following, and stopping.

The first step in the optimisation framework of [71] is to find $d\colon\, [t_0, t_1] \to \mathbb{R}$ and $s\colon\, [t_0, t_1] \to \mathbb{R}$ that minimise certain cost functions over a given time interval $[t_0, t_1]$. The parameters and the structure of the cost functions differ for each different task. For instance, the cost for the lateral offset in the case of high speed lane keeping task can be written as

$$C_d = k_1 \int_{t_0}^{t_1} \dddot{d}^2(\tau)\mathrm{d}\tau + k_2(t_1 - t_0) + k_3 d^2(t_1),$$

where $k_1, k_2, k_3 > 0$ are the weight coefficients that characterise the contribution of different cost terms. The first cost term $\int_{t_0}^{t_1} \dddot{d}^2(\tau)\mathrm{d}\tau$ is the integral of the lateral jerk (time-derivative of the acceleration) and is related to the passenger comfort, the second term $t_1 - t_0$ is the total time length, and the last term $d^2(t_1)$ is the squared lateral deviation from the path at the final time t_1.

Given the initial time t_0 and the initial offset $d(t_0)$ as well as its derivatives $\dot{d}(t_0)$, $\ddot{d}(t_0)$, the algorithm of [71] first prepares a set of different candidate time intervals T_j (representing $t_1 - t_0$) and candidate final lateral offsets d_j (representing $d(t_1)$), where j denotes the candidate index. For each candidate j, the integral cost $\int_{t_0}^{t_0+T_j} \dddot{d}^2(\tau)\mathrm{d}\tau$ is minimised by considering the boundary conditions

$$d(t_0+T_j) = d_j,\ \dot{d}(t_0+T_j) = 0,\ \ddot{d}(t_0+T_j) = 0.$$

For each j, the minimisation yields $d(t)$ as a quintic polynomial over $t \in [t_0, t_0 + T_j]$. Then, the total cost $C_{d,j} = k_1 \int_{t_0}^{t_0+T_j} \dddot{d}^2(\tau)\mathrm{d}\tau + k_2 T_j + k_3 d_j$ is computed to evaluate the performance of the associated trajectory.

An important point of the algorithm is that certain candidate trajectories can be discarded if they fail validity checks. Valid trajectories are the ones that avoid collision and stay within the drivable road areas. Noting that a similar optimisation approach (with different cost structure and parameters) is used for designing the arc length s as a function of time, validity checks can be conducted for the candidate pairs (d, s). The algorithm picks the valid trajectories for both d and s that yield the lowest costs.

Repeated Trajectory Optimisation

The trajectory optimisation for d and s is repeated frequently to take account of potential variations in the road environment. The duration of the planned trajectories can be longer than the time period between two consecutive optimisations. An important point of the algorithm of [71] is that the end times of the candidate trajectories in any optimisation overlap with the end times of candidate trajectories of previous optimisations. This ensures "temporal consistency", meaning that a computed optimal trajectory is followed without any modification if the environment does not change. A related goal is to ensure that the vehicle's acceleration does not face instantaneous jumps between two motion planning cycles. This is achieved if the end-point acceleration values of sequentially planned trajectories are equal. The work [78] proposes a re-planning scheme to achieve this. The scheme proposed in that work also accounts for the processing time needed for trajectory optimisation.

Practical Issues

Optimisation-based approaches with the Frenet frame representation can handle static/dynamic obstacles as well as stop signs, speed limits, and traffic lights commonly faced in the urban traffic [20]. For instance, it is mentioned in [20] that a red traffic light can be represented as an obstacle that is present for a certain time period, and furthermore, speed limits and stop signs can be considered as velocity constraints. Moreover, the work [22] discusses how a Frenet frame-based trajectory planner can be equipped in an autonomous driving system with sensing and control components.

It is discussed in [46] that Frenet frame-based planning approaches may have performance issues on roads with large curvatures (representing sharp turns). There, a new optimisation-based approach is proposed to conduct trajectory planning in the Cartesian frame.

7.3.1.3 Spatiotemporal Lattices

We next look at the motion planning framework introduced by [48]. This framework uses both state lattices and spatiotemporal lattices in construction of a search space with finitely many elements representing potential trajectories of an on-road vehicle for several seconds. Similarly to the approach of [71], this framework can generate maneuvers around the center line of a road lane that avoid hitting static/dynamic obstacles. One of the key features of the framework of [48] is that it takes account of the kinematic constraints of a vehicle. This allows generation of trajectories that the vehicle can follow exactly. In addition, the lattice structures proposed in [48] also enable expressing relatively complex maneuvers.

State lattices were used previously in planning problems in different settings (see, e.g., [54]). The lattice-based framework of [48] is different from those of the previous works in that it is designed to be used in a structured environment such as highways with moving traffic.

State Lattices

We start with a short description of the state lattices considered in [48]. Similarly to [71], first, the center line of a lane is considered as a curve that is parameterized with arc length. Cartesian coordinates, orientation, and curvature of the center line at arc length s are represented respectively by $x(s), y(s), \theta(s), \kappa(s)$. Any arbitrary point on the road that is not on the center line can be characterized similarly. In particular, a point at arc length s and at lateral offset d has the Cartesian coordinates

$$x(s,d) = x(s) + d\cos(\theta(s)), \tag{7.4}$$

$$y(s,d) = y(s) + d\sin(\theta(s)), \tag{7.5}$$

the orientation

$$\theta(s,d) = \theta(s), \tag{7.6}$$

and the curvature

$$\kappa(s,d) = (\kappa^{-1}(s) + d)^{-1}. \tag{7.7}$$

Construction of the state lattice is based on discretisation of the arc length and lateral offset. More specifically, the lattice is composed of vertices at uniformly-separated increasing arc lengths $s_0, s_1, \ldots, s_{N-1}$ and uniformly-separated increasing lateral offsets $d_0, d_1, \ldots, d_{M-1}$. Each vertex in the lattice corresponds to a different pair of arc length and lateral offset. The Cartesian coordinates of each vertex are obtained by using (7.4) and (7.5), and moreover, the orientation and the curvature are obtained through (7.6) and (7.7). An illustration of the state lattice is given in Fig. 7.4.

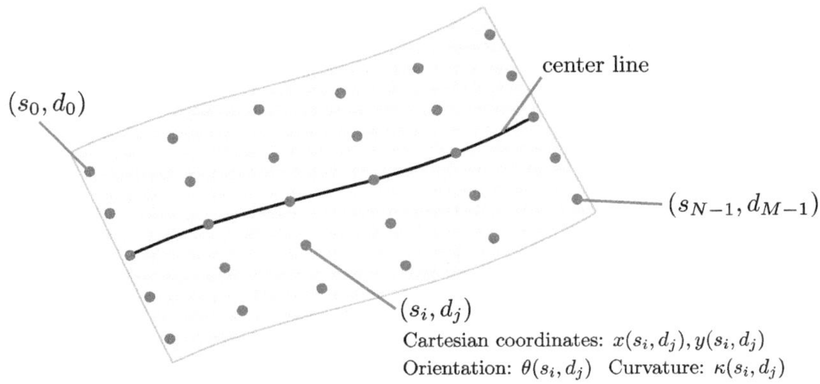

Figure 7.4: State lattice around a center line on a road segment.

The work [48] generates a path between two vertices in the state lattice by designing the curvature as a function of the arc length along the path. Each curvature function is defined as a cubic polynomial, and the coefficients of this polynomial are identified by considering the kinematic model of a vehicle. Given the known initial position (first vertex) and the desired final position (second vertex), the coefficients are obtained by solving equations with end-point constraints. The utility of this approach is that a vehicle with the considered model is guaranteed to be able to follow the generated path.

The works [14, 16] also take into account the constraints induced by the kinematic model of a vehicle. In [14], clothoids are used for representing potential paths of a vehicle, and moreover, probabilistic risk assessment methods are utilised for characterising the cost terms that are related to the safety of those paths. In [16], the authors tackle the problem of motion planning on general road networks composed of lanes, intersections, and roundabouts.

Spatiotemporal Lattices and Trajectories

Paths between the vertices of the state lattice do not contain any timing or velocity information, as the curvatures are characterized as functions of arc length instead of time. The work [48] supplements each generated path with time and velocity components. In particular, given a starting time, a starting velocity, and a constant acceleration value, the velocity profile of a vehicle moving on a path can be identified uniquely. Based on this observation, [48] extends the state lattice to a spatiotemporal lattice. Vertices in the spatiotemporal lattice contains not only the information of Cartesian coordinates, orientation, and curvature (x, y, θ, κ), but also time and velocity (t, v). Connecting a pair of vertices in the spatiotemporal lattice corresponds to a *trajectory*. Figure 7.5 illustrates an example spatiotemporal lattice, where a trajectory is indicated by two vertices with different spatial and temporal components.

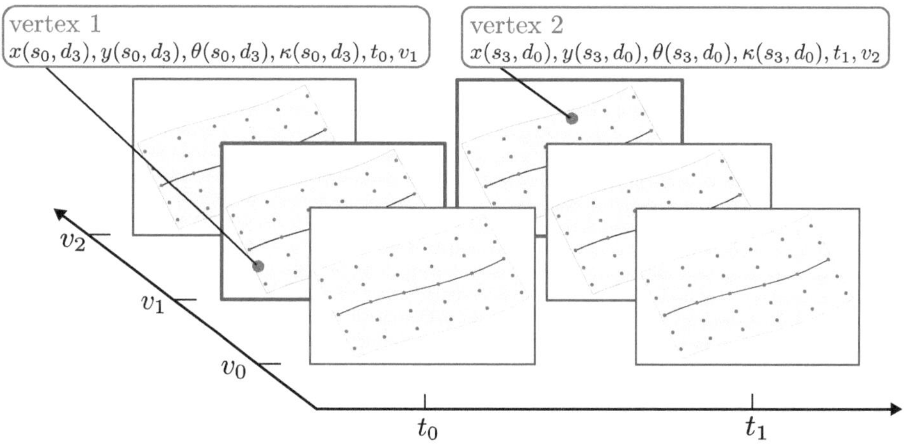

Figure 7.5: Illustration of spatiotemporal lattice as an extension of the state lattice with new temporal components (time and velocity). A trajectory in the spatiotemporal lattice is represented by a pair of vertices with different spatial and temporal components.

Costs Assigned to Paths and Trajectories

Once the spatiotemporal lattice is constructed, the framework of [48] assigns costs to pairs of vertices in the lattice. Then the sequence of vertices with the minimum sum of costs can be chosen as the best motion plan. The costs can penalize lateral acceleration to promote comfort, and moreover, they can penalize foreseen collisions with static/dynamic obstacles to ensure safety. An important aspect of the approach of [48] is that multiple trajectories can use the same path. To avoid repeated computations, the cost terms that depend only on position and curvature (but not time and velocity) are evaluated only once using the path information. These costs are often associated with efficiency, comfort, or safety. Physical interpretations and impacts of many different cost terms are provided in [73]. Moreover, a variety of common cost functions used in motion planning benchmarks are also discussed in [5].

Computational Issues

As a spatiotemporal lattice can have many vertices, direct search within the lattice can be slow. However, spatiotemporal lattices of [48] allow efficient re-parametrisations, which makes motion planning fast. The approach presented there is reported to be capable of repetitively planning the motion of a vehicle at every 0.1 seconds on a standard computer.

7.3.1.4 Trajectory Optimisation in Different Architectures

So far we have looked at works that propose motion planning techniques with the goal of finding trajectories that are optimal with respect to certain criteria. There are a few works where trajectory optimisation is combined with other concerns. For instance, [47] aims to ensure fail-safe operation. According to the architecture proposed in that work, in each planning cycle, the vehicle applies either the optimal trajectory or a certain collision-free emergency maneuver. The optimal trajectory is selected only if that trajectory guarantees existence of an emergency maneuver in future planning cycles. Otherwise, the vehicle executes the emergency maneuver. Typically, optimisation-based motion planners are considered to be separate components that interact with sensing, prediction, and local control components. However, we also note that there are approaches where sensing, prediction, planning, and control tasks are combined in a single component to enable the so-called end-to-end self-driving (see, e.g., [9, 60]).

7.3.2 Model Predictive Control Approach

Optimisation-based approaches discussed in Section 7.3.1 generate optimal trajectories as motion plans and pass them to local controllers. The role of a local controller is to provide control commands (e.g., steering angle and acceleration) to the vehicle so that the trajectories can be followed closely. The model predictive control (MPC) approach proposed in [50] provides combined trajectory generation and local control. In what follows, we present a short overview of this approach.

First, we describe the framework developed in [50] for generating optimal paths and trajectories to help a vehicle to move from an initial position to a desired destination in an environment with obstacles.

7.3.2.1 Visibility Graphs and Optimal Paths

The initial step for the path generation approach of [50] is to create a *visibility graph*. The nodes in a visibility graph are the initial position of the vehicle, the desired destination of the vehicle, and the corners of the polygons representing obstacles. The edges correspond to *visibility lines* between the nodes (see Fig. 7.6).

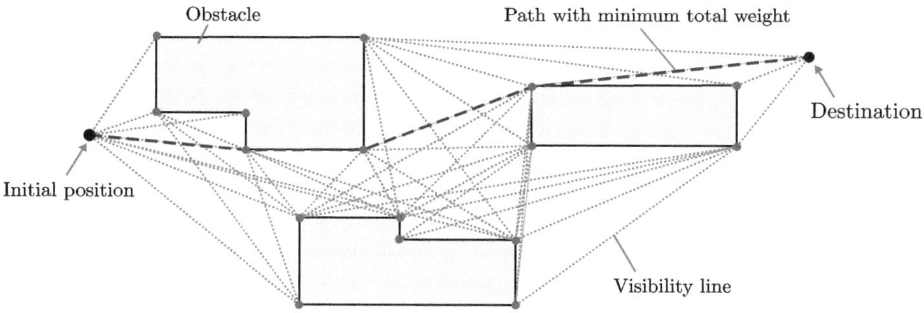

Figure 7.6: Path generation in the visibility graph.

The edges in the visibility graph are assigned positive weights [50]. The weights of the edges that are not connected to the initial position or the destination are proportional to the physical distance between the connected nodes. On the other hand, the edges that are connected to either the initial position or the destination are assigned weights based on both the physical distances and the so-called orientation gaps. To compute an orientation gap, one needs the information of the initial orientation of the vehicle and its required final orientation at the destination. Orientation gap for an edge connected to the initial position is the angle between the orientation of the edge and the initial orientation. Orientation gaps for the edges connected to the destination are computed similarly.

Once all edges are assigned weights, one can use a search algorithm (e.g., A*) to find the path with minimum total weight that connects the initial position to the destination [50].

7.3.2.2 Reference Trajectory

The obtained optimal path is then used for generating a reference trajectory. This is done by assigning a velocity profile that describes the velocity of the vehicle at different positions on the path. The trajectory is represented with the sequence $r(0), r(1), \ldots, r(T-1) \in \mathbb{R}^2$, where $r(0)$ is the initial position of the vehicle, and $r(t)$ represents the position at time $t\Delta$, where Δ is the sampling and the control input update period of the vehicle.

7.3.2.3 MPC Approach to Follow the Reference Trajectory

The goal of the model predictive control approach in [50] is to make the vehicle follow the reference trajectory closely. This is achieved by minimising a cost (related to the distance from the trajectory) in a rolling horizon fashion. In a simplified setting, the objective can be set to minimise

$$\sum_{i=1}^{h} \|p(t+i|t) - r(t+i)\|^2, \tag{7.8}$$

at each time t, where h is a positive integer denoting the optimisation horizon, $r(t+i) \in \mathbb{R}^2$ is the desired position on the reference trajectory at time $t+i$, and $p(t+i|t) \in \mathbb{R}^2$ is the prediction regarding the position of the vehicle at time $t+i$. The prediction is obtained at time t by using the vehicle dynamics

$$\begin{aligned} x(t+i+1|t) &= f(x(t+i|t), u(t+i)), \quad i \in \{0, 1, \ldots h-1\}, \\ p(t+i|t) &= g(x(t+i|t)), \end{aligned}$$

where $x(t+i|t) \in \mathbb{R}^n$ is the state variable with $x(t|t) = x(t)$ denoting the state measurement at time t, $u(t+i) \in \mathbb{R}^m$ is the control input, $f\colon \mathbb{R}^n \times \mathbb{R}^m \to \mathbb{R}^n$ describes the evolution of the state over one time step, and $g\colon \mathbb{R}^n \to \mathbb{R}^2$ is a function that maps the vehicle's state variable to vehicle position.

The cost in (7.8) is minimised with respect to the control inputs $u(t), u(t+1), \ldots, u(t+h-1)$. The optimisation problem in [50] incorporates multiple weighted cost terms and it also includes constraints on $p(t+i|t)$ and $u(t+i)$ so as to keep a certain distance from the obstacles and to satisfy actuation constraints (e.g., maximum acceleration).

The framework of [50] uses the MPC approach. The key idea in MPC is to solve the optimisation problem at time t, find optimal control inputs for the entire horizon (from t to $t+h-1$), and apply the first control input to the system (see, e.g., [11, 58] for more detailed discussions). The optimisation problem is solved again at the next time step $t+1$. This idea is illustrated in Fig. 7.7.

The utility of the MPC approach is that the optimisation problem is solved repeatedly with updated information about the system states. Uncertainties in the system and the environment as well as constraints on the inputs and the states can be explicitly taken into account [11, 58]. In particular, nonlinear model predictive control can be used for handling the planning and control needs when nonlinear vehicle models are considered (see, e.g., [28, 70]). In addition, robust MPC methods can be employed for operation under uncertainty. For instance, a robust tube-based MPC approach was developed in [39] to tackle scenarios where the obstacle information available to the ego-vehicle is subject to uncertainties. MPC-based approaches can also handle more complicated objectives involving passenger comfort and lane keeping performance (see [30, 44]).

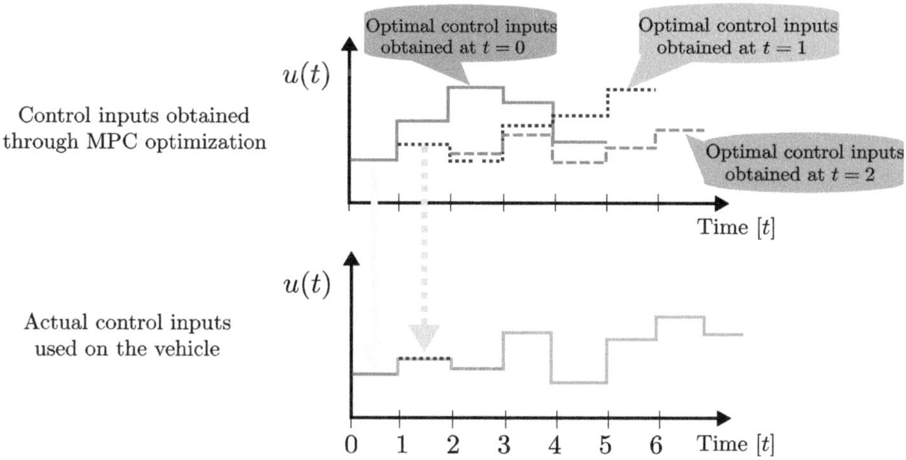

Figure 7.7: Optimal input values obtained with MPC approach at each time (top) and control inputs applied to vehicle (bottom).

MPC-based approaches are also used in other domains. For instance, [7] uses MPC in trajectory generation for robotic arms, and moreover, [76] uses an MPC-based approach to find optimal routes and schedules for a group of vehicles in mobility-on-demand systems.

7.3.3 Game Theory-Based Approaches

Let us consider the road situation in Fig. 7.8. While driving on the highway, we notice a vehicle trying to cut closely in front of us. In this situation, we can safely avoid the collision if we apply a sudden brake and let the other vehicle cut in. However, if we accelerate, we may force the other vehicle to decelerate and move to our lane behind us so that we could avoid the accident without a sudden brake.

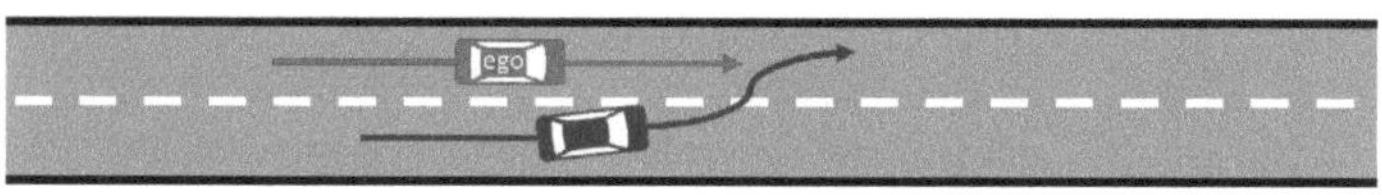

Figure 7.8: An example of a road situation where a vehicle (bottom) is cutting in front of our vehicle (top).

A game theory-based decision-making approach can naturally model the above inter-vehicle interactions in a human-like manner [61, 75]. Game theory-based approaches model the vehicles as players of the game, where the reward functions are designed based on several criteria such as the speed of the ego vehicle, the distance between vehicles, and the jerk-based smoothness of our movement. Then, the players decide their actions by optimizing their reward in response to the actions of others.

In this section, we focus on game theory-based approaches that model the vehicle interactions as a Stackelberg (or leader-follower) dynamic game [23,55, 57,61]. We first introduce the framework in [55,57]. Let us consider n vehicles, where the state (e.g., information of position and speed of the vehicle) of each vehicle $i \in \{1,\dots,n\}$ at time $t \geq 0$ is represented by $x_i(t)$. Then, we assume that the dynamics of the vehicle i is

$$x_i(t+\Delta) = f_{\theta_i}(x_i(t), u_i(t)),$$

where f_{θ_i} is the discrete-time dynamics function, Δ is the duration between discrete time steps, and u_i is the control input of the decision-maker of vehicle i. We assume that the dynamics function f_{θ_i} is dependent on a set θ_i of parameters such as the navigation path, target region, aggressiveness, and priority to the right of way. The parameters θ_j of the other vehicles $j \neq i$ are unknown to i.

The high-level overview of the algorithms in [55,57] is depicted in Fig. 7.9. First, the decision-maker of the ego vehicle i observes the states of each nearby vehicle j (step-1). Based on the observation, the ego vehicle i initiates the parameters in $\widehat{\theta}_{j,i}$, which is the estimation of θ_j (step-2). Then, the ego vehicle computes an optimal control input based on $f_{\widehat{\theta}_{j,i}}$ (step-3). In the next time step, the ego vehicle observes the other vehicles and updates the parameters in $\widehat{\theta}_{j,i}$, and repeats the process.

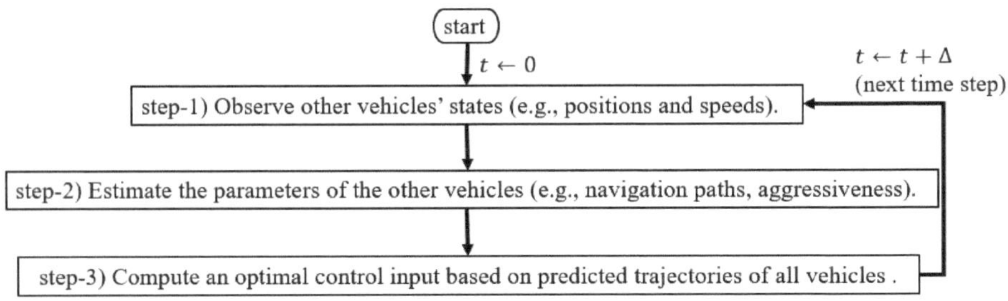

Figure 7.9: Overview of the game theory-based decision-making process proposed in [55,57].

We compute the optimal control input in step-3 by computing an equilibrium of a Stackelberg (or leader-follower) game: a strategic game in which the leader player makes her decision first and then the follower player decides her action sequentially. For simplicity, we consider a two-player Stackelberg game played between two players: l and f, which are the leader and follower, respectively. Let the reward of player $i \in \{l, f\}$ at each time step t be given by $R_i(u_i(t), u_j(t))$, where $j \neq i$ is the other player and R_i is a reward function of i computed using $x_i(t)$, $x_j(t)$, f_{θ_i}, and $f_{\widehat{\theta}_{j,i}}$. The reward function can be designed based on several criteria such as the ego vehicle's speed, the distance between vehicles, and the smoothness of the movement.

Algorithm 1 Algorithm for a two-player Stackelberg game equilibrium.

$u_l^* \leftarrow$ an arbitrary control input of the leader.
for each control input u_l of the leader **do**
    $u_f^*(u_l) \leftarrow$ an arbitrary control input of the follower.
    for each control input u_f of the follower **do**
        if $R_f(u_l, u_f) > R_f(u_l, u_f^*(u_l))$ **then**
            $u_f^*(u_l) \leftarrow u_f$
    if $R_l(u_l, u_f^*(u_l)) > R_l(u_l^*, u_f^*(u_l^*))$ or $u_f^*(u_l^*)$ is undefined **then**
        $(u_l^*, u_f^*) \leftarrow (u_l, u_f^*(u_l))$
return (u_l^*, u_f^*)

Then, at each time step, a Stackelberg equilibrium (u_l^*, u_f^*) is computed as in Algorithm 1. Figure 7.10 shows the decision tree corresponding to Algorithm 1 for the example situation in Fig. 7.8. Notice that the leader makes the decision first, then the follower selects the best control input based on the leader's decision. Figure 7.10 shows possible decisions of the players in terms of the control inputs associated with acceleration and braking actions. As the leader usually has the advantage, the computed equilibrium depends on which player is the leader. The leader and the follower of the game can either be fixed for all time steps [23, 61] or decided adaptively based on the observations at previous time steps [55, 57].

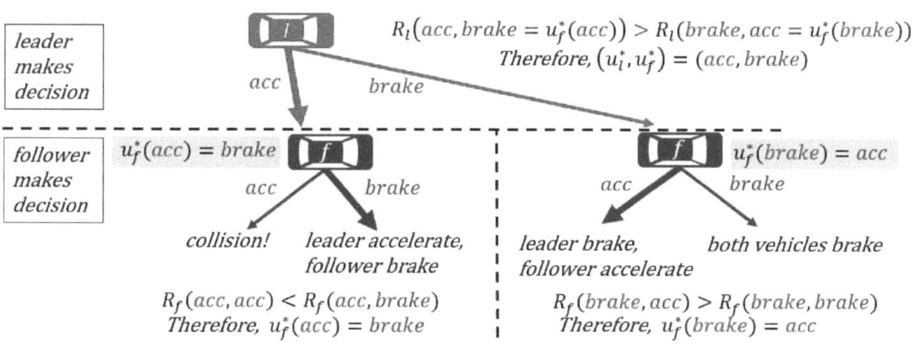

Figure 7.10: The decision tree corresponding to Algorithm 1 for the example situation in Fig. 7.8.

The main advantages of game theory-based decision-making approaches are that (1) they can model rational vehicle interactions in a human-like manner, and (2) they do not require previous training data as they are adaptive to changes (based on observation). However, as shown in Algorithm 1, the approaches require the control input set to be finite. Moreover, the complexity of the algorithms can be high if we consider a large set of actions and a game with many players [55].

7.3.4 Other Approaches

In this section, we looked at optimisation-based, model predictive control, and game theory-based motion planning approaches that are well-suited for autonomous vehicles. There are motion planning approaches such as sampling-based motion planning, which was originally designed for robots but can be applied to certain driving scenarios. Regardless of the state space and dynamic constraints, the sampling-based methods explore the state space and find an optimal path using random sampling. Examples of well-known sampling-based motion planning methods are Probabilistic Roadmaps (PRM) [38], Rapidly-exploring Random Trees (RRT) [45], and RRT* [37]. We refer interested readers to [40, 53] for the surveys of sampling-based motion planning approaches.

In recent years, machine learning techniques have had a significant impact on autonomous driving. Several machine learning paradigms, such as deep learning and reinforcement learning, have shown their efficiency both in academia and industry and have been continuously developed [29, 41]. However, one crucial issue of the learning-based approaches is the lack of formal guarantees and valid explanations for their reasoning. Although Explainable Artificial Intelligence has been widely studied, it is arguable whether the existing methods are adequate to provide trust in the decision-making for autonomous vehicles [26].

In the following sections, we introduce some formal verification techniques to ensure the safety of autonomous vehicles. We first introduce online verification techniques using reachability analysis methods, where the goal is to compute potential road occupancy areas of the ego vehicle and other road participants in Section 7.4. Then, in Section 7.5 we discuss the Responsibility-Sensitive Safety (RSS) framework: a framework that derives logical safety rules from the traffic regulations and common senses.

7.4 Online Verification Using Reachability Analysis

Formal verification can be combined with behavior and motion planners to ensure the safety of the autonomous vehicle. The safety of the vehicle is guaranteed for some future time interval if the vehicle's road occupancy area does not intersect with that of other road participants during the time interval. In order to verify the safety online, we can apply reachability analysis to predict all possible road occupancies of all road participants by considering all possible uncertainties such as sensor noise and environmental disturbances. In this section, we introduce two approaches for safety verification: the set-based and the probabilistic approaches.

7.4.1 Set-Based Reachability Analysis for Safety Verification

Set-based approaches verify the safety of autonomous vehicles by computing a set that contains all possible road occupancy areas of all road participants during a considered future time interval. In this section, we briefly introduce an approach to overapproximate the reachable set of the autonomous vehicle that is proposed in [3,4]. Let $\mathcal{X}$, $\mathcal{R}(0)$, and $\mathcal{U}$ be the state space, the bounded set of initial states, and the bounded set of inputs, respectively. The dynamics of the vehicle is given by a nonlinear function $\dot{x} = f(x,u)$, where $x(t) \in \mathcal{X}$ is the state vector and $u(t)$ is the input vector at time t. The solution of this system at time t is denoted by $\chi(t,x_{in},u)$, where x_{in} is the initial state vector. The reachable set during time interval t_f is

$$\mathcal{R}^e([0,t_f]) = \{\chi(t,x_{in},u) \mid t \in [0,t_f], x_{in} \in \mathcal{R}(0), u(t) \in \mathcal{U}\}.$$

The reachable set $\mathcal{R}^e([0,t_f])$ cannot be exactly computed in general. We, therefore, compute its overapproximation $\mathcal{R}([0,t_f]) \supseteq \mathcal{R}^e([0,t_f])$, which is a compact convex set that can be represented by a polytope or a zonotope. The overapproximated reachable set is computed by considering each time interval $[t_k,t_{k+1}]$, where $t_{k+1} - t_k$ is constant for all time step $k \geq 0$. The set $\mathcal{R}([t_k,t_{k+1}])$ is then computed from a linear system obtained by linearising the original nonlinear dynamics using a first-order Taylor expansion, then the linearisation errors are overapproximated using the Lagrange remainders. We consider a linearisation point $z^* = [x^*, u^*]^\mathsf{T}$, where x^* and u^* are points of state and input, respectively. One possible choice for z^* is the center of $\mathcal{R}([t_{k-1},t_k])$ and $\mathcal{U}$. Then, using the first-order Taylor expansion around the point z^*, the original differential equation of the ith coordinate is enclosed by

$$\forall t \in [t_k,t_{k+1}], \dot{x}_i(t) \in f_i(z^*,u^*) + \frac{\partial f_i(z,u^*)}{\partial z}\Big|_{z=z^*} (z - z^*) \oplus \mathcal{L}_i([t_k,t_{k+1}]), \quad (7.9)$$

where $\oplus$ is the Minkowski addition and $\mathcal{L}$ is the Lagrange remainder that covers all possible linearisation errors, which can be computed using interval arithmetic [4, 36]. Then, the overapproximated convex reachable sets are computed using the linearised system (7.9).

The set-based reachability analysis is an efficient methodology for the safety verification of autonomous vehicles. The approach has been demonstrated by test drives with an autonomous vehicle [4].

7.4.2 Probabilistic Approach Through Moment Propagation

Another approach to guarantee the safety of the autonomous vehicle is to use probabilistic verification. We focus on the method proposed in [56], which approximates the moments of a discrete-time polynomial stochastic system. These approximate moments are used to compute probability bounds for the state to be outside of given regions.

We consider a finite-dimensional discrete-time stochastic polynomial system

$$x(t+1) = \sum_{i=0}^{d_s} F_i(t)x^{[i]}(t), \quad t \in \mathbb{N}, \tag{7.10}$$

where $x(0) = x_{in}$ is a given initial state, $x(t) \in \mathbb{R}^n$ is the state vector, $x^{[i]}$ is the ith Kronecker power of x, and $F_i(t) \in \mathbb{R}^{n \times n^i}$ are randomly-varying coefficient matrices for all $i \in \{0, 1, \ldots, d_S\}$. We assume that the vector x_{in} and the matrices $F(t), t \in \mathbb{N}$, are all independent. Moreover, $F(t), t \in \mathbb{N}$, are identically distributed.

We use Carleman linearisation to obtain an infinite-dimensional linear system that describes the evolution of the Kronecker powers of the state vector $x(t)$. By defining

$$F(t) = \begin{bmatrix} F_0(t) & F_1(t) & F_2(t) & \cdots & F_{d_S}(t) \end{bmatrix}$$

and

$$y(k,t) = \begin{bmatrix} 1 & x(t)^{\mathsf{T}} & x^{[2]}(t)^{\mathsf{T}} & \cdots & x^{[k]}(t)^{\mathsf{T}} \end{bmatrix}^{\mathsf{T}},$$

the dynamical system (7.10) can be rewritten as

$$x(t+1) = F(t)y(d_S, t).$$

By using the *mixed-product property* of the Kronecker product, we obtain

$$\begin{aligned} x^{[j]}(t+1) &= F^{[j]}(t)y^{[j]}(d_S,t) \\ &= \sum_{k=0}^{jd_S} \left(\sum_{(i_l)_{l \le j} \in H_{j,k}} F_{i_1}(t) \otimes \cdots \otimes F_{i_j}(t) \right) x^{[k]}(t), \end{aligned}$$

for $j \in \mathbb{N}$, where

$$H_{j,k} = \left\{ (i_l)_{l \le j} \,\middle|\, \sum_{l=1}^{j} i_l = k \text{ and } i_l \le d_S \right\}.$$

In order to give a simpler description of the system, we introduce the matrices

$$A_{j,k}(t) = \sum_{(i_l)_{l \le j} \in H_{j,k}} F_{i_1}(t) \otimes \cdots \otimes F_{i_j}(t),$$

and

$$A(t;N,M)=\begin{bmatrix} A_{0,0}(t) & \cdots & A_{0,M}(t) \\ \vdots & \ddots & \vdots \\ A_{N,0}(t) & \cdots & A_{N,M}(t) \end{bmatrix}.$$

Then, for any $k \in \mathbb{N}$, we have

$$\begin{aligned} y(k,t+1) &= \begin{bmatrix}1 & x^{[1]}(t+1)^\mathsf{T} & \cdots & x^{[k]}(t+1)^\mathsf{T}\end{bmatrix}^\mathsf{T} \\ &= A(t;k,kd_S)\begin{bmatrix}1 & x^{[1]}(t)^\mathsf{T} & \cdots & x^{[kd_S]}(t)^\mathsf{T}\end{bmatrix}^\mathsf{T} \\ &= A(t;k,kd_S)y(kd_S,t). \end{aligned} \tag{7.11}$$

Note that the vector $y(k,t)$ contains all Kronecker powers $x^{[0]}(t), x^{[1]}(t), \ldots x^{[k]}(t)$ of the state $x(t)$. We now derive the deterministic system that describes the evolution of the moments of $x(t)$ by taking expectation in (7.11). This gives

$$\mathbb{E}\begin{bmatrix}1 & x^{[1]}(t)^\mathsf{T} & \cdots & x^{[k]}(t)^\mathsf{T}\end{bmatrix} = \mathbb{E}[y(k,t+1)] = E(k,kd_S)\mathbb{E}[y(kd_S,t)],$$

where $E(k,kd_S)$ is the expectation of $A(t;k,kd_S)$, which is independent of t thanks to the assumption that the matrices $F(t), t \in \mathbb{N}$, are all independent and identically distributed.

Using the first and second moments, $\mathbb{E}[x(t)]$ and $\mathbb{E}[x^{[2]}(t)]$, at a given time t, we can compute probability bounds for the state vector $x(t)$ using Chebyshev's inequality. However, to compute the second moment $\mathbb{E}[x^{[2]}(t)]$, we need the knowledge of matrices $E(2,2d_S)$, $E(2d_s,2d_S^2)$, ..., $E(2d_S^{t-1},2d_S^t)$ of exponentially increasing dimensions. As a result, the exact computation for the moments becomes infeasible when t is large. Therefore, we need an efficient approach to approximate the moments at a given time t. The approach in [56] considers the truncated system with a fixed truncation limit. We fix $N_T \in \mathbb{N}$ and define $\widetilde{x}_i(t) \in \mathbb{R}^{n^i}$, $i \in 1,\ldots,N_T$, by

$$\begin{bmatrix}1 & \widetilde{x}_1(t)^\mathsf{T} & \cdots & \widetilde{x}_{N_T}(t)^\mathsf{T}\end{bmatrix}^\mathsf{T} = E(N_T,N_T)^t\begin{bmatrix}1 & \mathbb{E}[x_{in}]^\mathsf{T} & \cdots & \mathbb{E}[x_{in}^{[N_T]}]^\mathsf{T}\end{bmatrix}^\mathsf{T}.$$

The vector $\widetilde{x}_i(t)$ represents an approximation of the moment $\mathbb{E}[x^{[i]}(t)]$ that is computed using only our knowledge of the first N_T moments of x_{in}. The truncated system allows us to iteratively compute approximations of the moments of $x(t)$ at consecutive time instants. The approach only requires an offline computation of the matrix $E(N_T,N_T)^t$. Efficient ways to compute the upper bounds of the truncation error are proposed in [56]. Figure 7.11 shows the over-approximated road occupancy with probability 0.9, which is computed using Chebyshev's inequality, of an autonomous vehicle modeled using kinematic bicycle model in [56].

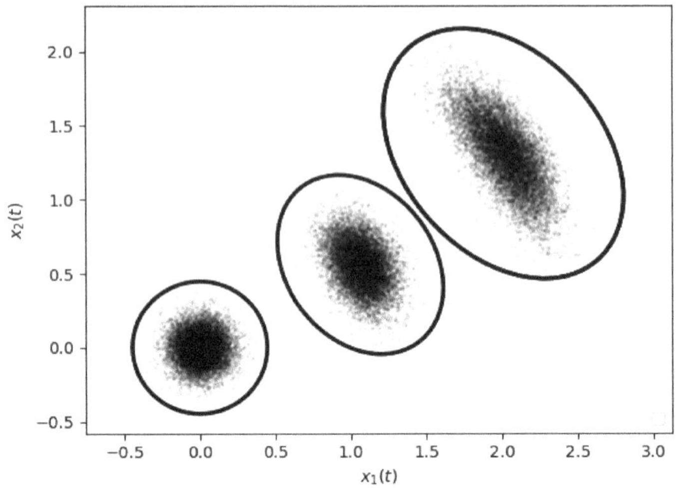

Figure 7.11: The over-approximated road occupancy region with probability 0.9 (ellipses), in comparison to 1000 runs of Monte Carlo simulation (dots), of an autonomous vehicle modeled using kinematic bicycle model in [56].

7.5 Responsibility-Sensitive Safety

7.5.1 What is RSS?

Responsibility-sensitive safety (RSS) is an approach to the ADS safety that attracts growing attention. Safety is a central theme in every aspect of the ADS research, including the approaches we discussed earlier in this chapter. RSS is unique among these approaches in that it provides *safety rules* that are rigorously formulated in mathematical terms. Unlike the algorithms and techniques we discussed earlier, RSS is not so much about *how to drive safely* as about *breaking down the ultimate goal* (namely safety) *into concrete and instantly-checkable conditions*. The goal of RSS is that those safety rules guarantee the ADS safety in the rigorous form of *mathematical proofs*. How far RSS has gone so far towards this ultimate goal of ADS safety proofs, and what are still needed for the ultimate goal, are questions that we will discuss later in Section 7.5.6.

The methodology and basic techniques of RSS are first introduced by researchers at Mobileye/Intel [64]. The safety rules derived in RSS have a variety of possible usages (as we will discuss in Section 7.5.5), giving RSS a multifaceted character. Moreover, its presentation to non-expert audience such as Mobileye/Intel's webpage[1]—an appeal to the general public is important given the social significance of the ADS safety—tends to emphasise RSS's

[1]https://www.mobileye.com/responsibility-sensitive-safety

ideological aspects. All these have somewhat blurred the technical substance of RSS. In this section, in contrast, we aim at a technical account of RSS from the viewpoints of ADS research, theory of dynamical systems, logic, and software science.

Remark 7.1 (RSS terminologies) Although there are already a number of research works on RSS as well as practical works (including the efforts to use it in international standards), comprehensive technical accounts on RSS are scarce. In particular, terminologies for the core concepts in RSS do not seem to be fixed in the literature.

In this section, we will use the terminologies listed in Table 7.1. Note that they may differ from the terminologies used in the literature (which vary from a paper to another). The notions in the table will be introduced in due course. Their correspondence to those used in the original paper [64] is also shown in the table. ▮

Table 7.1: RSS terminologies

In this section	in [64]	Meaning
RSS safety rule	(No explicit name)	A pair of an RSS safety condition and a proper response
RSS safety condition	Safe distance	An instantly-checkable condition that guarantees safety henceforth. Often stated in terms of a safety distance
Proper response	Proper response	A control strategy that realises safety
RSS responsibility principle	Safety rule, common sense rule	A high-level and informal principle used as assumptions on the behaviours of vehicles

7.5.2 Elements of RSS

7.5.2.1 RSS Safety Rules

The constructs central in RSS are *RSS safety rules*. Here is their semi-formal definition—a concrete example will appear in Section 7.5.4.

Definition 7.1 (RSS safety rule, RSS safety condition, proper response) An *RSS safety rule* is a pair $R = (C, P)$ of an RSS safety condition C and a proper response P, designed specifically for each driving scenario (straight road or crossing, other cars driving in the same direction or another, etc.). Here,

- an *RSS safety condition* C is a rigorous condition formulated in mathematical terms, which must be instantly checkable (it speaks only about the current values but not about future ones); and

- a *proper response* P is a control strategy that achieves safe driving, typically until the vehicle comes to a halt.

Here is the first fundamental requirement of an RSS safety rule.

Requirement 1 (safety assurance in RSS) *An RSS safety rule* $R = (C,P)$ *is required to satisfy the following:*

> Whenever the RSS condition C is satisfied, executing the proper response P from that moment leads to safe driving (i.e., driving with no collision).

By considering the beginning (at time 0) of the execution of P, the above requirement implies, in particular, that

> *Whenever the RSS condition C is satisfied, there is no collision at that moment.*

Ideally, this safety assurance should be mathematically proved, building on the rigorous formulation of the RSS safety condition C as well as that of the proper response P. Such a mathematical proof makes the requirement a *theorem*.

A crucial feature of an RSS safety rule $R = (C,P)$ is that it reduces the problem of safety *in the future* to a condition that can be verified *at present*, namely the RSS safety condition C. The RSS safety condition C ensures the safety of executions of the proper response P, much like a *precondition* in a program logic ensures the safety of executions of a program (see, e.g., [72]).

Such reduction of the future to the present is not easy. In fact, ensuring future safety is impossible without posing suitable assumptions about the behaviours of other vehicles. Figure 7.12 is a situation that is often used as an example in the literature (including [64]): a collision happens if the other vehicles come close to the subject vehicle (SV); and the SV can do nothing to avoid it.

7.5.2.2 RSS Responsibility Principles

It is therefore needed to formulate certain behavioural constraints that all traffic participants are expected to respect. Such constraints can be thought of as *contracts* in driving situations, when the participants are identified with multi-agent systems. Each traffic participant must respect these constraints; moreover, each agent can act assuming that all the other agents obey these constraints. In RSS, these behavioural constraints are derived from a high-level idea of *responsibility*.

Specifically, RSS expresses its idea of responsibility in the following informal form of *RSS responsibility principles*. (Note that the name “RSS responsibility principle” differs from those which appear in the literature—see Remark 7.1.)

Figure 7.12: The subject vehicle (SV) is surrounded by other vehicles. When they come close, there is nothing that the SV can do to avoid a collision.

Definition 7.2 (RSS responsibility principles [64])

1. Don't hit the car in front of you.
2. Don't cut in recklessly.
3. Right of way is given, not taken.
4. Be cautious in areas with limited visibility.
5. If you can avoid a crash without causing another one, you must.

Note that these principles indeed embody a natural idea of responsibility, or "duties of care", in driving. They are close to those rules commonly expected of human drivers.

The uses of the five RSS responsibility principles are twofold. Firstly, they can be used in establishing Requirement 1—more specifically, in a mathematical proof for the satisfaction of Requirement 1, one can derive some assumptions from the informal principles in Definition 7.2, and require them of other vehicles' behaviours.

The second use of the RSS responsibility principles is that, as a participant in a traffic scenario (that is thought of as a multi-agent system), the subject vehicle (SV) must itself respect those principles:

Requirement 2 (responsibility in RSS) *An RSS safety rule $R = (C,P)$ is required to satisfy the following:*

> *Let E be an arbitrary execution of the proper response P; assume that the execution E starts at a state in which the RSS safety condition C is true. Then the execution E respects the RSS responsibility principles (Definition 7.2).*

Requirement 2 is expected to be mathematically proven, too. In doing so, one has to turn the RSS responsibility principles (that are only informally stated, Definition 7.2) into certain rigorous mathematical conditions, in a way specific to the driving scenario in question.

7.5.3 The RSS Framework

The elements of RSS that we have described constitute what we call the *RSS framework*, depicted in Fig. 7.13. Each RSS safety rule $R = (C, P)$ is formulated for an individual driving scenario ("single-lane same-direction", "cut in", "crossing", etc.)—the design of the rule, as well as the proofs that it satisfies Requirements 1–2, is heavily dependent on the choice of a driving scenario. An example will be presented in Section 7.5.4.

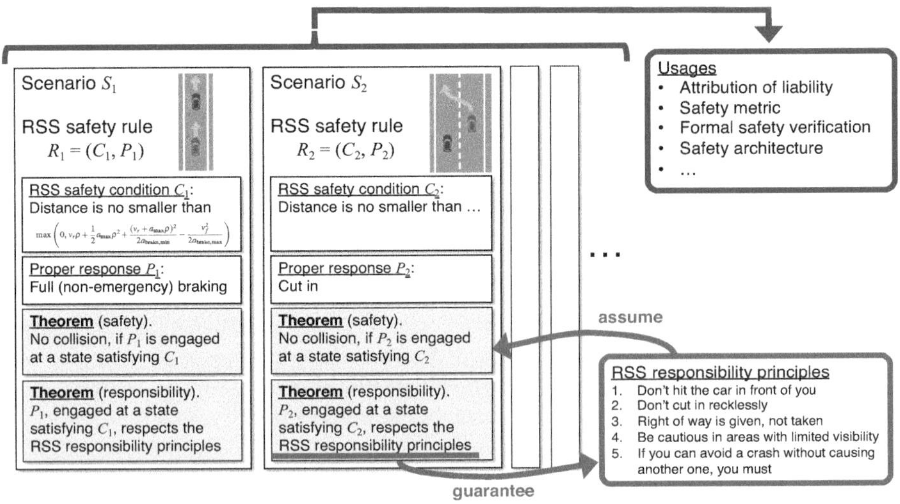

Figure 7.13: The RSS framework. An RSS safety rule is formulated for an individual driving scenario. Each rule is expected to satisfy safety (Requirement 1) and responsibility (Requirement 2). The latter *guarantees* that the subject vehicle acts responsibly; conversely, the safety proof assumes that the other vehicles act responsibly.

In the proof of safety, it is assumed that other vehicles respect the RSS responsibility principles (Definition 7.2)—otherwise safety is often unachievable, see Fig. 7.12. Conversely, it is needed to show that the SV respects the RSS responsibility principles, so that other vehicles can rely on it when they plan their behaviours. This is Requirement 2 (responsibility).

While concrete usages of the RSS framework are discussed later in Section 7.5.5, we can already see that conceptual benefits of RSS are significant.

For one, the RSS safety rules give a precise recipe for safe driving, in which (1) triggering conditions and control strategies are specified in mathematical terms, and (2) safety is given a mathematical proof. RSS can therefore be seen as a promising approach to the goal of *formal verification of ADS safety*.

It is, however, unrealistic to expect that RSS realises a world with zero traffic accidents. Even if the SV acts according to the RSS safety rules, other vehicles may not, especially those driven by human drivers. RSS is still useful in such situations with traffic accidents. The RSS responsibility principles explicate natural "contracts" that each traffic participant is expected to follow. These principles can therefore be used for identifying who is liable for an accident—namely the one who did not respect them.

The last argument can be turned upside down and yield the following:

> *a vehicle is not responsible for an accident as long as it respects the RSS responsibility principles.*

This is an answer to a major challenge that is currently hindering large-scale deployment of automated driving, namely

> *the difficulty of determining the boundary of the responsibilities of automated driving systems and their manufacturers.*

The last boundary of responsibilities is currently vague for ADS, which exposes the ADS manufacturers to the risk of unexpected and exceeding liabilities. This risk is a big burden when a company wants to run an ADS business, potentially blocking the deployment and development of the ADS technology. Clarification of the boundary of responsibilities is therefore pursued by many parties, including standardisation bodies such as ISO, UL, and SAE. Indeed, the use of RSS is often advocated in these standardisation efforts, notable among which is the efforts towards the IEEE 2846 standard.[2]

7.5.4 Example: An RSS Safety Rule for the Single-Lane Same-Direction Scenario

We exhibit an example of an RSS safety rule and its safety and responsibility proofs. The example is one of the first examples in the literature and is taken from [64].

7.5.4.1 The Scenario S_{slsd}

The driving scenario in question, denoted by S_{slsd}, has a single lane and two vehicles driving in the same direction. See Fig. 7.14: the vehicle behind is the one under our control (the subject vehicle (SV)); the SV drives behind another vehicle that is called the *principal other vehicle (POV)*.

[2]`https://sagroups.ieee.org/2846/`

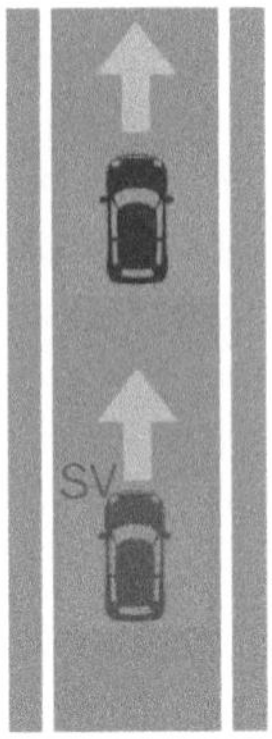

Figure 7.14: The single-lane same-direction scenario. The subject vehicle (SV) is following the other vehicle. The latter is called the *principal other vehicle (POV)*.

The goal is to avoid collision regardless of the behaviour of the principal other vehicle (POV). Here, however, we rule out some unrealistic but physically possible behaviours of the POV from our consideration—such as the POV being hit by a comet and suddenly coming to a halt. The worst case within our consideration is the POV engaging emergency braking and coming to a halt in a short moment. We want the SV to stop without colliding with the POV. We want the SV to do so by comfortable braking and not by emergency braking. Moreover, we have to take into account the *response time*, the time between the moment the POV starts braking and the moment the SV starts its response. All these require a suitable distance between the POV and the SV; the question is how large precisely the distance should be.

Note that the above restriction of the scenario's scope ("no comet") is similar to the use of the RSS responsibility principles, although there is no explicit RSS principle that corresponds to the "no comet" assumption. The argument here is "if a collision happens because of a comet, the SV is not held responsible", which is logically parallel to one that uses RSS responsibility principles such as "if a collision happens because of another vehicle's reckless cut in, the SV is not held responsible".

7.5.4.2 The RSS Safety Condition and the Proper Response

In [64], an RSS safety condition C_{slsd} and a proper response P_{slsd} are given as follows. They together constitute an RSS safety rule $R_{\mathrm{slsd}} = (C_{\mathrm{slsd}}, P_{\mathrm{slsd}})$.

The RSS Safety Condition C_{slsd}

The condition C_{slsd} is given by

$$x_f - x_r > \max\left(0,\, v_r\rho + \frac{1}{2}a_{\text{max}}\rho^2 + \frac{(v_r + a_{\text{max}}\rho)^2}{2a_{\text{brake,min}}} - \frac{v_f^2}{2a_{\text{brake,max}}}\right). \tag{7.12}$$

Here, the following are dynamic parameters that describe the current state of the driving situation.

- x_f, x_r are the positions of the front vehicle (POV) and the rear vehicle (SV), respectively.
- v_f, v_r are their velocities, respectively, modelled in the 1-dimensional lane coordinate.

Besides, the following are static parameters for the driving scenario—they do not change from one state to another. Their values are decided according to the traffic laws, regional customs, vehicle specs, etc.

- ρ is the maximum *response time* that the rear vehicle might take to initiate the required braking.
- a_{max} is the maximum (forward) acceleration rate of the rear vehicle.
- $a_{\text{brake,min}}$ is the maximum comfortable braking rate for the rear vehicle.
- $a_{\text{brake,max}}$ is the maximum emergency braking rate for the front vehicle. We assume $0 < a_{\text{brake,min}} < a_{\text{brake,max}}$.

An example of these parameter values is found in [74]: $\rho = 0.3$ s, $a_{\text{max}} = 2\ \text{m/s}^2$, $a_{\text{brake,min}} = 4\ \text{m/s}^2$, and $a_{\text{brake,max}} = 8\ \text{m/s}^2$.

The Proper Response P_{slsd}

A proper response is a control strategy that is expected to *avoid any collision in the future, no matter how other vehicles would behave* (within a prescribed range of possible behaviours—recall the "no comet" argument in the above). The proper response P_{slsd} in [64] is

> to engage the maximum comfortable braking, after a response time that is no bigger than ρ. The behaviour during the response time is arbitrary.

The response time is included since otherwise the control strategy would be unrealizable. The behaviour during the response time is constrained physically—we do not consider an instant acceleration to the speed of light, for example. Clearly, the worst case behaviour is the acceleration at the maximum rate a_{max}.

7.5.4.3 The Safety Proof

Theorem 7.1 (safety of R_{slsd})
Let s be a state of the driving scenario S_{slsd}, and let the positions and the velocities of the vehicles in s be denoted by x_f, x_r, v_f, v_r as described in Section 7.5.4.2.

Assume that the state s satisfies the RSS safety condition C_{slsd}. Consider an execution e of the proper response P_{slsd} that starts at s and ends when the SV comes to a halt or collides into the POV. Then, no collision occurs in this execution e of P_{slsd}.

The proof here follows the outline of the one presented in [64].

Proof 7.1 In the execution e of P_{slsd} in question, there are two arbitrary components: the behaviour of the SV during the response time, and the behaviour of the POV. It is however obvious that the worst choices in terms of safety are: (1) the SV accelerates at the maximum rate $a_{\max}$, at the beginning of e, for the longest possible response time ρ, and (2) the POV engages the maximum braking, namely at the rate $a_{\text{brake,max}}$, for the whole period of e. In the rest of the proof, we assume these behaviours of the SV and the POV without loss of generality.

We study the sign of the relative velocity of the SV with respect to the POV. To do so, we list up the possible patterns of the time-varying relationship between the velocities of the SV and the POV. See Fig. 7.15.

- In the case 1, the relative velocity of SV is always positive, therefore the *inter-vehicle distance* between the SV and the POV becomes minimum when the SV comes to a halt.
- In the cases 2–3, the relative velocity of SV turns from negative to positive. Therefore the inter-vehicle distance becomes minimum either in the beginning (at time 0) or when the SV comes to a halt.
- In the case 4, the relative velocity of SV is always negative, therefore the inter-vehicle distance becomes minimum in the beginning (at time 0).

It is easy to see that the four cases are exhaustive (the assumption $0 < a_{\text{brake,min}} < a_{\text{brake,max}}$ is crucial here). It therefore suffices to ensure no collision at time 0 and at the time when the SV comes to a halt.

A necessary and sufficient condition for avoiding collision at time 0 is, obviously,

$$x_f - x_r > 0. \tag{7.13}$$

A condition for avoiding collision at the time when the SV comes to a halt is derived as follows.

- The SV travels the distance $v_r\rho + \frac{1}{2}a_{\max}\rho^2$ during the response time, at the end of which its velocity reaches $v_r + a_{\max}\rho$. The subsequent braking phase takes the time $\frac{v_r + a_{\max}\rho}{a_{\text{brake,min}}}$, during which the SV travels the distance $\frac{1}{2}(v_r + a_{\max}\rho)\frac{v_r + a_{\max}\rho}{a_{\text{brake,min}}}$.

- To compute the distance that the POV travels until the SV comes to a halt, note first that we can restrict to the cases 1–3 in Fig. 7.15—this is because the minimum inter-vehicle distance is at time 0 (not when the SV comes to a halt) in the case 4. We can easily see that, in the cases 1–3, the POV comes to a halt earlier than the SV does. Therefore the travelled distance for POV is $\frac{1}{2}v_f\frac{v_f}{a_{\text{brake,max}}}$.
- We conclude that, in the cases 1–3 (the case 4 can be ignored as argued above), the positions of the SV and the POV when the SV comes to a halt are

$$x_r + v_r\rho + \frac{1}{2}a_{\max}\rho^2 + \frac{(v_r + a_{\max}\rho)^2}{2a_{\text{brake,min}}} \quad \text{and} \quad x_f + \frac{v_f^2}{2a_{\text{brake,max}}},$$

respectively. There is no collision at that moment if and only if

$$x_f - x_r > v_r\rho + \frac{1}{2}a_{\max}\rho^2 + \frac{(v_r + a_{\max}\rho)^2}{2a_{\text{brake,min}}} - \frac{v_f^2}{2a_{\text{brake,max}}}. \tag{7.14}$$

The RSS safety condition (7.12) implies both (7.13–7.14), and thus ensures that there is no collision at the beginning or at the end of an execution of P_{slsd}, which in turn ensures that there is no collision at *any* moment during the execution. This concludes the proof.

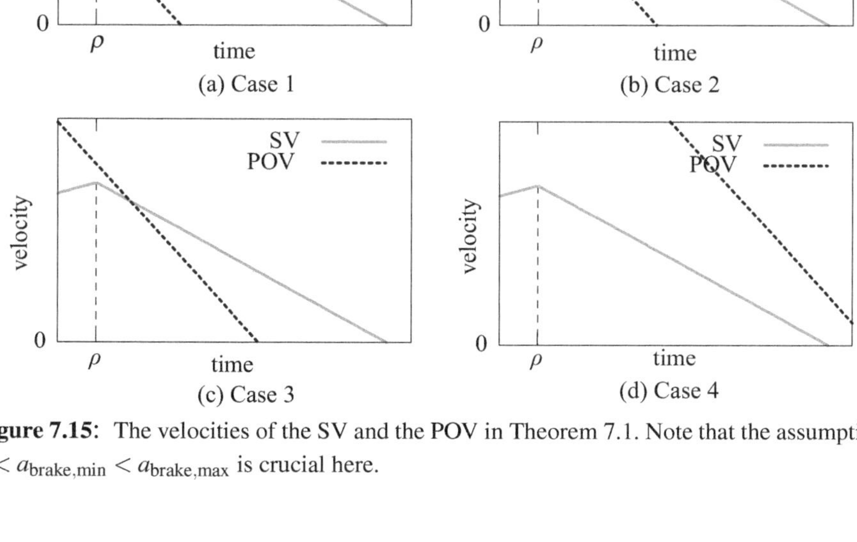

Figure 7.15: The velocities of the SV and the POV in Theorem 7.1. Note that the assumption $0 < a_{\text{brake,min}} < a_{\text{brake,max}}$ is crucial here.

7.5.4.4 The Responsibility Proof

The following is not a "theorem" in a rigorous sense—this is because the RSS responsibility principles (Definition 7.2) are not formally defined conditions. We nevertheless provide an argument for it; such arguments are of great practical values, when it comes to such matters as explainability to the public and attribution of liability.

Theorem 7.2 (responsibility of R_{slsd})
Assume the same setting as in Theorem 7.1. Then the execution of the proper response P_{slsd} satisfies the RSS responsibility principles.

Proof 7.2 (an informal argument) The principles 1 and 5 are ensured by safety (Theorem 7.1). The other principles (2–4) do not apply to the current driving scenario S_{slsd}.

7.5.5 Usages of RSS

Some usages of the RSS framework (Fig. 7.13) have been already hinted above. Here we go into their details, providing some pointers to related scientific works and ongoing practical efforts at the same time.

7.5.5.1 Attribution of Liability

Attribution of liability is one of the first applications that were pursued for RSS. We already discussed it briefly in Section 7.5.3: when a collision happened, the traffic participant that did not comply with the RSS responsibility principles is held liable. The validity of this reasoning is supported by the mathematical elements of RSS: the RSS safety rules, especially their safety and responsibility requirements, ensure that there is no collision as long as all the traffic participants adhere to the RSS responsibility principles.

Here, the roles of RSS safety rules are twofold. For one, an RSS safety rule can be seen as a mathematical incarnation of the RSS responsibility principles, one that enables a rigorous safety proof (the safety theorems in Fig. 7.13). Another important role is as an *evidence of acting responsibly*. The RSS responsibility principles are informal conditions; in contrast, RSS safety rules are rigorous and mathematically checkable. Demonstration of the compliance with RSS safety rules is therefore a strong evidence of acting responsibly, via the responsibility requirements for those rules (the responsibility theorems in Fig. 7.13).

A comprehensive case study of such use of RSS is found in [65], where the RSS framework is applied to NHTSA pre-crash scenarios.

7.5.5.2 As a Safety Metric

Safety metrics for ADS are an active topic of research, see e.g. [2, 66, 68, 69, 77]. The question here is how to evaluate the safety of a trajectory of a vehicle. A naive metric can be given by the minimum distance—the minimum distance to obstacles and other traffic participants exhibited during the trajectory—but its limitations are obvious, too, such as the insensitivity to the vehicle speed. A safety metric that has been commonly used is the *time to collision (TTC)*.

It is straightforward to derive another safety metric from the last usage of RSS (Section 7.5.5.1). This *RSS safety metric* measures the degree with which a trajectory satisfies RSS safety rules: the margin with which the inequalities in the RSS safety conditions are satisfied is its safety score. Violation of the RSS safety conditions results in a negative safety score.

Multiple variations and extensions are possible for the above (informal and rudimentary) notion of RSS safety metric. Several of such are presented in [2]; among them are a metric that takes into account whether a proper response was engaged when needed.

7.5.5.3 Formal Verification of ADS Safety

Another obvious usage of RSS is for *formal verification* of the ADS safety. By proving that a vehicle's control complies with RSS safety rules, one can conclude the safety of the vehicle via the safety theorems of the RSS safety rules. Here, the "compliance with RSS safety rules" means the following.

Definition 7.3 (compliance with RSS safety rules) Let $\mathcal{R} = (R_1, R_2, \ldots, R_n)$ be a list of RSS safety rules, where $R_i = (C_i, P_i)$ for each $i \in \{1, 2, \ldots, n\}$. We say that a vehicle's trajectory *complies with* $\mathcal{R}$ if, at each moment in the trajectory, either

- the RSS safety condition C_i is true for some $i \in \{1, 2, \ldots, n\}$, or
- the proper response P_i is being executed for some $i \in \{1, 2, \ldots, n\}$. Additionally, we require that this execution of P_i started in a state in which the corresponding RSS safety condition C_i was true.

Theorem 7.3 (safety of an RSS-compliant trajectory)
Let $\mathcal{R} = (R_1, R_2, \ldots, R_n)$ be a list of RSS safety rules, where $R_i = (C_i, P_i)$ for each $i \in \{1, 2, \ldots, n\}$. Assume further that each rule R_i satisfies the safety requirement (Requirement 1). If a vehicle's trajectory T complies with $\mathcal{R}$ (Definition 7.3), then T exhibits no collision.

Proof 7.3 Consider an arbitrary moment of the trajectory T. If some RSS safety condition C_i is true at that moment, then there is no collision at that moment—this follows from the remark immediately after Requirement 1. If some proper response

P_i is being executed at that moment, the safety guaranteed in Requirement 1 implies that no collision occurs then.

Towards the goal of mathematically proving safety of an ADS, according to Theorem 7.3, it suffices to ensure that every trajectory of the ADS complies with some list of RSS safety rules in the sense of Definition 7.3. However, there are two major challenges in doing so.

- An ADS controller is a complex system, involving a number of numeric optimisation algorithms and statistical machine learning. Therefore, proving its compliance with RSS safety rules is hard.
- Compliance in the sense of Definition 7.3 requires a comprehensive list $\mathcal{R}$ of RSS safety rules—otherwise there will be moments that are not covered by any C_i or P_i. Obtaining such a comprehensive list will take an enormous effort: a safety proof for a single safety rule is nontrivial already for a simple scenario (Section 7.5.4); and the number of driving scenarios to be covered is huge.

7.5.5.4 Safety Architecture

A promising "workaround" to the last two challenges in formal verification (Section 7.5.5.3) is the use of RSS in a *safety architecture*.

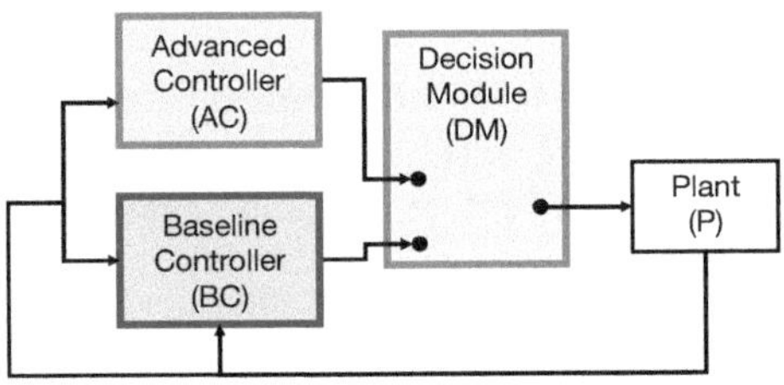

Figure 7.16: The simplex architecture.

A prototypical safety architecture is the *simplex architecture* shown in Fig. 7.16 [15,63]. Here the plant represents the driving system to be controlled, the *advanced controller* (AC) is a complicated controller that pursues not only safety but also performance (such as comfort, progress, and fuel efficiency); the *baseline controller* (BC) is a simpler controller with a strong emphasis on safety; and the *decision module* (DM) switches between the two controllers. The DM tries to use the AC as often as possible for its superior performance. However, when the DM finds that the current situation is safety critical, it switches to the BC whose behaviours are more predictable and easier to analyse.

A notable feature of safety architectures such as the simplex architecture is that it enables formal verification of a system that contains a black-box component (such as the AC in the simplex architecture). It does so by "wrapping" the black-box component with safety-centric components that closely monitor the black-box component and overtake the control when it is needed.

The elements of RSS map naturally to the simplex architecture:

- The DM makes use of RSS safety conditions. When they are about to be violated, the DM switches the control from the AC to the BC.
- The BC implements proper responses. It thus executes a control strategy whose safety is guaranteed.
- If it so happens that the BC restores the RSS safety condition, then the DM switches the control back to the AC.

The way the resulting *RSS-supervised controller* operates reflects the argument in Section 7.5.5.3: it lets the AC control as long as some RSS safety condition C_i is true; once it comes close to violating all the RSS safety conditions, it executes a proper response P_i to maintain safety. The possibility of switching back to the AC is advantageous for performance.

Such use of RSS in a safety architecture successfully addresses the two challenges in formal verification (Section 7.5.5.3).

- The complexity of the ADS controller becomes no problem since it is confined to the AC—the safety of the whole architecture is ensured exclusively by the properties of the DM and the BC.
- On the difficulty of obtaining a comprehensive list of RSS safety rules, a safety architecture offers a *proactive* and *best-effort* alternative to formal verification. Even if a rule list $\mathcal{R} = (R_1, R_2, \ldots, R_n)$ does not cover all possible driving scenarios, $\mathcal{R}$ can still be implemented in a safety architecture and ensure the safety of those scenarios which are covered by $\mathcal{R}$. (For those driving scenarios which are not covered, we just let the AC do its best to maintain safety). After all, formal verification is about analysing a given controller (Section 7.5.5.3), while safety architectures is about modifying the controller (identified with the AC) and making it safe.

Such use of RSS in a safety architecture is advocated increasingly often. In fact, some recent papers such as [51, see Fig. 1] present RSS in the format of safety architectures, although they may not explicitly refer to the term "safety architecture".

7.5.6 Current State and Future Directions

7.5.6.1 Making a Rule Set Comprehensive

Towards the ultimate goal of ADS safety proofs, one important area of RSS that requires further work is the formulation of RSS rules. As we already discussed, RSS rules are formulated and proved correct in a scenario-specific manner, and there are a huge number of driving scenarios.

While a comprehensive RSS rule set is obviously important and desired, the path to it might seem endless. We nevertheless believe that efforts in this direction are worthwhile.

- One reason is the use of the rules in a safety architecture (Section 7.5.5.4). Even if the rule list $\mathcal{R}$ at hand is not comprehensive yet (and hence not enough to prove the overall safety), the "best-effort" usage in a safety architecture ensures safety at least for the scenarios that are already covered by $\mathcal{R}$.
- Another reason is that RSS rules are irrevocable, with their correctness being mathematically established. Once derived, they can be used for the coming dozens of years or even longer—they can be seen as common assets of humankind. With more efforts thrown in for the formulation of RSS rules, the rule set grows monotonically.

7.5.6.2 Tool Support for Using, Deriving, and Verifying RSS Rules

Tool support for RSS is pursued actively in recent years. One example is an implementation of some RSS rules: it is offered as a library [25] that can be used in combination with simulation environments such as Baidu Apollo. This implementation is for *using* RSS rules.

Another area that calls for tool support is the derivation of RSS rules. As we discussed in Section 7.5.5.4, RSS rule derivation requires systematic and organised efforts, much like formal verification by theorem proving. Therefore the task needs tool support—much like tool support for theorem proving is given by *proof assistants* such as Coq [19] and Isabelle/HOL [49]—otherwise the efforts will be messed up with human errors.

Yet another area that calls for tool support is the (formal) verification of RSS rules. Existing correctness proofs for RSS rules (such as the ones in Section 7.5.4) are mathematical yet not formalised or mechanised; it is not hard to imagine human errors in such proofs, especially when we work with complex driving scenarios. The problem of giving formalised and mechanised proofs to RSS rules is investigated in [59]. Their trial is based on a rigorous notion of signal; they argue that none of the existing *automated* verification tools is suited for the verification problem. We believe that the use of theorem provers that allow human interaction should be pursued. An example of such provers is KeYmaera X [24].

7.5.6.3 Permissive, and Thus Practical, RSS Rules

The scope of RSS rules is principally safety. They can sacrifice other practical performance metrics such as comfort, progress, and fuel efficiency. It is therefore desired that RSS rules are less conservative and more permissive.

One work in this direction is [52]; it proposes an extension of RSS in which the balance between safety and other performance metrics is proactively adjusted. Other promising directions seem to include the following.

- *Fine-grained formalisation* of the RSS responsibility principles, so that the formalisation respects different traffic circumstances (urban, rural, or highway), different regions and countries, and different driving cultures.
- *RSS that is aware of intentions and knowledge*. Responsibilities of other vehicles will be totally different depending on whether the SV's turn signal is blinking or not. Moreover, when SV makes a maneuver with its turn signal blinking, it is also desired that the SV makes sure that other vehicles are aware of the turn signal.
- *Enlarging the variety of proper responses*. For example, [17] proposes an extension of the RSS rule in Section 7.5.4 in which swerves (in addition to braking) are allowed as evasive maneuvers [17]. This makes the corresponding safety condition much weaker, and the RSS rule more permissive.

7.5.6.4 Perceptual Uncertainties

In the presence of perceptual uncertainties (such as errors in position measurement and object recognition), the perceived values that RSS safety conditions depend on can be erroneous. Making RSS rules tolerant of such perceptual uncertainties is pursued in [62]; the methodology used there is more generally formulated in [43] and used in combination with the modelling and verification framework Event-B [1].

A big source of perceptual uncertainties in modern ADS is its *statistical machine learning* components (neural networks to be specific). One way to lessen such ML-related uncertainties is to enrich the output of an ML component. For example, the use of DNNs' confidence scores is proposed in [6]. In [13], it is proposed to look at inconsistencies between perceptual data of different modes.

7.5.6.5 Logical Formalization of RSS

A recent development that potentially contributes a lot to many directions discussed above is about *formalization* of RSS. This amounts to devising a framework in which (1) properties and statements are written down in some formal language and (2) reasoning steps in proofs must conform with symbolic *derivation rules*. See Chapter 9 for basics of logical formalization. Informally speaking, logical formalization means a transition from *informal* pen-and-paper proofs to *formal* proofs written in a computer. The latter can be mechanically and automatically checked for their validity, using software called *proof assistants*.

Specifically, the work [31] introduces the first formalization of logical reasoning for RSS by devising a formal system called dFHL (differential Floyd–Hoare logic). dFHL is primarily an extension of a classic Floyd–Hoare logic [33] with

- ordinary differential equations for reasoning about continuous-time dynamics, and
- so-called *safety conditions* that, in addition to standard *pre-* and *post-conditions* in Floyd–Hoare logic that express what should hold *before* and *after* dynamics, assert what should hold *throughout* dynamics.

The formalization in [31] allows one to apply RSS to a wider class of more complicated scenarios, beyond the simple single-lane same-direction scenario in Fig. 7.14. One example of such a scenario is the pull over scenario in Fig. 7.17, where the ego car should reach a specified position on the shoulder by changing lanes and avoiding collision at the same time.

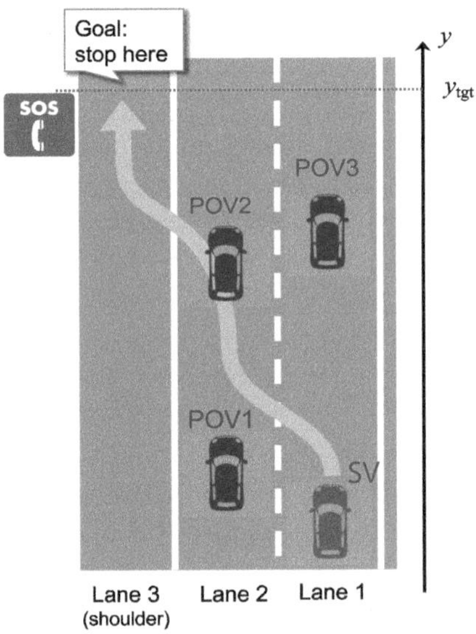

Figure 7.17: The pull over scenario.

In the formalization of RSS in [31], as a benefit of the use of dFHL, one can prove *goal achievement* in addition to collision avoidance. This is suited for many scenarios such as Fig. 7.17, allowing one to prove e.g. safe arrival at the destination position on the shoulder.

Logical formalization is expected to aid many future directions around RSS. It is a theoretical foundation for tool support for deriving and verifying RSS rules (Section 7.5.6.2), thereby accelerating the development of more RSS rules (Section 7.5.6.1). The formalization in [31] has been further extended to more specialized frameworks for rule derivation and usage, namely for safety architectures [21] and for intersection scenarios [32].

7.6 Conclusions

In this chapter, we provided an overview to decision making architectures in automated driving systems. In particular, we explained different motion planning techniques and discussed recent approaches to safety guarantee.

Motion planning for automated driving is an important problem that comes with several technical challenges. In particular, many urban driving scenarios require motion planners to generate trajectories that provide safety and passenger comfort while satisfying constraints with respect to the dynamical limitations of a vehicle. What makes achieving these tasks particularly difficult is that the information available to motion planners is subject to uncertainties. Specifically, vehicle dynamics are not precisely known, sensors provide noisy data, and the driving environment often includes other participants with unmodeled behavior.

In this chapter, we explained several motion planning architectures that tackle different aspects of these challenges. We looked at optimisation-based and model predictive motion planning approaches that can allow maximising the passenger comfort while avoiding obstacles in different driving scenarios. Automated driving systems are expected to work in environments with other automated driving agents and human-driven vehicles. Game theory is a natural and useful framework to model potential interactions between different driving agents in the same environment. In this chapter, we discussed leader-follower dynamic games to model vehicle interactions. Such interaction models are shown to be especially useful in making driving decisions in intersections and roundabouts.

Guaranteeing safety is a critical hurdle in automated driving systems. Motion planners need to take account of sensor noise and environmental disturbances when generating trajectories. In this chapter, we explained two reachability analysis methods where the goal is to compute potential road occupancy areas of the ego vehicle and other road participants. The first method is a set-based overapproximation method, whereas the second one is a probabilistic method that allows computation of probability bounds for the vehicles to be outside of given regions. In addition, we provided a discussion on the RSS framework. This framework takes existing traffic regulations and common-sense as starting points to derive logical rules, which in turn facilitate formal proofs of safety. We discussed the five main RSS principles, an example of an RSS rule, and their use in the attribution of liability as well as verification of safety. We then explained a safety architecture that incorporates RSS. Furthermore, we discussed the current position of RSS and highlighted important future research directions.

References

[1] Abrial, J-R., Butler, M., Hallerstede, S., Hoang, T.S., Mehta, F. and Voisin, L. 2010. Rodin: An open toolset for modelling and reasoning in Event-B. STTT, 12(6): 447–466.

[2] Altekar, N., Elli, M., Weast, J., Chen, Y., Wishart, J., Como, S., et al. 2020. Driving safety performance assessment metrics for ADS-equipped vehicles. SAE International Journal of Advances and Current Practices in Mobility, 2(5): 2881–2899, April 2020.

[3] Althoff, M. and Dolan, J.M. 2011. Set-based computation of vehicle behaviors for the online verification of autonomous vehicles. *In*: 2011 14th International IEEE Conference on Intelligent Transportation Systems (ITSC), pages 1162–1167. IEEE.

[4] Althoff, M. and Dolan, J.M. 2014. Online verification of automated road vehicles using reachability analysis. IEEE Transactions on Robotics, 30(4): 903–918.

[5] Althoff, M., Koschi, M. and Manzinger, S. 2017. Commonroad: Composable benchmarks for motion planning on roads. *In*: Proceeding IEEE Intelligent Vehicles Symposium, pages 719–726.

[6] Angus, M., Czarnecki, K. and Salay. R. 2019. Efficacy of pixel-level OOD detection for semantic segmentation. CoRR, abs/1911.02897.

[7] Mahdi Ghazaei Ardakani, M., Olofsson, B., Robertsson, A. and Johansson, R. 2015. Real-time trajectory generation using model predictive control. *In*: Proceeding IEEE International Conference on Automation Science and Engineering, pages 942–948.

[8] Bast, H., Delling, D., Goldberg, A., Müller- Hannemann, M., Pajor, T., Sanders, P., et al. 2015. Route planning in transportation networks. *In*: Kliemann, L., Sanders, P. (eds.) Algorithm Engineering. Lecture Notes in Computer Science, Vol. 9220. Springer, Cham. pages 19–80, 2016.

[9] Bojarski, M., Del Testa, D., Dworakowski, D., Firner, B., Flepp, B., Goyal, P., et al. 2016. End to end learning for self-driving cars. arXiv preprint arXiv:1604.07316.

[10] Brechtel, S., Gindele, T. and Dillmann. R. 2011. Probabilistic MDP-behavior planning for cars. *In*: 2011 14th International IEEE Conference on Intelligent Transportation Systems (ITSC), pages 1537–1542. IEEE.

[11] Camacho, E.F. and Alba, C.B. 2013. Model Predictive Control. Springer.

[12] Castellano, E., Cetinkaya, A., Ho Thanh, C., Klikovits, S., Zhang, X. and Arcaini, P. 2021. Frenetic at the SBST 2021 tool competition. *In*: Proc. 2021 IEEE/ACM 14th International Workshop on Search-Based Software Testing, pages 36–37.

[13] Chow, J., Richmond, V., Wang, M., Guajardo, U., Jackson, D., Arechiga, N., et al. 2020. Certified Control: A New Safety Architecture for Autonomous Vehicles, 2020. preprint.

[14] Chu, K., Lee, M. and Sunwoo, M. 2012. Local path planning for off-road autonomous driving with avoidance of static obstacles. IEEE Transactions on Intelligent Transportation Systems, 13(4): 1599–1616.

[15] Crenshaw, T.L., Gunter, E., Robinson, C.L., Sha, L. and Kumar, P.R. 2007. The simplex reference model: Limiting fault-propagation due to unreliable components in cyber-physical system architectures. *In*: 28th IEEE International Real-Time Systems Symposium (RTSS 2007), pages 400–412.

[16] da Silva, J.A.R., Pach, I., Wolf, D.F. and Grassi, V. 2021. Sparse road network model for autonomous navigation using clothoids. IEEE Transactions on Intelligent Transportation Systems, 2021 in IEEE Transactions on Intelligent Transportation Systems, 23(2): 885–898, February 2022.

[17] de Iaco, R., Smith, S.L. and Czarnecki, K. 2020. Safe swerve maneuvers for autonomous driving. *In*: 2020 IEEE Intelligent Vehicles Symposium (IV), pages 1941–1948.

[18] Delling, D., Sanders, P., Schultes, D. and Wagner, D. 2009. Engineering route planning algorithms. *In*: Algorithmics of Large and Complex Networks, pages 117–139. Springer.

[19] The Coq development team. 2004. The Coq proof assistant reference manual. LogiCal Project, 2004. Version 8.0.

[20] Ding, W., Zhang, L., Chen, J. and Shen, S. 2019. Safe trajectory generation for complex urban environments using spatio-temporal semantic corridor. IEEE Robotics and Automation Letters, 4(3): 2997–3004.

[21] Eberhart, C., Dubut, J., Haydon, J. and Hasuo. I. 2023. Formal verification of safety architectures for automated driving. *In*: IEEE Intelligent Vehicles Symposium, IV 2023, Anchorage, AK, USA, June 4–7, 2023, pages 1–8. IEEE. Preprint available as CoRR abs/2308.10365.

[22] Levinson, J., Askeland, J., Becker, J., Dolson, J., Held, D., Kammel., S., et al. 2011. Towards fully autonomous driving: Systems and algorithms. *In*: Proceeding IEEE Intelligent Vehicles Symposium, pages 163–168.

[23] Fisac, J.F., Bronstein, E., Stefansson, E., Sadigh, D., Shankar Sastry, S. and Dragan, A.C. 2019. Hierarchical game-theoretic planning for autonomous vehicles. *In*: 2019 International Conference on Robotics and Automation (ICRA), pages 9590–9596. IEEE.

[24] Fulton, N., Mitsch, S., Quesel, J-D., Völp, M. and Platzer, A. 2015. KeYmaera X: An axiomatic tactical theorem prover for hybrid systems. *In*: Amy P. Felty and Aart Middeldorp, (eds). CADE, volume 9195 of LNCS, pages 527–538. Springer.

[25] Ga.mann, B., Oboril, F., Bürkle, C., Liu, S., Yan, S., Elli, M.S., et al. 2019. Towards standardization of AV safety: C++ library for responsibility sensitive safety. *In*: 2019 IEEE Intelligent Vehicles Symposium, IV 2019, Paris, France, June 9-12, 2019, pages 2265–2271. IEEE.

[26] Glomsrud, J., Ødegårdstuen, A., Clair, A. and Smogeli, O. 2019. Trustworthy versus explainable AI in autonomous vessels. *In*: International Seminar on Safety and Security of Autonomous Vessels (ISSAV), Espoo, Finland, 09.

[27] Gonzālez, D., Pérez, J., Milans V., and Nashashibi, F. 2015. A review of motion planning techniques for automated vehicles. IEEE Transactions on Intelligent Transportation Systems, 17(4): 1135–1145.

[28] Götte, C., Keller, M., R.smann, C., Nattermann, T., Haß., C., Glander, K-H., et al. 2016. A real-time capable model predictive approach to lateral vehicle guidance. *In*: Proceeding IEEE 19th International Conference on Intelligent Transportation Systems, pages 1908–1913.

[29] Grigorescu, S., Trasnea, B., Cocias, T. and Macesanu, G. 2020. A survey of deep learning techniques for autonomous driving. Journal of Field Robotics, 37(3): 362–386.

[30] Gutjahr, B., Gröll, L. and Werling, M. 2016. Lateral vehicle trajectory optimization using constrained linear time-varying MPC. IEEE Transactions on Intelligent Transportation Systems, 18(6): 1586–1595.

[31] Hasuo, I., Eberhart, C., Haydon, J., Dubut, J., Bohrer, R., Kobayashi, T., et al. 2023. Goal-aware RSS for complex scenarios via program logic. IEEE Transactions on Intelligent Vehicles, 8(4): 3040–3072, 2023. Preprint available as CoRR abs/2207.02387.

[32] Haydon, J., Bondu, M., Eberhart, C., Dubut, J. and Hasuo, I. 2023. Formal verification of intersection safety for automated driving. *In*: 26th IEEE International Conference on Intelligent Transportation Systems, ITSC 2023, Bilbao, Spain, September 24–28, 2023. IEEE. To appear, preprint available as CoRR abs/2308.10365.

[33] Hoare, C.A.R. 1969. An axiomatic basis for computer programming. Communications of the ACM, 12: 576–580, 583.

[34] Hubmann, C., Schulz, J., Becker, M., Althoff, D. and Stiller, C. 2018. Automated driving in uncertain environments: Planning with interaction and uncertain maneuver prediction. IEEE Transactions on Intelligent Vehicles, 3(1): 5–17.

[35] International Organization for Standardization. 2021. ISO/FDIS 21448 Road vehicles—safety of the intended functionality.

[36] Jaulin, L., Kieffer, M., Didrit, O. and Walter, E. 2001. Applied Interval Analysis, ed: Springer, London.

[37] Karaman, S. and Frazzoli, E. 2011. Sampling-based algorithms for optimal motion planning. The International Journal of Robotics Research, 30(7): 846–894.

[38] Kavraki, L.E., Kolountzakis, M.N. and Latombe, J-C. 1998. Analysis of probabilistic roadmaps for path planning. IEEE Transactions on Robotics and Automation, 14(1): 166–171.

[39] Khaitan, S., Lin, Q. and Dolan, J. 2021. Safe planning and control under uncertainty for self-driving. IEEE Transactions on Vehicular Technology.

[40] Kingston, Z., Moll, M. and Kavraki, L.E. 2018. Sampling-based methods for motion planning with constraints. Annual Review of Control, Robotics and Autonomous Systems, 1: 159–185.

[41] Kiran, B.R., Sobh, I., Talpaert, V., Mannion, P., Al Sallab, A.A., Yogamani, S. and P.rez, P. 2021. Deep reinforcement learning for autonomous driving: A survey. IEEE Transactions on Intelligent Transportation Systems.

[42] Kobayashi, S. 2019. Differential Geometry of Curves and Surfaces. Springer, 2019.

[43] Kobayashi, T., Salay, R., Hasuo, I., Czarnecki, K., Ishikawa, F. and Katsumata, S-Y. 2021. Robustifying controller specifications of cyber-physical systems against perceptual uncertainty. *In*: Aaron Dutle, Mariano M. Moscato, Laura Titolo, César A. Muñoz and Ivan Perez, (eds). NASA Formal Methods—13th International Symposium, NFM 2021, Virtual Event, May 24–28, 2021, Proceeding, volume 12673 of Lecture Notes in Computer Science, pages 198–213. Springer.

[44] Lattarulo, R. and Rastelli, J.P. 2021. A hybrid planning approach based on (MPC) and parametric curves for overtaking maneuvers. Sensors, 21(2): 595.

[45] LaValle, S.M. and Kuffner Jr., J.J. 2001. Randomized kinodynamic planning. The International Journal of Robotics Research, 20(5): 378–400.

[46] Li, B. and Zhang, Y. 2000. Fast trajectory planning in Cartesian rather than Frenet frame: A precise solution for autonomous driving in complex urban scenarios. IFAC-PapersOnLine, 53(2): 17065–17070.

[47] Magdici, S. and Althoff. M. 2016. Fail-safe motion planning of autonomous vehicles. *In*: Proceeding IEEE 19th International Conference on Intelligent Transportation Systems, pages 452–458.

[48] McNaughton, M., Urmson, C., Dolan, J.M. and Lee, J-W. 2011. Motion planning for autonomous driving with a conformal spatiotemporal lattice. *In*: Proc. IEEE International Conference of Robotics and Automation, pages 4889–4895.

[49] Nipkow, T., Paulson, L.C. and Wenzel, M. 2002. Isabelle/HOL: a proof assistant for higher-order logic, volume 2283. Springer Science and Business Media.

[50] Nolte, M., Rose, M., Stolte, T. and Maurer, M. 2017. Model predictive control based trajectory generation for autonomous vehicles—an architectural

approach. *In*: Proceeding IEEE Intelligent Vehicles Symposium, pages 798–805.

[51] Oboril, F. and Scholl, K-U. 2020. Risk-aware safety layer for AV behavior planning. *In*: 2020 IEEE Intelligent Vehicles Symposium, IV 2020, Las Vegas, United States, October 19–22, 2020, page 7. IEEE.

[52] Oboril, F. and Scholl, K-U. 2021. RSS+: Pro-active risk mitigation for AV safety layers based on rss. *In*: 2021 IEEE Intelligent Vehicles Symposium, IV 2021. IEEE.

[53] Paden, B., Ĉáp, M., Yong, S.Z., Yershov, D. and Frazzoli, E. 2016. A survey of motion planning and control techniques for selfdriving urban vehicles. IEEE Transactions on Intelligent Vehicles, 1(1): 33–55.

[54] Pivtoraiko, M., Knepper, R.A. and Kelly, A. 2009. Differentially constrained mobile robot motion planning in state lattices. Journal of Field Robotics, 26(3): 308–333.

[55] Pruekprasert, S., Dubut, J., Zhang, X., Huang, C. and Kishida, M. 2019. A game-theoretic approach to decision making for multiple vehicles at roundabout. arXiv preprint arXiv:1904.06224.

[56] Pruekprasert, S., Takisaka, T., Eberhart, C., Cetinkaya, A. and Dubut. J. 2020. Moment propagation of discrete-time stochastic polynomial systems using truncated carleman linearization. IFACPapersOnLine, 53(2): 14462–14469.

[57] Pruekprasert, S., Zhang, X., Dubut, J., Huang, C. and Kishida, M. 2019. Decision making for autonomous vehicles at unsignalized intersection in presence of malicious vehicles. *In*: 2019 IEEE Intelligent Transportation Systems Conference (ITSC), pages 2299–2304. IEEE.

[58] Rawlings, J.B., Mayne, D.Q. and Diehl, M.M. 2017. Model Predictive Control: Theory, Computation and Design. Nob Hill Publishing.

[59] Roohi, N., Kaur, R., Weimer, J., Sokolsky, O. and Lee, I. 2018. Self-driving vehicle verification towards a benchmark. CoRR, abs/1806.08810, 2018.

[60] Sadat, A., Casas, S., Ren, M., Wu, X., Dhawan, P. and Urtasun, R. 2020. Perceive, predict and plan: Safe motion planning through interpretable semantic representations. *In*: Proceeding European Conference on Computer Vision, pages 414–430.

[61] Sadigh, D., Sastry, S., Seshia, S.A. and Dragan, A.D. 2016. Planning for autonomous cars that leverage effects on human actions. *In*: Robotics: Science and Systems, volume 2, pages 1–9. Ann Arbor, MI, USA.

[62] Salay, R., Czarnecki, K., Elli, M.S., Alvarez, I.J., Sedwards, S. and Weast. J. 2020. PURSS: Towards perceptual uncertainty aware responsibility sensitive safety with ML. *In*: Hu.scar Espinoza, José Hernández-Orallo, Xin Cynthia Chen, Seán S. ÓhÉigeartaigh, Xiaowei Huang, Mauricio Castillo-Effen, Richard Mallah and John Mc-Dermid, (eds). Proceeding of the Workshop

on Artificial Intelligence Safety, Co-located with 34th AAAI Conference on Artificial Intelligence, SafeAI@AAAI 2020, New York City, NY, USA, February 7, 2020, volume 2560 of CEURWorkshop Proceeding, pages 91–95. CEUR-WS.org.

[63] Seto, D., Krogh, B., Sha, L. and Chutinan. A. 1998. The simplex architecture for safe online control system upgrades. *In*: Proceeding of the 1998 American Control Conference. ACC (IEEE Cat. No. 98CH36207), volume 6, pages 3504–3508.

[64] Shalev-Shwartz, S., Shammah, S. and Shashua, A. 2017. On a formal model of safe and scalable self-driving cars. CoRR, abs/1708.06374.

[65] Shashua, A., Shalev-Shwartz, S. and Shammah, S. 2018. Implementing the RSS model on NHTSA pre-crash scenarios. Technical report, Mobileye.

[66] Silberling, J., Wells, P., Acharya, A., Kelly, J. and Lenkeit, J. 2020. Development and application of a collision avoidance capability metric. *In*: WCX SAE World Congress Experience. SAE International, April 2020.

[67] Underwriters Laboratories. 2020. UL4600: Standard for Evaluation of Autonomous Products, 1 edition, April 2020.

[68] Wang, L., Fernandez Lopez, C. and Stiller, C. 2020. Realistic single-shot and long-term collision risk for a human-style safer driving. *In*: 2020 IEEE Intelligent Vehicles Symposium, IV 2020, Las Vegas, United States, October 19–22, 2020, page 8. IEEE.

[69] Weng, B., Jagannatha Rao, S., Deosthale, E., Schnelle, S. and Barickman, F. 2020. Model predictive instantaneous safety metric for evaluation of automated driving systems. *In*: 2020 IEEE Intelligent Vehicles Symposium, IV 2020, Las Vegas, United States, October 19–22, 2020, page 8. IEEE.

[70] Werling, M. and Liccardo, D. 2012. Automatic collision avoidance using model-predictive online optimization. *In*: *Proceeding* IEEE 51st IEEE Conference on Decision and Control, pages 6309–6314.

[71] Werling, M., Zieglerand, J., Kammel, S. and Thrun, S. 2010. Optimal trajectory generation for dynamic street scenarios in a Frenet frame. *In*: Proceeding IEEE International Conference on Robotics and Automation, pages 987–993.

[72] Winskel, G. 1993. The Formal Semantics of Programming Languages. The MIT Press.

[73] Xu, W., Wei, J., Dolan, J.M., Zhao, H. and Zha, H. 2012. A real-time motion planner with trajectory optimization for autonomous vehicles. *In*: Proceeding IEEE International Conference on Robotics and Automation, pages 2061–2067.

[74] Xu, X., Wang, X. and Wu, X. 2020. Calibration and evaluation of responsibility sensitive safety model on autonomous car—following maneuvers using naturalistic driving study data. Transportation Research Board (TRB) Annual Meeting.

[75] Yu, H., Tseng, H.E. and Langari, R. 2018. A human-like game theory-based controller for automatic lane changing. Transportation Research Part C: Emerging Technologies, 88: 140–158.

[76] Zhang, R., Rossi, F. and Pavone, M. 2016. Model predictive control of autonomous mobility-on-demand systems. *In*: Proceeding IEEE International Conference on Robotics and Automation, pages 1382–1389.

[77] Zhao, H., Zhang, Y., Meng, P., Shi, H., Li, E., Lou, T., et al. 2020. Safety score: A quantitative approach to guiding safety-aware autonomous vehicle computing system design. *In*: 2020 IEEE Intelligent Vehicles Symposium, IV 2020, Las Vegas, United States, October 19–22, 2020. IEEE.

[78] Ziegler, J., Bender, P., Dang, T. and Stiller, C. 2014. Trajectory planning for Bertha—A local, continuous method. *In*: Proceeding IEEE Intelligent Vehicles Symposium Proceeding, pages 450–457.

Chapter 8

Formal Modelling

Stefan Klikovits
Johannes Kepler University, Linz, Austria

Fuyuki Ishikawa
National Institute of Informatics, The University of Electro-Communications and SOKENDAI, Tokyo, Japan

Tsutomu Kobayashi
Japan Aerospace Exploration Agency (JAXA), Tsukuba, Japan

8.1 System Modelling

The use of *models* is widespread in everyday life and in most (if not all) professions. Models serve as vital aids for task completion, since their *abstraction* capabilities remove unnecessary details and present a clear view on the underlying information. For instance, we might look at Fig. 8.1, which shows a map of Tokyo's public transport network on the left, and a satellite view of the same area on the right.

If our goal is to take the train from one place to another, it would be significantly harder to do so if the only information we had was the satellite view—even though, with enough patience and zooming we might be able to identify the necessary train stops. Evidently the subway map is a much more useful, as this model was created specifically for this purpose and applies several improvement steps: First, the model abstracts over (i.e., removes) unnecessary details (houses, streets, smaller rivers). Second, other aspects are

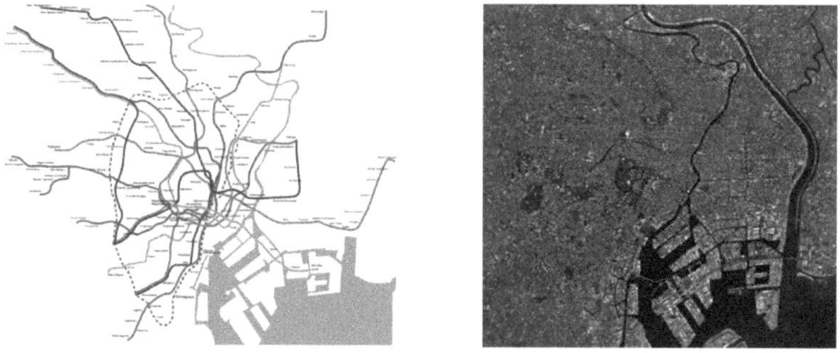

Figure 8.1: Left: Tokyo's public transport network; right: satellite view of the same area. (Image credit: Cnes—Spot Image; via Wikimedia Commons).

aggregated and simplified, such as the division in only land and water (instead of buildings, parks, etc.) and the rough drawing of shorelines and rivers using straight edges. Third, the important aspects, i.e., train stations, are emphasised using labels. Finally, the model adds train lines and uses colours as identifiers, to make the task of route planning easier.

Similarly, in most (large) projects, we use cost, time and risk models to plan ahead and help make decisions. These models abstract away certain details and allow us to take a focused *viewpoint* on a given aspect. In software engineering, the *V-model* [31] (see Fig. 8.2) is a widely spread template, offering a structured view of different project phases. Due to their genericity, however, such models need to be extended and refined with other types of details. For instance, the system's high-level *specifications* should ideally be captured in a *requirement* model, and the system architecture can be expressed as a software *architecture model*. Each model thereby abstracts away specific details, so that its own *model purpose* (i.e., the task(s) for which the model was created) can be efficiently carried out.

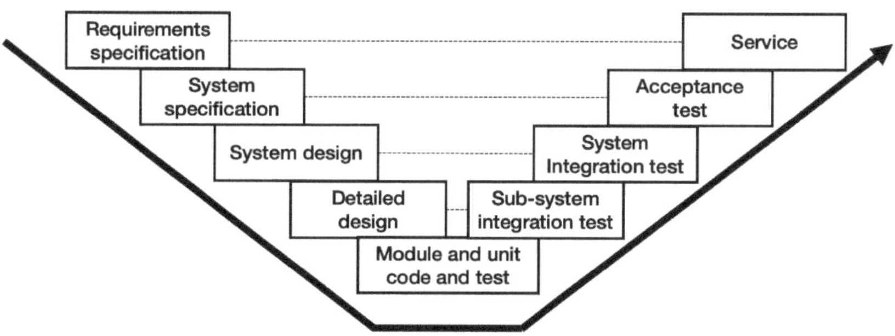

Figure 8.2: The V-model, as defined in [31].

Due to their abstraction-based *system views*, models are a powerful tool and enable clearer analysis and communication. However, to avoid ambiguities, they have to be expressed in a language whose concepts are clear to all stakeholders. These *modelling languages* are typically based on *formalisms*, which provide the (mathematical) basis for the languages. Figure 8.3 provides a framework that expresses the relation between several concepts of the modelling world. On the left, we see viewpoints as the high-level concepts, which typically express the *stakeholder*'s interests of a system or project. These might be for instance a system's *architecture*, safety, communication flows, physical resources, etc. Viewpoints are supported by formalisms. Roughly speaking, a formalism consists of an abstract syntax and a matching semantics, such that the abstract syntax defines a "vocabulary" and *grammar* and the *semantics* define the "meaning" of each abstract syntax concept. Well-known formalisms are *automata*, *state charts*, various logic systems, or *graphs*, for instance.

Modelling languages build upon these formalisms and offer a user-friendly interface to the underlying formalism by adding a *concrete syntax*. For instance, in the public transport example, the abstract information is defined as a graph formalism with nodes (subway stops) and edges (subway lines). A concrete syntax defines how these are represented on a map (e.g., circles for stops and coloured lines for subway routes). In the *systems engineering* world we typically use common modelling languages such as *Event-B*, *Simulink*, *Modelica*, *SysML* and *AADL*.

As cyber-physical system (CPS) models tend to be quite complex, especially when considering the various viewpoints involved, we typically employ modelling tools that implement languages and help in the model creation, editing and execution processes. They often also offer *consistency checks* and make sure that each part of the model (specified in concrete syntax) adheres to the rules defined in the formalism (abstract syntax and semantics).

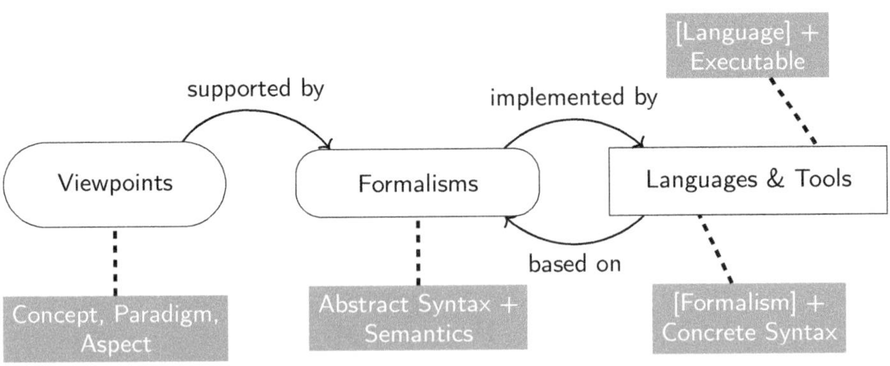

Figure 8.3: Formal framework of viewpoints, formalisms, languages and tools; based on Broman et al. [9].

In the past, a plethora of modelling formalisms, languages and tools have been defined and developed to support the many different viewpoints involved in systems engineering and beyond. Nonetheless, the creation of a system model is a non-trivial process. Modelling languages are often very generic, aiming to be versatile enough to support many types of systems and a multitude of viewpoints. Often, it requires months of training and/or years of practical experience to become efficient with their use.

Another type of difficulties lies in managing models of different abstraction levels. *Stepwise refinement*, from abstract models to concrete ones, has been one of the key principles in software engineering as we cannot construct the code or the detailed models in an one-shot activity for complex software systems. This is especially the case for CPS with increasing complexity and enlarged scopes of the models including interaction between the software system and environments. Rigorous mechanism is required to ensure that concrete models should comply with what were defined and checked in abstract models.

In the rest of this chapter, we introduce two approaches to systems modelling that facilitate the modelling process, thereby lowering the entry bar and helping in the effective use of system models. *Domain specification*, the first method, enables the definition of models using system vocabulary, thereby avoiding generic modelling languages and *system-agnostic* formalisms. This approach often allows for faster development on a higher abstraction level, such that domain experts may more easily participate in the modelling processes. The second approach describes *modelling through* refinement. It is a rigorous approach that enables developers to start system development on a high level and systematically lower the model's abstraction level while at the same time asserting consistency and validity properties.

8.2 Domain-Specific Modelling

As introduced above, we notice that modelling formalisms such as automata or state charts look quite different from the systems that they represent. Conceptually, we might even compare them to the calculi that provide the formal computational foundations on which modern general-purpose programming languages (GPPLs) are built. Just as when using these calculi, when using modelling formalisms it is often very difficult to recognise a given system at all, after its conversion into such a rather mathematical description. Another analogy would be to compare the modelling of a system using a formalism directly to writing a software in form of processor instructions and assembly code. While it is certainly valid to do so, very few software developers would voluntarily choose to "step down" to this level of verbosity, and instead use a modern GPPLs such as C, Java or Python, and also use integrated development environments (IDEs) and other tools to facilitate their development. Similarly, most model engineers will aim to choose an appropriate modelling language and corresponding tooling.

Nonetheless, from experience we know that in software engineering we often further facilitate our work by relying on frameworks, or using/creating reusable *libraries* and application programming interfaces (APIs). Next to the reduced development effort and repetition, this *domain-specification* also extends the generic *syntax* and programming concepts with *domain vocabulary*, which allows users to think in terms of resulting application rather than underlying GPPL details. For instance, a library such as Python's `Flask`[1] provides means to easily create web servers. In doing so, the library bridges the gap between the programming language's generic syntax and the target domain, extending it by the necessary vocabulary used in the web application itself (URL routing, HTTP methods, port definitions). Using this vocabulary, libraries not only increase development speed and reduce repetition, but they also allow non-experts—i.e., users with little knowledge of the technical details—to create and implement complex code. Indeed, `Flask` users do not need to know how to implement a HTTP method themselves, it is enough to specify which one they want to use.

The latter is a vital advantage, and raises a rather important question: *Can we facilitate systems modelling in a similar way?* For instance, it would be beneficial to conceptualise a way for modellers to develop system models using a language, syntax and vocabulary that is familiar to them, without having to learn the details of the underlying formalisms. The short answer is: ***Yes!*** Similarly to GPPLs, many modelling languages can be further customized and extended using domain concepts. This section introduces three domain-specification methods that allow bridging the gap between generic modelling languages and the systems they model. Then, the second part takes a closer look at CREST, a modelling language that builds on top of these concepts to provide convenient means for facilitated modelling of small-scale CPSs. It thereby focuses on ease of use and *reuse* of well-known modelling concepts.

8.2.1 Domain-specification Approaches

While it is certainly possible to combine different approaches, in general, we distinguish between three domain-specification methods: 1. *libraries*, 2. *profiles* and *extensions*, and 3. *domain-specific languages (DSLs)*. Similar to APIs and software frameworks, they are positioned at the *domain-specific* level of the domain adaptation hierarchy [20], as shown in Fig. 8.4. The figure also positions *general purpose* (e.g., general modelling languages and tools) and *system-agnostic* (i.e., formalisms) concepts. Furthermore, calculi, GPPLs, and APIs & frameworks are displayed for illustrative comparison.

[1]`https://flask.palletsprojects.com`

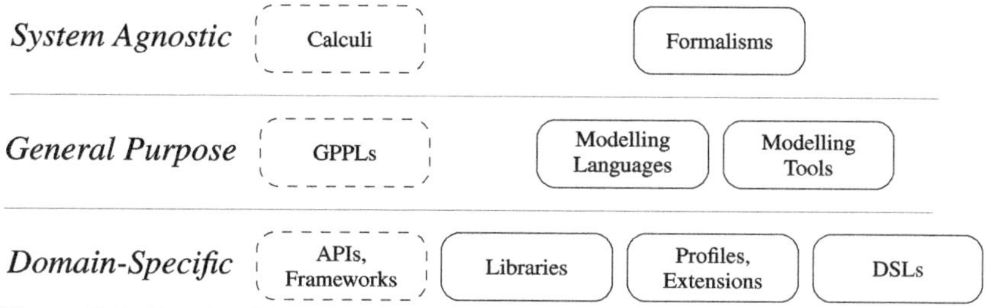

Figure 8.4: Levels of domain adaptation in systems modelling. Adapted from: [20].

8.2.1.1 Libraries

The first, rather straightforward approach to *domain-specification* is the use of model, object and template *libraries*. While their primary goal is typically the avoidance of repetition and increased speed of development, libraries can serve as a convenient step towards domain-specification. By abstracting over low-level and general purpose *modelling concepts*, modelling with libraries becomes more accessible to domain experts, facilitates discussions and avoids the need to "mentally switch" between the modelling and system views. Furthermore, it is easy to replace "modelling vocabulary" with domain-specific terms from the system. It is noteworthy though, that libraries (typically) do not change the base language's syntax or semantics.

Modelling libraries can be obtained from several sources, including the developers of the modelling language themselves, third party providers—e.g., OEMs aiming conveniently extend a language's support to their products—or the system's developers and users. Common modelling languages and tools supporting object and template libraries include Ptolemy II, Simulink, LabVIEW and Modelica[2]. When starting to develop a library, there are few "true downsides" beyond the slightly heightened testing and distribution effort. Note however, that the need for quality increases with the number of clients and users.

8.2.1.2 Extensions, Profiles

While libraries primarily operate on the model-level, there is sometimes the need to actually extend the base language itself. As the name suggests, *language extensions* build on an existing modelling language but introduce new features. In contrast to libraries though, these extensions usually add new syntactic concepts which typically also define their own meaning and semantics. The BPMN language [25] for instance extends the *Petri nets* formalism and introduces additional concepts for the analysis of business workflows. These additions carry new meaning beyond the standard semantics of Petri nets. Some

[2]https://www.modelica.org/libraries

languages such as the Architecture Analysis & Design Language (AADL) [30] even provide built-in extension opportunities. AADL's *annex* mechanism for instance allows users to extend architecture models, and define e.g., discrete [5] and continuous [19] behaviour, or model error extensions [13].

Closely related to language extensions are *profiles*, most widely known through its arguably most common application in the Unified Modeling Language (UML) [27]. These profiles are a more powerful means of altering languages, often adapting the base language's semantics. The Systems Modeling Language (SysML) [28] for instance is a well-known adaptation building on some of UML's diagrams and extending them. In the process, some of UML's semantics are changed and concretised so that the new language can be efficiently used for e.g., CPS systems modelling. Similarly, *MARTE* [26] is a profile for modelling of real-time and embedded systems; and the before-mentioned AADL itself is defined as UML profile. It has to be said though, that despite providing domain-specific concepts, SysML, MARTE and AADL are still rather generic and not aimed at specific applications. As a concrete example, we might look at RobotML [15] which describes the use of a UML profile dedicated to robotic applications.

Extensions and profiles are powerful mechanisms for language adaptation often significantly extending the capabilities of the base language. Nonetheless, these modifications go hand-in-hand with a non-negligible price. As the base-language is modified, developers have to be careful not to introduce inconsistencies and assert the correctness of the new/altered concepts. From a user's viewpoint, extensions do not always facilitate the language's use, since the need to learn the original language still persists. Furthermore, as profiles are capable of changing language semantics, it might be that pre-existing knowledge of the original language can lead to confusion in the best case and when unaware even lead to significant problems in the system models.

8.2.1.3 Domain-specific Languages

A third approach to domain-adaptation is the development of a new language that is dedicated to modelling specifically the systems within a target domain. The concept of DSLs is best known to software engineers from languages such as SQL, Makefile or GraphViz's `dot` language. In fact, modelling DSLs are closely related to software DSLs, and, given the tight integration of software engineering and modelling, it is sometimes even hard to distinguish whether a DSL is part of the "modelling" or "software engineering" world. Thus, for instance, we might find languages that define system architecture annotations (modelling concepts) directly into a software's API definitions (programming parts). Similarly, build tools such as Maven and Ant have evolved far beyond their original purpose and can support in the software release process and thus tightly integrate into a deployment model.

External DSLs

Typically, when talking about DSLs we mean *external DSLs*, i.e., languages that are—just like general-purpose languages—created as standalone products. They have a dedicated syntax and semantics, and in many cases even some kind of editor, syntax highlighting, language processor or execution engine and validator. As the effort of developing such DSLs is significant [35], it is nowadays usual to employ *language workbenches* [18] such as Xtext [6], MetaEdit+ [34] or MPS [10]. The advantage is that the workbenches generate many required or nice-to-have artefacts automatically based on a description of the DSL grammar alone. For instance, it is de-facto standard that for textual languages they create an IDE with auto-completion and syntax checker, and help in the creation of language transformation engines, code generators and other useful features. The languages created by these tools are often based on or allow simplified interfacing with the Eclipse Modeling Framework (EMF) [32], a de-facto standard when it comes to model description frameworks. This further enables easy integration with graphical modelling frameworks (e.g., GMF) to specify a graphical syntax and many other model-transformation engines.

Despite these advantages, it has to be said that the development of a DSL is still difficult and requires good knowledge of language grammars, parsers and similar technologies. At the end, however, users obtain a "clean" language that is geared directly to their system and application domain, and vitally avoids not needed (potentially confusing) concepts of other domains.

Internal DSLs

In contrast to external DSLs, *internal DSLs* are languages built inside another language. They use their *host language*'s syntax and semantics to embed advanced features. To this extent, they often "use, and abuse, the host language", by relying on features such as meta-programming, reflection and language tricks. Especially interpreted languages such as Python and JavaScript, but also other common languages such as C++ and Java, are used as hosts. The advantage of internal DSLs is that given a widespread host, its language libraries can be readily used which avoids development effort on the side of language developers and users alike. Furthermore, such DSLs integrate directly into existing development environments and provide language tooling, such as IDEs, static analysers and similar.

There are many examples of internal DSLs, but a very common one is *SystemC* [7]. This embedded systems modelling language uses C++ as its host. It relies on macros and library functions to define its syntax and builds its semantics directly into the library. A major advantage is that SystemC's modelling syntax is pure C++ code and therefore can be compiled, analysed and executed using widespread tools. Other examples for internal DSLs include testing frameworks such as *JUnit* and extensions for programming languages such as *Ruby on Rails*. Section 8.2.2 provides an example for the use of an internal modelling DSL using Python as a host.

Embedded DSLs

In contrast to the above, *embedded languages* sit between internal and external DSLs. Though self-contained in theory, these languages "live" inside another *host language*, due to their conceptual proximity. For instance, programmers might be aware of various forms of code documentation, which use a language's comment function to document function and package information using keywords, etc. While independent in theory, some of them might also access and parse some information of their host languages—e.g., by automatically extracting data types or architecture information from package data. Embedded modelling DSLs are often also defined as annotations in graphical languages, providing extra information.

The advantage of embedded DSL is that they are easily created and very flexible. It has to be said that, despite being convenient, for more powerful approaches such as the automatic data extraction it is still necessary to develop a clear language syntax and semantics as well as parsers to automatically process information.

8.2.2 CREST—A Case Study

CREST [20] is a *domain-specific modelling language* that has been created for the design, development, testing and verification of small-scale CPSs. Despite its focus on simplicity and easy-to-learn features, CREST incorporates well-known concepts from existing languages and builds on solid formal foundations to create a clear semantics. In fact, CREST consists of three components each serving a specific purpose:

1. A *formalism* that clearly defines CREST's semantics.
2. CREST diagrams, a *graphical DSL* for the visualisation of components and resource flows.
3. `crestdsl`, an *internal DSL* built in Python, that enables *simulation* and *verification*.

Here, we detail the development phases of domain-specific modelling languages using CREST as an example.

8.2.2.1 Language Purpose Evaluation

The creation of a DSL typically requires non-negligible planning and development effort. This means that before starting, it is necessary to clearly

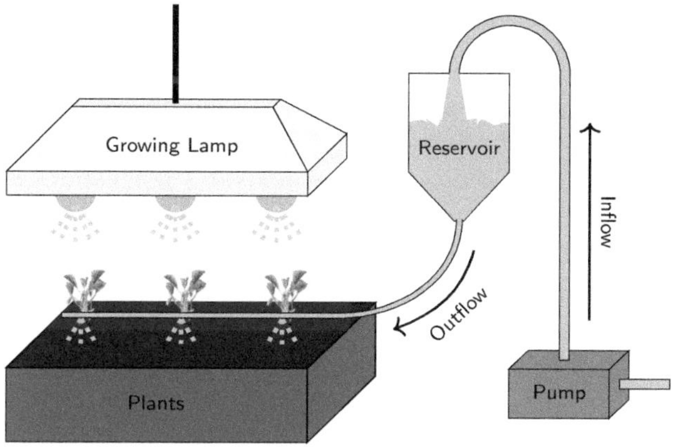

Figure 8.5: Schema of the plant system's components.

identify several key data points such as the language's *purpose*, the target *audience* and environment *setting*.

In CREST's case, the language purpose is to support users during the creation of CPSs that are dominated by the flow of resources such as electricity, light and heat. Examples of such installations can be home and office automation systems, or automated plant growing applications. A simplified schema of such an application is shown in Fig. 8.5.

The modelling intent, i.e., the reason for employing system models spans three modelling activities. First, the *model creation* helps users during conceptualisation, design and building of the CPS. Second, to evaluate the designed system's behaviour a *simulation* engine should allow exploratory analysis. Finally, the *verification* of the models should rule out conceptual flaws in the system design.

The target audience consists of model creators who search for an easy-to-use solution for small scale systems on the one hand, and non-experts and modelling newcomers searching for a low-entry bar solution that will help them simulate and verify their systems, on the other. In either case, the focus on simplicity and clarity is predominant, while not restricting the expert modellers in advanced modelling use cases is another. Furthermore, it is assumed that CPS creators and thus CREST's users have some pre-existing knowledge in a technical field such as IT, programming, engineering or are willing to learn.

Finally, some more information about usage environment—closely linked to the target applications and users—is identified. Thus, it is assumed that users of CREST do not have the financial or temporal means to acquire a costly tool or spend a long time on becoming expert users of *open-source* modelling solutions. Additionally, we suppose that CREST is a self-contained application and the need for integration with other modelling frameworks is limited.

8.2.2.2 Language Development

After the evaluation and exclusion of various existing candidate languages, the decision was made to develop a new language, rather than adapt an existing one. Nonetheless, to increase the familiarity and flatten the learning curve, CREST aims to *reuse* well-known concepts from existing languages. To this extent, CREST combines architectural and behavioural system aspects, as well as *resource flows* and *timed semantics*. The guiding principles during CREST's creation can be summarised in three points:

- *Reuse* well-known formalisms and languages.
- Combine the concepts in a formally sound manner.
- Focus on *usability* and comprehensibility.

The preliminary evaluation of existing language candidates for the CREST project pointed out the benefits and disadvantages of various design choices. Following the first, reuse principle, the CREST was then designed to reuse various concepts such as *hierarchical entities* and *ports*—similar to architecture description languages, *reactive dataflow*—inspired by *synchronous formalisms*, continuous variable evolution—known from hybrid models, and support for non-determinism.

The concepts were formally defined and combined in the CREST formalism, such that their meaning is unambiguous and allows formal verification. Furthermore, the two CREST languages were developed to enable convenient modelling and discussion.

CREST diagrams is a *graphical language* visualising a *system model*. Fig. 8.6 shows a model of a plant growing lamp, as used in the automated plant growing system. Without going into the details of the language, the example shows that the model specifies typed **input** (Σ▭) and **output** (▭▷) ports, **states** and guarded **transitions** ((Off) —guard→ (On)), as well as **update** functions (- ➝) which propagate continuous value updates in form of resource flows and also apply transformation functions (e.g., fahrenheit_to_celsius, light_electricity_zero). Furthermore, the modules ("entities") are hierarchically organised, meaning that components behaviour can be compartmentalised, if needed. Each element in the diagram is either directly defined in the language semantics or, as is the case with **influences** (⟶) and **actions** (···►), can be transformed into other syntactic elements (in this case *updates*) which map onto the semantics. This asserts that no elements with unclear semantics are used.

Next to the visual language, CREST defines a second syntax geared at efficient model creation, simulation and verification. `crestdsl` is an internal DSL defined within Python. It uses Python objects as entities and enables the use of classes as templates for defining these objects. Object attributes are used to define states and ports and annotated methods specify state transitions, updates and actions. Listing 8.1 shows an example of such a `crestdsl` class

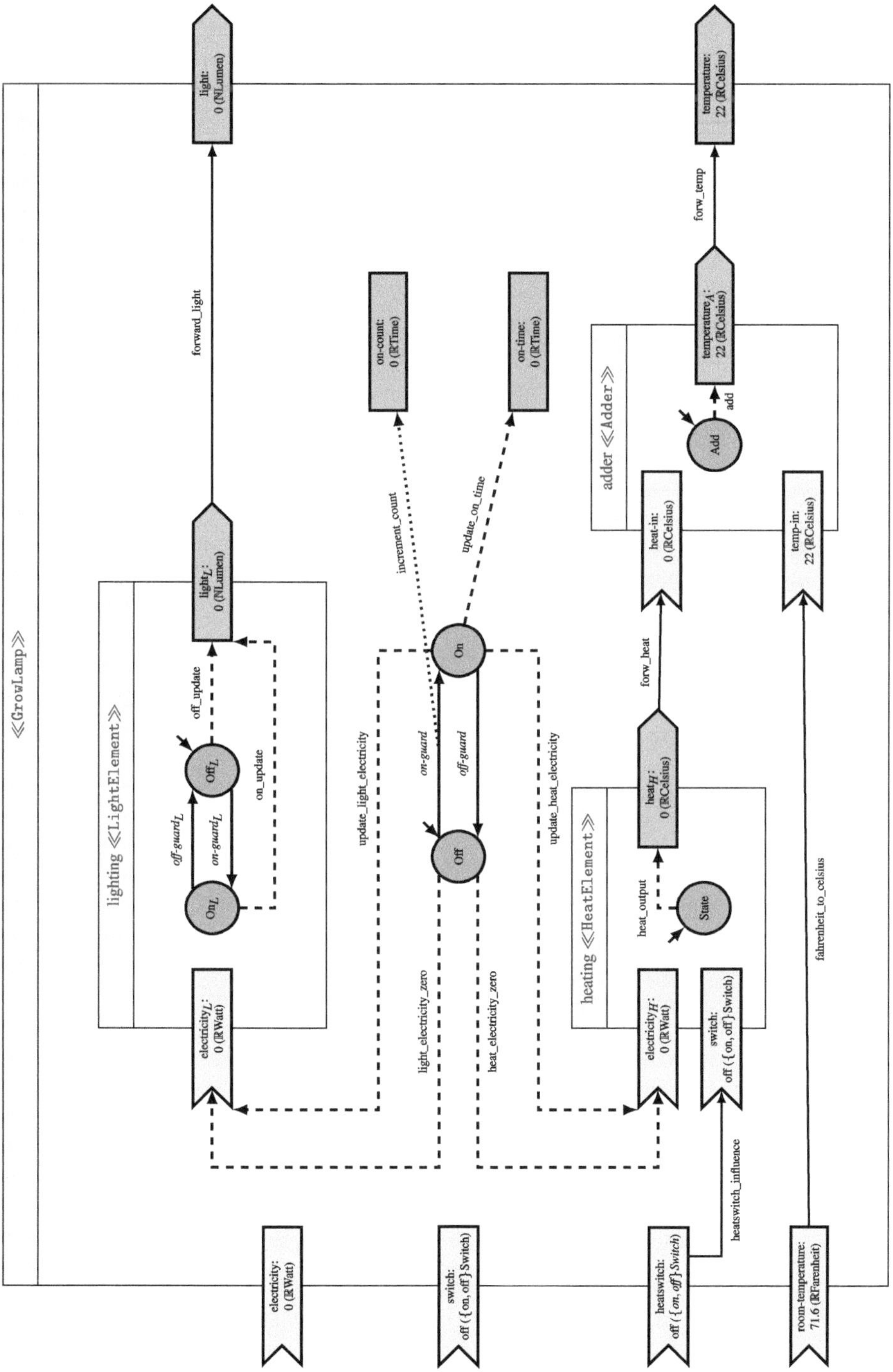

Figure 8.6: A CREST diagram of a growing lamp entity with submodules, ports, and resource flows as arrows.

that declares an entity type that can be used to instantiate models. The implementation is based on pure Python code and uses meta-programming and reflection to intercept class and object creation. This allows us to dynamically alter an object's definition, access the methods' source codes and also modify the class composition.

Furthermore, the use of Python as a host language asserts that existing IDEs, analysers and runtimes can be used. Additionally, due to the popularity, there exists a plethora of libraries that naturally interface with the model basis. Thus, for instance, the simulator is implemented using the Z3 [12] SMT solver, which defines a native Python interface. `crestdsl` also provides plotting interfaces, such that it is possible to generate and visualise CREST diagrams directly from the `crestdsl` model. These extensions define a `matplotlib` plotter which uses the widespread Python plotting library, but also a web user interface (UI) generator which uses JavaScript and HTML to create interactive CREST diagrams.

Listing 8.1: (partial) entity template with transitions and update functions.

```
class GrowingLamp(crest.Entity):
    lighting = LightElement()
    heating = HeatElement()
    electricity = crest.Input(resource=watt, value=100)
    switch = crest.Input(resource=switch, value="off")
    on = crest.State()
    off = crest.State()
    current = off

    @crest.transition(source=off, target=on)
    def on_guard(self):
        return self.electricity.value < 100 or\
                    self.switch.value == "off"

    @crest.transition(source=on, target=off)
    def off_guard(self):
        return self.electricity.value >= 100 and \
                    self.switch.value == "on"

    @crest.update(state=on, target=on_time)
    def output_when_off(self, dt):
        return self.on_time.value + dt

    ...

# instantiation
growinglamp = GrowingLamp()
```

Summarising, one can say that CREST is a good example for the use of DSL techniques for a specific target domain, namely the modelling of small-scale CPSs. The implementation effort for the language was roughly one person-year. Since then, CREST has been shown to be usable, adaptable and extensible even by non-expert modellers. By reusing common techniques and concepts, CREST caters to users with pre-existing modelling knowledge, while its implementation in Python renders `crestdsl` usable by anyone who is familiar with programming.

Nonetheless, it has to be remarked that—as is common with DSLs—CREST has been created for a specific target audience and domain. While it is capable of being adapted to various use cases, the language and implementation reaches its limits when it comes to large-scale models, safety-critical systems and high performance simulations. It serves however as a good example for users who might have similar needs and language requirements.

8.3 Modelling with Stepwise Refinement

The usefulness of most modern software systems is due to their functionalities of interacting with their *environment*. Therefore, when we check the correctness of software, investigating only the software itself is not enough; we need to check that the software and its environment satisfy particular *requirements* together [36]. This claim is sometimes formally expressed as

$$E \parallel S \models R,$$

where E is the set of assumptions on the environment, S is the set of specifications of the software, and R is the set of requirements that should be satisfied. Note that the environment can be hostile to the software; we should design the software to prepare for the worst behaviour of the environment.

The explosion of *Ariane 5*, a satellite launcher of the *European Space Agency* launched in 1996, is an example of an accident caused by the mismatch of the behaviour of software and its environment [8, 23]. The analysis team found that an overflow of variables in the inertial reference systems caused the failure, despite that the same software perfectly worked for the previous version (Ariane 4). It turned out that the overflow happened because the software was not ready for the trajectory of Ariane 5, which was significantly different from Ariane 4's.

While it seems natural to take the environment into account, doing it thoroughly is difficult in practice because the system becomes quite complex; in addition to the software and the environment's behaviour, we need to consider interactions between them. For such systems, it is crucial to construct high-quality models and do reasoning in an early phase before programming because even *understanding* the system is difficult. In [1], Abrial advocates:

> "Our task is not limited to the software part alone, because what we intend to build is a system within which the piece of software we will construct is just one component among many others. In doing this as engineers, we are not supposed to instruct a computer; rather, we are supposed to instruct ourselves".

The challenge here is to rigorously construct and verify models while dealing with the *complexity* of the target system. One of the promising approaches to the complexity is *stepwise refinement* to gradually construct complex concrete models from abstract ones. The specific difficulty is to ensure the *consistency* between the abstract and concrete models. In other words, concrete models should comply with, thus not disposing the effort of, what were defined and checked in abstract models. Rigorous mechanisms are necessary to support this process. In the rest of this section, we first introduce *Event-B* [2], which was proposed to tackle this challenge. We then describe our methods for improving the *maintainability* and *reusability* of system models specified in Event-B by *refactoring refinements*. We also give a summary of good practices of refinement found from our previous study.

8.3.1 Modelling in Event-B

Event-B is a formal method designed for modelling of and reasoning on complex software systems. The modelling language is based on and , and it supports *event-driven modelling* of the behaviour of the target system. The language and the *Rodin platform* [3, 17], the modelling environment for Event-B, are designed to facilitate *theorem proving* — mathematically proving that certain properties hold (see also Chapter 9.1) — for verifying system models.

One of the most notable features of Event-B is the support of modelling through stepwise refinement, which enables developers to model the target system on multiple *abstraction levels*. The modelling process in Event-B proceeds in an incremental manner; the modeller starts from constructing an *abstract* version of the model and gradually makes the model *concrete*. This mechanism enables modellers to distribute the complexity over multiple steps.

For instance, suppose that we are constructing a model of a parking lot system.[3] There is an indoor parking lot with ten slots connected to the outside with a narrow corridor (M_2 of Fig. 8.7). The software to be developed is a controller of traffic lights located at both ends of the corridor. The main requirements which the controller should satisfy are that (1) the traffic light outside should be red if vehicles are leaving or the parking lot is full, (2) the traffic light inside should be red if vehicles are entering the building, and (3) at least one of two traffic lights should be red to keep the corridor one-way or the other. Note that we assume that vehicles strictly follow traffic lights.

To model this system with the stepwise refinement mechanism (Fig. 8.7), the modeller can construct an *abstract model* first. For instance, in M_0, by *abstracting away* details about the traffic lights and the internal structure of the building (the corridor and the parking lot), only changes in the number of vehicles inside the building are modelled. The core concern here is that the abstract model M_0 should

[3]This example is a modified version of the one in [2, Chapter 2].

be consistent with the target system. Therefore, vehicles' behaviour should not be random; although traffic lights are not modelled yet, vehicles find whether there are no more vehicles than the building capacity when entering the building as if there were traffic lights.

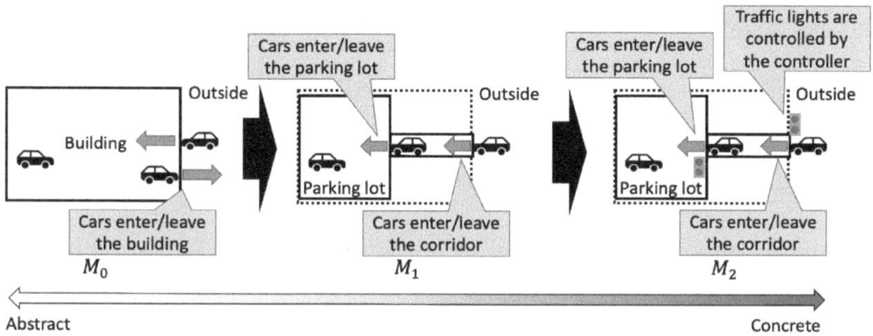

Figure 8.7: Modelling a parking lot system with stepwise refinement. M_0: an abstract model that abstracts away traffic lights and the internal structure of the building. M_1: an intermediate model that abstracts away traffic lights. M_2: the final (concrete) model.

Next, a more *concrete model* is newly created. For example, in M_1, traffic lights are still abstracted away, but the structure inside the building is introduced. The behaviour of vehicles also becomes more concrete; entering and leaving the corridor and the parking lot are modelled. To make the model consistent with the target system, again, although traffic lights are not modelled, vehicles entering the corridor find whether other vehicles are coming from the opposite side as if there were traffic lights.

The modeller repeats this process until the constructed model becomes concrete enough. For instance, in M_2, the behaviour and the requirements of traffic lights are modelled. In this model, the behaviour of vehicles in M_2 is simply following the traffic lights.

The modeller may go further, for example, by introducing sensors that detect vehicles and send signals to the controller as more concrete models $M_3, M_4, \ldots$.

This mechanism of stepwise refinement enables modellers to focus on limited aspects of the target system in each step and construct complicated models rigorously and incrementally.

Event-B has tool support for seamless integration of modelling and verification. The *Rodin platform* [3, 17], an extensible modelling environment for Event-B, gives various kinds of feedback about the correctness of models so that modellers can focus on modelling and "instructing themselves". Rodin has an on-the-fly mechanism of (1) generating proof obligations (POs) that serve as justifications of the system's correctness (2) attempting to prove the generated POs automatically. In case Rodin fails to prove some POs, the situation indicates that either the model has faulty parts or the POs were too complicated

```
machine M_A

variables v_A
invariants I_A(v_A)
events
  event e_A

    when G_{e_A}(v_A)
    then v_A :| T_{e_A}(v_A, v'_A)
  end
  ...
```

```
machine M_C
refines M_A
variables v_C
invariants I_C(v_A, v_C)
events
  event e_C
    refines e_A
    when G_{e_C}(v_C)
    then v_C :| T_{e_C}(v_C, v'_C)
  end
  ...
```

Figure 8.8: Generic abstract machine M_A and concrete machine M_C (simplified[4]).

for Rodin to prove automatically. The modeller can use an interface for *interactive proof* to investigate the details of proof attempt — that provides the reason for proof failure — and either find faulty parts of the model or manually finish the proof. In this manner, Rodin helps modellers in constructing models and doing reasoning of them.

Event-B has been applied to industrial projects. One of the most notable applications is the FP7 *Deploy project* [14, 29], where Event-B is used to construct models of automotive systems, railway systems, space systems, enterprise systems, and the instruction set architecture of a microprocessor. In addition, the application to *cyber-physical systems (CPSs)* has been studied. For instance, in the FP7 Advance project [4], Event-B is combined with methods for multi-simulation and testing to validate CPSs. Moreover, the language and the set of POs of Event-B are extended to model hybrid systems [16].

In Event-B, dynamic properties of a target system are specified as a component called *machine!in Event-B* (Fig. 8.8). An Event-B machine contains definitions of *variables* v, *invariants* $I(v)$, and *events* e. The *safety* of the target system (i.e., the property that should always hold) is specified as invariant predicates, and phenomena of state transitions are specified as events. A definition of event e contains *guard* predicates $G_e(v)$ that define when e is allowed to occur, and definitions of the state transition of e in terms of *before-after predicates* $T_e(v, v')$, which describes the relationship between the before-state v and the after-state v'. The *event-driven* description of behaviour is suitable for modelling systems, including the environment of software, which tend to have concurrent behaviour.

8.3.1.1 Example

Snippets of Event-B machines M_0 and M_1 for the parking lot system (Fig. 8.7) are shown in Listings 8.2 and 8.3, respectively.

[4]We focus on only basic constructs of Event-B models in this article, but more things are specified in general development in Event-B. For example, events are *parametric* and static properties of the target system are specified as components called *contexts*. See [2, 17] for details of the Event-B language.

Listing 8.2: M_0: the most abstract machine of the parking lot system.

```
machine M_0
variables n_bldg
invariants
  @SAF_bldg_capacity: n_bldg ≤ 10
event enter_bldg_m0
  when
    @not_full: n_bldg < 10
  then
    @enters: n_bldg :| n'_bldg = n_bldg + 1
end
event leave_bldg_m0
...
```

Listing 8.3: M_1: the intermediate machine of the parking lot system.

```
machine M_1 refines M_0
variables n_parked, n_entering, n_leaving
invariants
  @SAF_one_way: n_entering = 0 ∨ n_leaving = 0
  @n_bldg_gluing: n_bldg = n_parked + n_entering + n_leaving
event enter_bldg_m1 refines enter_bldg_m0
  when
    @not_full: n_parked + n_entering < 10
    @no_vehicles_leaving: n_leaving = 0
  then
    @enters: n_entering :| n'_entering = n_entering + 1
end
event enter_parking_m1
...
```

Due to the abstraction, M_0 has only one variable representing the number of vehicles in the building (n_{bldg}). The invariant `@SAF_bldg_capacity` states the safety condition about the number of vehicles in the building. This condition should be satisfied after every possible occurrence of the specified events. The event `enter_bldg_m0` is about a vehicle's entrance to the building that can happen only when the building has available slots (the guard `@not_full`). The number of vehicles in the building after the event (n'_{bldg} of the action `@enters`) is one more than that before the event (n_{bldg} of the action `@enters`). Because the guard `@not_full` of the event `enter_bldg` ensures that the event can only occur when the building is not full yet, the invariant `@SAF_bldg_capacity` is preserved by the state transition (`@enters`).

M_1 is declared as a refinement of M_0. In such a case, we call M_0 and M_1 an *abstract machine* and a *concrete machine*, respectively. Similarly, we use abstract/concrete variables, abstract/concrete events, etc. In M_1, there are variables of the number of vehicles entering, leaving, and being in the parking lot. The invariants' clause has a safety invariant `@SAF_one_way` and a special invariant (*gluing invariant*) `@n_bldg_gluing` that describes the relationship between an abstract variable (n_{bldg}) and concrete ones (n_{parked}, $n_{entering}$, and $n_{leaving}$), which is used for checking the consistency between M_0 and M_1. The event of entering the building (`enter_bldg_m1`) is rewritten using the new variables; guard predicates (`@not_full` and `@no_vehicles_leaving`) are more specific than `enter_bldg_m0`'s.

As a more concrete machine, a modeller may want to construct M_2' (Listing 8.4). In M_2', variables of traffic lights and invariants related to them are introduced. The one-way property is now expressed with the vocabulary of traffic lights (`@SAF_TL_one_way`), and the requirements of the traffic lights are specified in terms of the number of vehicles (`@SAF_light_outside`, `@SAF_light_inside`). Now that variables of traffic lights are introduced, how the vehicles judge whether they may enter can be specified; they simply see the colour of the traffic light (e.g., the guard `@light_outside_grn`).

Listing 8.4: M_2': the concrete machine of the parking lot system (inconsistent).

```
machine M2' refines M1
variables n_parked, n_entering, n_leaving, l_outside, l_inside
invariants
  @SAF_TL_one_way: l_outside = red ∨ l_inside = red
  @SAF_light_outside: l_outside = green ⟹ (n_leaving = 0 ∧ n_parked + n_entering < 10)
  @SAF_light_inside: l_inside = green ⟹ n_entering = 0
  ...
event enter_bldg_m2 refines enter_bldg_m1
  when
    @light_outside_grn: l_outside = green
  then
    @enters: n_entering :| n'_entering = n_entering + 1
end
...
```

Although M_2' may look fine at a glance, it is inconsistent. How can we find the inconsistency and the reason for it?

8.3.1.2 Checking Consistency of Event-B Machines

The primary consistency of a machine to be checked is *invariant preservation*, which states that a particular invariant holds after every occurrence of an event (every behaviour of the software and the environment) given that the invariant holds beforehand. Therefore, the proof obligation (PO) for invariant preservation (INV PO) is generated for every pair of an invariant predicate and an event. Specifically, for event e and invariant predicate i in machine M, the preservation of i by e (e/i/INV) [5] is:

[5] We present a simplified (and incomplete) version of the PO formula here. See [2,17] for the complete definition of INV PO, where predicates of axioms, the abstract invariant, and witnesses are also concerned.

$$I(v) \wedge G_e(v) \wedge T_e(v,v') \Longrightarrow i[v'/v](v'),$$

where $[v'/v]$ is the substitution of variables v by v'.

For instance, the PO `enter_bldg_m2/@SAF_light_outside`/INV generated for M_2' (Listing 8.4) is as follows:

$$
\begin{array}{ll}
(l_{outside} = green \Longrightarrow & \\
\quad (n_{leaving} = 0 \wedge n_{parked} + n_{entering} < 10)) \wedge & (\texttt{@SAF\_light\_outside}) \wedge \\
l_{outside} = green \wedge & (\texttt{@light\_outside\_grn}) \wedge \\
n'_{entering} = n_{entering} + 1 \wedge & (\texttt{@enters}) \wedge \\
l'_{outside} = l_{outside} \wedge & (\text{implicit}) \wedge \\
n'_{leaving} = n_{leaving} \wedge & (\text{implicit}) \wedge \\
n'_{parked} = n_{parked} \wedge & (\text{implicit}) \wedge \\
\ldots & \ldots \\
\Longrightarrow & \Longrightarrow \\
(l'_{outside} = green \Longrightarrow & (\texttt{@SAF\_light\_outside} \\
\quad (n'_{leaving} = 0 \wedge n'_{parked} + n'_{entering} < 10)). & \text{with variables substituted})
\end{array}
$$

The automatic prover of Rodin fails to prove this. In fact, we cannot prove this since $n_{entering} < 10 \wedge n'_{entering} = n_{entering} + 1 \Longrightarrow n'_{entering} < 10$ is not valid (counterexample: $n_{entering} = 9$). This feedback indicates that the traffic lights should turn red after the tenth vehicle enters to preserve the invariant `@SAF_light_outside`. Then, the modeller can replace the event with two events with a case splitting (Events `enter_bldg_m2_grn` and `enter_bldg_m2_red` in Listing 8.5) to resolve this.

Listing 8.5: M_2: the concrete machine of the parking lot system (modified, consistent).

```
...
event enter_bldg_m2_grn refines enter_bldg_m1
  when
    @light_outside_grn: l_outside = green
    @far_from_full: n_parked + n_entering < 9
  then
    @enters: n_entering :| n'_entering = n_entering + 1
end
event enter_bldg_m2_red refines enter_bldg_m1
  when
    @light_outside_grn: l_outside = green
    @almost_full: n_parked + n_entering = 9
  then
    @enters: n_entering :| n'_entering = n_entering + 1
    @turn_red: l_outside :| l'_outside = red
end
...
```

8.3.1.3 Checking Consistency of Refinement

Another primary consistency of Event-B machines is the consistency of refinement. As we demonstrated in Section 8.3.1.1, a machine created in an arbitrary manner can be declared as a refinement of another machine. Therefore, we should check the consistency between behaviour of the abstract machine and that of the concrete machine. Specifically, we need to check the consistency between the abstract machine's events and the concrete machine's events because events are units of the behaviour.

The consistency between an abstract event e_A in an abstract machine M_A and a concrete event e_C in a concrete machine M_C is defined with the *simulation relation*. The intuition of the simulation is that "e_C does not do what e_A does not do", i.e., it implies that the invariant preserved by e_A is also preserved by e_C. Therefore, if all events of a concrete machine M_C simulate corresponding events of the abstract machine M_A, then it is guaranteed that M_C carries over the invariant of M_A, and we do not need to check that M_C preserves the abstract invariant. In this manner, the refinement mechanism realizes the *divide-and-conquer* of verifying model consistency; we can focus on discussing abstract requirements in abstract machines and concrete requirements in concrete machines.

The primary proof obligations for verifying simulation are POs of *guard strengthening* (GRD) and *action simulation* (SIM). The guard strengthening PO states that the guard of a concrete event should imply the guard of the abstract event. The action simulation PO states that the state transition of the concrete event is more deterministic than or equivalent to that of the abstract event.

The GRD PO and the SIM PO are generated for every pair of an abstract event and a concrete event.[6] Specifically, for an event e_A of the abstract machine M_A and an event e_C of the concrete machine M_C, the guard strengthening (e_C/e_A/GRD) [7]is:

$$I_A(v_A) \wedge I_C(v_A, v_C) \wedge G_{e_C}(v_C) \Longrightarrow G_{e_A}(v_A),$$

and the action simulation (e_C/e_A/SIM) [7] is:

$$I_A(v_A) \wedge I_C(v_A, v_C) \wedge W(v_A, v'_A, v_C, v'_C) \wedge G_{e_C}(v_C) \wedge T_{e_C}(v_C, v'_C) \Longrightarrow T_{e_A}(v_A, v'_A),$$

where $W(v_A, v'_A, v_C, v'_C)$ is *witness*, which should be given by the modeller to specify how the state change by the concrete event can be seen as a state change in the abstract machine.

For example, the guard of `enter_bldg_m2_grn` of M_2 should be stronger than or equivalent to that of `enter_bldg_m1` of M_1. The PO for this requirement (`enter_bldg_m2_grn/enter_bldg_m1`/GRD) is as follows:

[6]To be precise, for dividing the proof obligations as much as possible, GRD is generated for each pair of *a guard predicate* of an abstract event and a concrete event. SIM is generated for each pair of *a before-after predicate* of an abstract event and a concrete event.

[7]We present a simplified (and incomplete) version of PO formulas here. See [2, 17] for details.

$(l_{outside} = green \Longrightarrow$	
$\quad (n_{leaving} = 0 \wedge n_{parked} + n_{entering} < 10)) \wedge$	(`@SAF_light_outside`) $\wedge$
$l_{outside} = green \wedge$	(`@light_outside_grn`) $\wedge$
$n_{parked} + n_{entering} < 9 \wedge$	(`@far_from_full`) $\wedge$
$\ldots$	$\ldots$
$\Longrightarrow$	$\Longrightarrow$
$n_{parked} + n_{entering} < 10 \wedge$	(`@not_full`) $\wedge$
$n_{leaving} = 0.$	(`@no_vehicles_leaving`)

Also, the state transition of `enter_bldg_m1` of M_1 should be more deterministic than or equivalent to that of `enter_bldg_m0` of M_0. With a witness $n'_{bldg} = n'_{parked} + n'_{entering} + n'_{leaving}$, the PO for this requirement (`enter_bldg_m1/enter_bldg_m0/SIM`) is as follows:

$n_{bldg} = n_{parked} + n_{entering} + n_{leaving} \wedge$	(`@n_bldg_gluing`) $\wedge$
$n'_{bldg} = n'_{parked} + n'_{entering} + n'_{leaving} \wedge$	(witness) $\wedge$
$n'_{entering} = n_{entering} + 1 \wedge$	(`@enters` of M_1) $\wedge$
$n'_{parked} = n_{parked} \wedge$	(implicit)
$n'_{leaving} = n_{leaving} \wedge$	(implicit)
$\ldots$	
$\Longrightarrow$	
$n'_{bldg} = n_{bldg} + 1.$	(`@enters` of M_0)

Those formulas are valid.

8.3.2 Automatic Refinement Refactoring

Software *maintenance* — involving removing faults, improving artefacts, and coping with changes in the environment or requirements — is one of the primary software development activities. It is also known to be one of the most costly activities [24]. Therefore, improving the *maintainability* of software artefacts during and after the development is essential.

Another essential quality of software artefacts is *reusability*; reusing software artefacts for reducing the cost and faults is one of the central goals of software engineering.

In developing formal models, improving maintainability and reusability is particularly crucial because formal models are complex and costly to develop. For instance, according to reports of industrial application of Event-B, modellers of an automotive system [29, Chapter 3] and a satellite system [29, Chapter 5] had to evolve models and deal with changes in requirements after constructing the initial version; improving maintainability is crucial in such situations. Moreover, development reports about systems of automotives, satellites, and trains [29, Chapter 7] mention the importance and difficulty of reusing constructed Event-B models.

In the rest of this section, we introduce our *refinement refactoring* method [22] for improving the maintainability and reusability of Event-B machines.

8.3.2.1 Refinement Strategy

The refinement mechanism described in Section 8.3.1 supports gradually adding aspects of the target system and is thus suitable for constructing comprehensive models of complex target systems.

For every sequence of machines constructed with the refinement mechanism, there is the *refinement strategy*, that is, the order of introducing elements of target systems. Formally, we define the refinement strategy as follows:

Definition 8.1 (refinement chain) A *refinement chain* $(M_i)_{i\in\{0,\dots,n\}}$ is a sequence of machines such that M_{j+1} refines M_j for every $j \in \{0,\dots,n-1\}$.

Definition 8.2 (refinement strategy) For a refinement chain $c = (M_i)_{i\in\{0,\dots,n\}}$, the refinement strategy of c is $\mathrm{str}(c) = (\mathrm{var}(M_{j+1}) \setminus \mathrm{var}(M_j))_{j\in\{0,\dots,n-1\}}$, where $\mathrm{var}(M)$ is the set of variables of the machine M.

For instance, $\mathrm{str}((M_0, M_1, M_2)) =$ $(\{n_{bldg}\}, \{n_{parked}, n_{entering}, n_{leaving}\}, \{l_{outside}, l_{inside}\})$.

8.3.2.2 Refinement Refactoring

The refinement mechanism is quite flexible; there may be various refinement strategies for constructing a single machine with refinement. Because a refinement strategy reflects the elements of the target system focused in each step, different refinement strategies result in different machines.

For instance, assume that we follow another refinement strategy $(\{n_{bldg}\}, \{n_{parked}, n_{entering}, n_{leaving}, l_{inside}\}, \{l_{outside}\})$ for constructing M_2. In a refinement chain with this strategy, the intermediate machine is an abstraction of M_2 that abstracts away the traffic light outside. A possible refinement chain with the strategy can be (M_0, M_1', M_2), where the new intermediate machine M_1' is as shown in Listing 8.6. The difference from M_2 is that the guard `@light_inside_red` is specified in M_1' instead of `@light_outside_grn` of M_2. Specifying the guard `@light_inside_red` is reasonable because the event is about the case where the light outside is green (`@light_outside_grn` of `enter_bldg_m2_grn`), and at least one of the lights must be red (`@SAF_TL_one_way` of M_2).

Listing 8.6: M_1': another intermediate machine of the parking lot system constructed by following the refinement strategy $(\{n_{bldg}\}, \{n_{parked}, n_{entering}, n_{leaving}, l_{inside}\}, \{l_{outside}\})$.

```
machine M1' refines M0
variables n_parked, n_entering, n_leaving, l_inside
invariants
  @SAF_light_inside: l_inside = green ⟹ n_entering = 0
  ...
event enter_bldg_m1'_grn refines enter_bldg_m0
  when
    @light_inside_red: l_inside ≠ green
    @far_from_full: n_parked + n_entering < 9
    @no_vehicles_leaving: n_leaving = 0
  then
    @enters: n_entering :| n'_entering = n_entering + 1
end
...
```

The event `enter_bldg_m1'_grn` of M_1' preserves the safety invariant `@SAF_light_inside` because of the guard `@light_inside_red`; since the event is about situations where the light inside is red, the safety invariant (about the case where the light inside is green) is irrelevant for this event.

In this example, from a refinement chain (M_0, M_1, M_2), we constructed a new refinement chain (M_0, M_1', M_2) that has a different refinement strategy. We call such a process *refinement refactoring*.

Refinement refactoring is effective for improving the understandability, maintainability, and reusability of existing machines. We are often interested in a limited set of aspects of the target system when trying to understand and maintain a machine. In that case, the refinement refactoring gives us an intermediate machine written only with variables of the aspects of interest. Moreover, the refinement refactoring enables us to extract the reusable part of the target system as an intermediate machine, which can be used as an abstract machine of a new machine. Thus, refinement refactoring facilitates flexible uses of existing formal system models, which tend to be difficult to understand, maintain, and reuse.

The core of our refinement refactoring method is *splitting* a refinement, that is, for a given refinement chain (M_A, M_C), constructing the intermediate machine M_B that forms a new refinement chain (M_A, M_B, M_C).

The main challenge in splitting refinement is that PO formulas generated for M_B should be valid; in concrete, invariants in M_B should be preserved, M_B should refine M_A, and M_B should be refined by M_C. For example, in the event `enter_bldg_m1'_grn` of M_1' (Listing 8.6), the new guard `@light_inside_red` is necessary for preserving the invariant `@SAF_light_inside`. However, unlike other predicates of the event, the predicate of the guard `@light_inside_red` is not specified in the original concrete machine (M_2). Therefore, we need to construct this additional predicate for consistency.

We found that *Craig interpolation* [11], defined in the following theorem, is related to such additional predicates.

Theorem 8.1 (Craig's interpolation theorem [11])
For every valid formula with the form $P \Longrightarrow Q$, there exists a formula I (an interpolant *of $P \Longrightarrow Q$) such that*

- $P \Longrightarrow I$,
- $I \Longrightarrow Q$, *and*
- $\mathrm{sym}(I) \subseteq \mathrm{sym}(P) \cap \mathrm{sym}(Q)$, *where* $\mathrm{sym}(\phi)$ *is the set of non-logical symbols that appear in the formula* ϕ.

The intuition of the interpolant I of the formula $P \Longrightarrow Q$ is that I describes why $P \Longrightarrow Q$ is valid in the vocabulary common in P and Q.

In our automatic refinement splitting method, we obtain additional predicates specified in the intermediate machine as Craig's interpolants of predicates constructed from POs as follows:

1. We transform the formula of a PO ($\phi \Longrightarrow \psi$) into another formula $\phi' \Longrightarrow \psi'$ such that $(\phi \Longrightarrow \psi) \Longleftrightarrow (\phi' \Longrightarrow \psi')$ and $\mathrm{sym}(\phi') \cap \mathrm{sym}(\psi')$ is a subset of variables of the intermediate machine.
2. We obtain an additional predicate ρ that is necessary to prove the PO as an interpolant of $\phi' \Longrightarrow \psi'$.

The intuition of ρ is a description of the reason why the PO formula is valid in the vocabulary available for the intermediate machine.

For example, the formula of the PO `enter_bldg_m2_grn/@SAF_light_inside/INV` of M_2 is:

$(l_{outside} = red \vee l_{inside} = red) \wedge$	(`@SAF_TL_one_way`) $\wedge$
$(l_{inside} = green \Longrightarrow (n_{entering} = 0)) \wedge$	(`@SAF_light_inside`) $\wedge$
$l_{outside} = green \wedge$	(`@light_outside_green`) $\wedge$
$n'_{entering} = n_{entering} + 1 \wedge$	(`@enters` of M_2) $\wedge$
$l'_{inside} = l_{inside}$	(implicit) $\wedge$
...	...
$\Longrightarrow$	$\Longrightarrow$
$(l'_{inside} = green \Longrightarrow (n'_{entering} = 0))$.	(`@SAF_light_inside` with variables substituted)

This can be transformed into the following equivalent formula:

$$
\begin{aligned}
&(l_{outside} = red \vee l_{inside} = red) \wedge \\
&(l_{inside} = green \Longrightarrow (n_{entering} = 0)) \wedge \\
&l_{outside} = green \wedge \\
&\ldots \\
&\Longrightarrow \\
&(l'_{inside} = green \Longrightarrow (n'_{entering} = 0)) \vee \\
&\neg\, (l'_{inside} = l_{inside}) \vee \\
&\neg\, (n'_{entering} = n_{entering} + 1).
\end{aligned}
\tag{8.1}
$$

The premise part and the consequence part of Equation 8.1 have common variables and constants $\{l_{inside}, green, n_{entering}\}$, which can be specified in M'_1. We can obtain the predicate of the guard `@light_inside_red` ($l_{inside} \neq green$) as an interpolant of Equation 8.1.

The refinement refactoring method is implemented as a plug-in[8] of the Rodin platform.

8.3.3 Design Exploration of Refinement

As we described earlier, in refinement in Event-B, a modeller constructs two arbitrary models and checks whether they are in a refinement relationship (*posit and prove* style). Therefore, this refinement mechanism has a large flexibility (compared with e.g., generating the concrete model by converting the abstract model with a rule-based converter).

However, the flexibility results in a large space of refinement strategies and makes the strategy's decision difficult. Despite the importance of this problem, most existing studies provide only domain-specific guidelines and examples of good refinement design; they lack explicit discussions of how to evaluate refinement designs.

Which traits do good refinement strategies have? We presume that *good refinement strategies reduce complexities well* because the refinement mechanism is introduced to reduce complexities. In particular, we focus on the complexity of machines themselves and the complexity of proofs.

The rest of this section summarises[9] our analysis of the complexities of various refinement strategies using our automatic refinement refactoring method. Specifically, we compare different refinement strategies for constructing the same most concrete machine.

[8] Available at `http://research.nii.ac.jp/slicenmerge/`

[9] See our paper [21] for details of the analysis.

8.3.3.1 Evaluation Settings

For evaluating the machine complexity, we counted each machine's numbers of model constructs (variables, invariants, and events) in refinement chains. We assume that the fewer newly introduced constructs, the less complex the machine is because the modeller can focus on less elements in the machine. Therefore, we consider that good refinement strategies distribute the introduction of constructs well.

For evaluating the proof complexity, we counted the numbers of POs generated for each machine. We also counted the numbers of POs that were not automatically proved by Rodin's standard automatic prover because the numbers reflect the amount of the burden of manual proofs.

We used a machine of a train system [2, Chapter 17] (M_{Tr}) and a machine of an autonomous satellite flight formation system [33] (M_{Sat}) constructed by expert Event-B modellers as materials of the most concrete machines. In other words, we investigated strategies of refinement chains with the form $(\ldots, M_{Tr})$ and $(\ldots, M_{Sat})$.

We focused on two characteristics of refinement strategies, which are *granularity* and *order*. For instance, a strategy $(\{a,b,c\})$ that has only one step of introducing three variables and another strategy $(\{a\},\{b\},\{c\})$ that is about introducing a variable every step are different in granularity. Strategies $(\{a\},\{b\},\{c\})$ and $(\{c\},\{b\},\{a\})$ are different in order. In the evaluation, we changed the granularity and the order of existing refinement strategies and compared the results. Specifically, we conducted the evaluation with four refactoring operations: *merging*, *splitting*, *swapping*, and *reversing* some steps of the original strategy.

8.3.3.2 Findings

From the results, we found the following best practices of refinement structure designs:

Finding 1. Using fine-grained refinement strategies is effective for reducing complexity in general. Decomposing refinement reduced the complexities per refinement step. At the same time, it increased the sum of the number of POs generated for all machines in refinement chains. The increase is due to (1) additional refinement POs (e.g., GRD and SIM) generated for new refinement steps and (2) additional POs generated for interpolant predicates. The additional POs of both kinds were simple enough for automatic provers; it should be because new refinement steps divide proofs into a larger number of simpler proofs, and interpolant predicates work as lemmas that divide proofs too.

Changing the order of introducing variables may change the variance of the number of introduced invariants and generated POs. This is because of dependencies between invariants and variables. For instance, suppose that the most concrete machine has variables v, w and invariants $i(v), j(v,w)$. By following the refinement strategy $(\{v\},\{w\})$, we can distribute the introduction of invariants because we can introduce $i(v)$ in the first step and $j(v,w)$ in the second step. On the contrary, if we follow another strategy $(\{w\},\{v\})$, we need to introduce both in the second step because we cannot specify either $i(v)$ or $j(v,w)$ in the first step since both need the variable v. We found that introducing "key" variables used in many invariants in early steps distributes invariants and POs well (**Finding 2**).

References

[1] Abrial, J-R. 2009. Faultless systems: Yes we can! Computer, 42(9): 30–36.

[2] Abrial, J-R. 2010. Modeling in Event-B: System and software engineering. Cambridge University Press.

[3] Abrial, J-R., Butler, M., Hallerstede S., Hoang, T.S., Mehta, F. and Voisin. L. 2010. Rodin: An open toolset for modelling and reasoning in Event-B. International Journal on Software Tools for Technology Transfer, 12(6): 447–466.

[4] Advance Project. FP7 Advance project. https://web.archive.org/web/20191125075302/http://www.advance-ict.eu/index.html.

[5] Ahmad, E., Larson, B.R., Barrett, S.C., Zhan, N. and Dong, Y. 2014. Hybrid annex: An AADL extension for continuous behavior and cyber-physical interaction modeling. *In*: ACM SIGAda Ada Letters, volume 34, pages 29–38. ACM.

[6] Bettini, L. 2016. Implementing domain-specific languages with Xtext and Xtend. Packt Publishing Ltd.

[7] Black, D.C., Donovan, J., Bunton, B. and Keist, A. 2010. System C: From the Ground Up. Springer, Secaucus, NJ, USA.

[8] Inquiry Board. 1996. ARIANE 5 flight 501 failure report. https://hal.inria.fr/file/index/docid/73613/filename/RR-3079.pdf.

[9] Broman, D., Lee, E.A., Tripakis, S. and Törngren, M. 2012. Viewpoints, formalisms, languages, and tools for cyber-physical systems. *In*: Proceeding of the 6th International Workshop on Multi-Paradigm Modeling, MPM '12, page 49–54, New York, NY, USA, 2012. Association for Computing Machinery.

[10] Campagne, F. 2014. The MPS language workbench: volume I, Fabien Campagne.

[11] Craig, W. 1957. Three uses of the Herbrand-Gentzen theorem in relating model theory and proof theory. The Journal of Symbolic Logic, 22(3): 269–285.

[12] De Moura, L. and Bjørner, N. 2008. Z3: An Efficient SMT Solver. *In:* 14th International Conference on Tools and Algorithms for the Construction and Analysis of Systems, volume 4963 of Lecture Notes in Computer Science, pages 337–340. Springer.

[13] Delange, J. and Feiler, P. 2014. Architecture fault modeling with the AADL error-model annex. *In*: 2014 40th EUROMICRO Conference on Software Engineering and Advanced Applications, pages 361–368. IEEE.

[14] Deploy Project. FP7 Deploy project. http://www.deploy-project.eu/.

[15] Dhouib, S., Kchir, S., Stinckwich, S., Ziadi, T. and Ziane, M. 2012. Robotml, a domain-specific language to design, simulate and deploy robotic applications. *In*: International Conference on Simulation, Modeling, and Programming for Autonomous Robots, pages 149–160. Springer.

[16] Guillaume, D., Yamine, A.-A., Neeraj, K.S. and Marc, P. Event-B hybridation: A proof and refinement-based framework for modelling hybrid systems. ACM Transactions on Embedded Computing Systems, 20(4), May 2021.

[17] Jastram, M. (editor). Rodin User's Handbook. https://stups. hhu-hosting. de/handbook/rodin/current/html/.

[18] Fowler, M. 2005. Language workbenches: The killer-app for domain specific languages. https://www.martinfowler.com/articles/languageWorkbench. html.

[19] Bedin França, R., Rolland, J-F., Filali Amine, M., Bodeveix, J-P. and David Chemouil, D. 2007. Assessment of the AADL Behavioral Annex. Journées FAC, page 13.

[20] Klikovits, S. and Didier Buchs, D. 2021. Pragmatic reuse for DSML development. Software and Systems Modeling (SoSyM), 20: 837–866.

[21] Kobayashi, T. and Ishikawa, F. 2018. Analysis on strategies of superposition refinement of Event-B specifications. *In*: Jing Sun and Meng Sun, (eds). Formal Methods and Software Engineering, pages 357–372, Cham, 2018. Springer International Publishing.

[22] Kobayashi, T., Ishikawa, F. and Honiden. S. 2019. Consistency preserving refactoring of refinement structures in Event-B models. Formal Aspects of Computing, 31(3): 287–320, June 2019.

[23] Le Lann, G. 1997. An analysis of the Ariane 5 flight 501 failure—a system engineering perspective. *In*: Proceeding of the 1997 International Conference on Engineering of Computer-Based Systems, ECBS'97, page 339–346, USA. IEEE Computer Society.

[24] Lientz, B.P., Swanson, E.B. and Tompkins, G.E. 1978. Characteristics of application software maintenance. Communications of the ACM, 21(6): 466–471, June 1978.

[25] Object Management Group. Business Process Model and Notation (BPMN) Version 2.0, 2011. OMG Document Number: formal-2011-01-03.

[26] Object Management Group. UML Profile for MARTE: Modeling and Analysis of Real-Time Embedded Systems (OMG MARTE) Version 1.1, 2011. OMG Document Number: formal/11-06-02.

[27] Object Management Group. Object Management Group: Unified Modeling Language (UML) Version 2.5.1, 2017. OMG Document Number: formal/17-12-05.

[28] Object Management Group. OMG Systems Modeling Language (OMG SysML) Version 1.5, 2017. OMG Document Number: formal-2017-05-01.

[29] Romanovsky, A. and Martyn Thomas, M. 2013. Industrial deployment of system engineering methods. Springer.

[30] Society of Automotive Engineers. 2017. Society of Automotive Engineers: Architecture Analysis and Design Language (SAE AADL) Version 2.2. SAE Standard: AS5506C.

[31] Ian Sommerville, I. 2007. Software Engineering. Addison-Wesley.

[32] Steinberg, D., Budinsky, F., Merks, E. and Paternostro, M. 2008. EMF: Eclipse modeling framework. Pearson Education, 2008.

[33] Tarasyuk, A., Pereverzeva, I., Troubitsyna, E. and Latvala, T. 2015. The formal derivation of mode logic for autonomous satellite flight formation. *In*: SAFECOMP 2015, volume 9337, pages 29–43, Heidelberg, Springer.

[34] Tolvanen, J-P. and Rossi, M. 2003. Metaedit+: Defining and using domain-specific modeling languages and code generators. *In*: Companion of the 18th annual ACM SIGPLAN Conference on Object-oriented Programming, systems, Languages, and Applications, pages 92–93. ACM.

[35] van Deursen, A., Klint, P. and Visser, J. 2000. Domain-specific languages: An annotated bibliography. ACM SIGPLAN Notices, 35(6): 26–36, June 2000.

[36] Zave, P. and Jackson, M. 1997. Four dark corners of requirements engineering. ACM Transactions on Software Engineering and Methodology, 6(1): 1–30, January 1997.

Chapter 9

Theorem Proving at Work

Tsutomu Kobayashi
Japan Aerospace Exploration Agency (JAXA), Tsukuba, Japan

Stefan Klikovits
Johannes Kepler University, Linz, Austria

Rose Bohrer
Worcester Polytechnic Institute, Worcester, USA

Akihisa Yamada
National Institute of Advanced Industiral Science (AIST), Tokyo, Japan

Ichiro Hasuo
National Institute of Informatics and SOKENDAI, Tokyo, Japan

9.1 Introduction: Theorem Proving for Safety Assurance

Automotive systems are complex systems. They can exhibit unexpected behaviours, and those unexpected behaviours are often unsafe and can endanger human lives. There are a number of mechanical and control-theoretic methods for automotive safety accumulated over time. However, these traditional measures are rarely enough for modern automotive systems that are heavily *software-controlled*.

Therefore attention is growing towards *safety assurance methods for software and ICT systems* applied to automotive systems. The current chapter discusses one of the major families of those methods, namely *theorem proving*.

Before introducing theorem proving, let us discuss its position in the broader landscape of safety assurance methods for software. See Fig. 9.1.

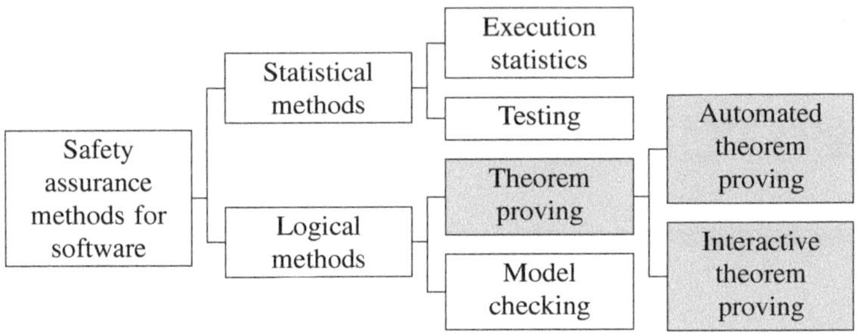

Figure 9.1: Safety assurance methods for software.

The first classification is into *statistical* and *logical* methods. In statistical methods, the target system is analysed by collecting its individual executions and summarising their characteristics. For example, if a piece of software has been executing for 30 years without a single dangerous behaviour, this fact gives confidence to the safety of the software. When we say *testing*, it usually means that the input data to the target system is carefully chosen for the purpose of safety assurance, and that the execution statistics is collected using those input data (called *test inputs*). A common criterion for the choice of test inputs is *coverage* (how big a variety of system behaviours are covered). When an error is rare, one can see testing as a *search* problem for a test input that causes an error. This view is called *search-based testing*.

Statistical methods for safety assurance are predominant in industry practise. However, they also come with limitations. One such limitation is the *absence of absolute guarantees*. Guarantees by statistical methods are always *empirical*: "from what we have observed so far, it is likely that the system is safe". Moreover, such guarantees often come with threshold values, such as ε in "the error rate was less than ε so the system is deemed safe". The validity of such thresholds ("is ε small enough?") is an issue that is always debatable.

Another major limitation of statistical methods is their *explainability*. It is usually very hard to extract, from the execution of a statistical method (e.g., a test result), a structured argument that explains why the system should be safe. This is a big problem in the automotive domain where *absolute safety* is usually not realisable—cars do crash in the real world, often for the reason that the ego vehicle cannot help (e.g., a DUI vehicle running into the ego vehicle). Facing such an unavoidable accident, one's goal will be to effectively argue that the accident is not their fault. This argument must be logically structured and most statistical methods are incapable of generating it.

This turns us to *logical methods* where we use logic to give mathematical proofs to safety claims. A mathematical proof can be seen as an ultimate correctness guarantee—following well-established logical reasoning steps, proofs can never go wrong. A sequence of reasoning steps in a safety proof also serves as a good *explanation* why its conclusion (namely safety) should be true. Moreover, proofs can explicate the *assumptions* they rely on. This is a very important feature in the automotive domain where most safety claims are conditional. In other words, safety assurance by proofs can explicate *operational design domains (ODDs)* in which safety is ensured.

Logical methods are commonly classified to *theorem proving* and *model checking*. Theorem proving is much like writing proofs as we imagine—formulating rigorous statements and explicating logical steps that derive them. Writing such proofs can be automated (algorithmic search) or manual, as we discuss later. Model checking refers to a specific way of automating proof search, namely via (1) translation of the original problem into some graph-like formalism (typically some *automata*) and (2) applying a graph algorithm to conduct search.

There has been a large body of literature about upgrading both approaches from conventional software to *cyber-physical systems* such as automotive systems. Model checking emphasises automation by graphs, and these graphs must be finite for graph algorithms to operate on. This poses a big challenge when continuous physical quantities come into the picture (which commonly lead to an infinite state space). In contrast, theorem proving can easily accommodate continuous values, although it can come with penalties such as greater computational costs or human efforts. Theorem proving has the advantage of explicating logical steps, too, for better explainability. Therefore, in this chapter, we focus on theorem proving.

Theorem proving can be further classified to *automated theorem proving* and *interactive theorem proving*, as shown in Fig. 9.1. In this chapter, we will discuss a few approaches and tools from both families, with an eye on accommodation of continuous dynamics and iterative model refinement (both are important for automotive applications). These approaches are summarised in Table 9.1.

Table 9.1: Theorem proving approaches and tools discussed in the chapter

Approach/tool	Expressivity	Automation	Proof construction
Isabelle/HOL	higher-order logic	little (mostly interactive)	bottom-up
SAT solvers	propositional formulas	fully automatic, complete	—
SMT solvers	first-order formulas	fully automatic, incomplete	—
KeYmaera X	hybrid systems	partly automated	bottom-up
Event-B	event-driven systems	partly automated	top-down

9.1.1 Interactive Theorem Proving

In interactive theorem proving, a proof is written by a human prover. The difference from the usual pen-and-paper mathematical proofs is that they are written in a piece of software called a *proof assistant*. A proof assistant can help a human prover with many functionalities, the most important of which being *proof check* (checking if each logical reasoning step in a proof is valid). For example, checking the validity of a 1,000-page pen-and-paper proof is unrealistic; in contrast, a big proof in a proof assistant can be validated mechanically.

Therefore, interactive theorem proving is really about *manual* theorem proving. It is called interactive since proofs are written in the interaction between a human prover and a proof assistant.

In this chapter, after reviewing the landscape of different proof assistants (Section 9.3.1), we review a well-known proof assistant called Isabelle/HOL (Section 9.3.2).

9.1.2 Automated Theorem Proving

Interactive theorem proving is costly—it is not uncommon that a single proof takes months or years to complete. Therefore *automation* of proof search has been actively sought. There are some no-go results for automated theorem proving—the most famous one being *Gödel's incompleteness theorem*—and in general proof search is an expensive computational task. Nevertheless, the rise of *SAT* and *SMT solvers* is making automated theorem proving an attractive approach to many real-world problems.

SAT solvers prove propositional formulas (cf. Section 9.2) and they are *complete* (they can prove every formula that is semantically true). While the worst-case complexity is not good (NP-complete), it has been observed that modern SAT solvers can efficiently handle big formulas in real-world problems. We review SAT solvers in Section 9.4.2.

SMT solvers prove first-order formulas (cf. Section 9.2) and they are usually *incomplete* (they may not prove every formula that is semantically true). Still, their enhanced expressivity—they can prove formulas about real numbers, for example—is a great advantage and it has opened up many new applications. SMT solvers are reviewed in Section 9.4.3.

9.1.3 Accommodation of Continuous and Hybrid Dynamics

Accommodation of continuous and hybrid dynamics—where *hybrid* means the combination of discrete/digital and continuous/physical—may not sound easy for theorem proving approaches, since the latter are traditionally developed for safety assurance of (pure) *software*. However, the study of *hybrid systems theorem-proving* initiated in 2000s has identified many commonalities in software science and *control theory*—a scientific field that traditionally studies continuous dynamics. Thanks to its findings, we now see multiple approaches and tools that can efficiently prove theorems about continuous or hybrid dynamics.

In Section 9.5.1, we review one such tool called KeYmaera X. It is an interactive proof assistant in principle, but comes with many automation techniques, especially around the use of symbolic computer algebra engines.

9.1.4 Model Refinement

A big burden for theorem proving in practice is the problem of *modelling*. In mathematics, to state a theorem and prove it, one has to *define* all the notions used in the theorem. In safety assurance of automotive and other systems by theorem proving, therefore, one has to *define* what the target system is, describing all its behaviours and the principles behind them. This amounts to mathematical *modelling* of the target system.

Such modelling is especially hard for automotive systems: (1) they have complex physical components; (2) they typically come with third-party components whose internal working is not known; (3) modern ADAS and ADVs include machine-learning components (such as neural networks for object recognition) whose behaviours are hard to define; and so on. Note that Simulink models—"mathematical models" commonly used in automotive industry—are usually not fit for theorem proving. Simulink models, and also other *digital twins*, are principally constructed for the purpose of numeric simulation and not for the logical analysis in theorem proving.

As a countermeasure to this problem of modelling, in Section 9.6, we review a modelling and theorem-proving framework called Event-B. Its most notable feature is *model refinement*: one can start with a coarse-grained model of the target system and prove its safety; for stronger assurance, one can refine the model to include more details about the system and prove its safety. Moreover, the safety proof for the refined model can be done in an *incremental* manner, in the sense that the safety proofs for previous (coarser-grained) models play the role of key lemmas.

A mathematical proof obtained this way can be seen as a *top-down* one (cf. Table 9.1). The conventional direction of constructing proofs is *bottom-up*, in that unquestionable facts get composed to derive more complex facts, all the way towards the ultimate claim. In Event-B, in contrast, one *decomposes* the ultimate claim gradually into simpler claims, using increasingly detailed system models.

The problem of modelling is a general one in the application of software science methods to automotive systems. Model refinement by Event-B is one answer. Another answer, in the specific problem of safety of automated driving vehicles, is the methodology of RSS [53]. It addresses the problem of modelling by letting RSS rules cut a clear line between what should be modelled and what should not. See Section 7.5, and also recent papers [19, 32, 33].

9.2 Formal Logic

In this section, we review the common foundation of the theorem proving approaches we describe later, namely *formal logic*.

To ensure some correctness property of a system, an ultimate means is a mathematical proof. It is the record of a series of reasoning steps from the assumptions (such as system requirements, a system model, environmental contracts, etc.) all the way to the conclusion (namely the desired correctness property). Being a mathematical proof, there should not be any gap in the course of these reasoning steps, giving a very strong degree of guarantee. A mathematical proof offers a high level of *explainability*, too, explicating *why* the correctness property should hold in the course of its reasoning steps.

Such use of mathematical proofs, however, is easier said than done. The most common form of mathematical proofs are the pen-and-paper one. Checking the validity of such pen-and-paper "proof candidates" incurs the following challenges.

- Written in a natural language, the check must rely on human efforts. But human checkers—much like human provers—are never free from mistakes and oversights.

- Mathematical proofs of system correctness are very often of the least interesting and the most tedious kind. They are typically painfully long. Moreover, their technical substance is not so much in enlightening mathematical insights as in massive case distinctions. Humans easily get bored of such unexciting proofs, leading to mistakes.

Formal logic, which is almost synonymous with *symbolic logic*, is a scientific discipline that offers a solution. Formal logic provides a formal language in which mathematical proofs are written; more specifically, this formal language defines: (1) what symbolic expressions can be used for proofs, and (2) what symbolic *proof rules*, that tell which reasoning steps are valid, can be used. In this way, formal logic enables *mechanical* and *automated solution* of our problem of checking the validity of proof candidates: symbolic expressions can be processed by software; and the validity of proof candidates, amounting to the compliance with symbolic proof rules, can be effectively checked by an algorithm.

In the rest of Section 9.2, we describe two basic examples of such formal languages for writing proofs, namely *propositional logic* and *first-order logic*. Our exposition is brief and its emphasis is more on intuition than on mathematical precision. Interested readers are referred to classic textbooks such as [55, 57] as well as online lecture notes such as [31].

9.2.1 Propositional Logic

Our first example is *propositional logic*, one of the most basic logical languages that is about what true statements can be formed using common logical connectives such as $\wedge$ ("and"), $\vee$ ("or"), $\rightarrow$ ("implies"), and $\neg$ ("not").

The first component of propositional logic is a formal syntax in which statements of our interest—those which are composed using the aforementioned logical connectives and the truth of which we want to prove—are expressed. These statements, expressed symbolically, are called *(well-formed) formulas*.

Here is a formal definition. We fix a set of *atomic propositions*; its elements are denoted by $p, q, \ldots$ in what follows. The set of *propositional formulas* is defined by the following *BNF notation*:

$$\phi \ ::= \ p \mid \top \mid \bot \mid \neg\phi \mid \phi \wedge \phi \mid \phi \vee \phi \mid \phi \rightarrow \phi \tag{9.1}$$

where p ranges over atomic propositions. BNF notations such as (9.1) describe inductive definitions of syntax: the one in (9.1), for example, can be read as (1) each atomic proposition p is a propositional formula, (2) $\top$ and $\bot$ are both propositional formulas, and (3) combinations of propositional formulas with the connectives $\wedge, \vee, \rightarrow, \neg$ are also propositional formulas. Although $\phi \wedge \phi$ in (9.1) uses the same symbol ϕ twice, they do not have to be instantiated by the same propositional formula. See the example below.

Example 31 *Let p and q be atomic propositions. Then p and q are propositional formulas, too, by the first case of (9.1). By the last case of (9.1), $p \rightarrow q$ is a propositional formula, too; so is $p \wedge (p \rightarrow q)$ by the fifth case. Using the last case again, $(p \wedge (p \rightarrow q)) \rightarrow q$ is a propositional formula.*

The second component of propositional logic is a set of symbolic proof rules that one can use to derive propositional formulas in reasoning steps inside a proof. There are different styles of proof rules: the Hilbert style, natural deduction, sequent calculus, etc. The one presented here follows the natural deduction style. See [55].

Our proof rules for propositional logic derive *judgments*

$$\phi_1, \ldots, \phi_n \vdash \psi \tag{9.2}$$

instead of propositional formulas themselves. This use of judgments is for syntactic convenience. In a judgment such as (9.2), the sequence $\phi_1, \ldots, \phi_n$ of *assumption* propositional formulas is followed by a *consequent* propositional formula ψ, delimited by a special symbol $\vdash$. Carrying assumptions around explicitly, in this manner, is useful when we formulate proof rules. In what follows, sequences $\phi_1,\ldots,\phi_n$ of propositional formulas are often denoted by Γ, Δ, etc.

Let us now introduce *proof rules* for propositional logic. A proof rule consists of zero or more judgments $\Gamma_1 \vdash \phi_1,\ldots,\Gamma_n \vdash \phi_n$ called *premises*, and a judgment $\Delta \vdash \psi$ called the *conclusion*, and is often written in the following form, where NAME is the name of the rule:

$$\frac{\Gamma_1 \vdash \phi_1 \quad \ldots \quad \Gamma_n \vdash \phi_n}{\Delta \vdash \psi}\ \textsc{Name}$$

The rule is such that if all premise judgments $\Gamma_1 \vdash \phi_1,\ldots,\Gamma_n \vdash \phi_n$ have been already derived, then one can further derive the conclusion judgment $\Delta \vdash \psi$. Table 9.2 shows some of the proof rules of propositional logic.

Table 9.2: Some proof rules for propositional logic

Proof rule	Informal description
$\dfrac{}{\Gamma,\phi,\Delta \vdash \phi}\ \textsc{As}$	Assuming ϕ, one can infer ϕ
$\dfrac{}{\Gamma \vdash \top}\ \top\mathrm{I}$	One can always infer $\top$ ($\top$-introduction)
$\dfrac{\Gamma \vdash \bot}{\Gamma \vdash \phi}\ \bot\mathrm{E}$	If one has contradiction ($\bot$), then one can infer anything (ϕ) ($\bot$-elimination)
$\dfrac{\Gamma \vdash \phi \quad \Gamma \vdash \psi}{\Gamma \vdash \phi \wedge \psi}\ \wedge\mathrm{I}$	If one has both ϕ and ψ, then one can infer $\phi \wedge \psi$
$\dfrac{\Gamma \vdash \phi \wedge \psi}{\Gamma \vdash \phi}\ \wedge\mathrm{E1} \qquad \dfrac{\Gamma \vdash \phi \wedge \psi}{\Gamma \vdash \psi}\ \wedge\mathrm{E2}$	If one has $\phi \wedge \psi$, then one can infer both ϕ and ψ
$\dfrac{\Gamma,\phi \vdash \psi}{\Gamma \vdash \phi \to \psi}\ \to\mathrm{I}$	If assuming ϕ one has ψ, then on can infer $\phi \to \psi$
$\dfrac{\Gamma \vdash \phi \quad \Gamma \vdash \phi \to \psi}{\Gamma \vdash \psi}\ \to\mathrm{E}$	If one has both ϕ and $\phi \to \psi$, then one can infer ψ
$\dfrac{\Gamma \vdash \phi}{\Gamma \vdash \phi \vee \psi}\ \vee\mathrm{I1} \qquad \dfrac{\Gamma \vdash \psi}{\Gamma \vdash \phi \vee \psi}\ \vee\mathrm{I2}$	If one has either ϕ or ψ, then one can infer $\phi \vee \psi$
$\dfrac{\Gamma \vdash \phi \vee \psi \quad \Gamma,\phi \vdash \rho \quad \Gamma,\psi \vdash \rho}{\Gamma \vdash \rho}\ \vee\mathrm{E}$	If one has $\phi \vee \psi$, and assuming either ϕ or ψ has ρ, then one can infer ρ
$\dfrac{\Gamma,\phi \vdash \bot}{\Gamma \vdash \neg\phi}\ \neg\mathrm{I}$	If assuming ϕ one has $\bot$, then one can infer $\neg\phi$
$\dfrac{\Gamma \vdash \neg\neg\phi}{\Gamma \vdash \phi}\ \neg\neg\mathrm{E}$	Double negation elimination

Formal proofs are built by stacking appropriate proof rules.

Definition 9.1 A *formal proof* is a tree whose nodes are judgments, such that for each node, there is an instance of a proof rule whose conclusion is the node itself and whose premises are the children of the node. A formula ϕ is said to be a *theorem* if there is a formal proof whose root is the judgment $\vdash \phi$.

Example 32 *The formula* $(p \wedge (p \rightarrow q)) \rightarrow q$ *is a theorem in propositional logic. It is witnessed by the following formal proof.*

$$\dfrac{\dfrac{\dfrac{\dfrac{}{p \wedge (p \rightarrow q) \vdash p \wedge (p \rightarrow q)}\,\mathrm{As}}{p \wedge (p \rightarrow q) \vdash p}\,\wedge\mathrm{E1} \quad \dfrac{\dfrac{}{p \wedge (p \rightarrow q) \vdash p \wedge (p \rightarrow q)}\,\mathrm{As}}{p \wedge (p \rightarrow q) \vdash p \rightarrow q}\,\wedge\mathrm{E2}}{p \wedge (p \rightarrow q) \vdash q}\,{\rightarrow}\mathrm{E}}{\vdash (p \wedge (p \rightarrow q)) \rightarrow q}\,{\rightarrow}\mathrm{I}$$

It is emphasised that the proof rules we have introduced are purely symbolic entities. A major benefit of being so is that their application can be mechanised: for example, in the application of the $\wedge$I rule, what we have to check are (1) if the two premises and the conclusion have the same assumptions Γ, and (2) if the consequent of the conclusion is the $\wedge$-combination of those of the two premises. This symbolic check can be done without thinking of any meanings such as those in the "informal description" column of Table 9.2.

Here arise questions about the relationship between the symbolic machinery of proof rules and the "truth" of propositional formulas. Specifically,

- Do the proof rules derive only true formulas?
- Conversely, is every true formula derivable using the proof rules?

The first property is called *soundness* of proof rules, while the latter is called their *completeness*. To answer these questions, we need to be first able to talk about the *truth* (or the *meaning*) of propositional formulas; the theory of formal logic about doing so is called *semantics*. In contrast, the symbolic machinery that we have described belongs to *syntax*. See, e.g. [31].

As for propositional logic, the semantics of formulas is commonly defined in terms of two truth values (true and false). Our proof rules in Table 9.2 enjoy both soundness and completeness.

9.2.2 First-Order Logic

Our second example is *first-order logic*, which additionally accommodates quantifiers, namely $\forall$ ("for all") and $\exists$ ("there exists"). This requires the extension of the syntax by *terms*: terms represent objects, while *(first-order) formulas* represent statements about properties of such objects.

We first fix sets of *variable symbols* (denoted by $x, y, \ldots$), *constant/function symbols* (such as $0, 1, 2, \ldots, +, -, \times, \ldots$), and *predicate symbols* (such as $=$,

Table 9.3: Additional proof rules for first-order logic.

Proof rules	Informal description
$\dfrac{\Gamma \vdash \phi(x)}{\Gamma \vdash \forall x.\ \phi(x)}\ \forall\mathrm{I}$ where Γ does not mention x	If one has $\phi(x)$ without assuming anything on x, then one can infer $\forall x.\ \phi(x)$
$\dfrac{\Gamma \vdash \forall x.\ \phi(x)}{\Gamma \vdash \phi(s)}\ \forall\mathrm{E}$	If one has $\forall x.\ \phi(x)$, then one can infer $\phi(s)$ for any term s
$\dfrac{\Gamma \vdash \phi(s)}{\Gamma \vdash \exists x.\ \phi(x)}\ \exists\mathrm{I}$	If one has $\phi(s)$ for some term s, then one can infer $\exists x.\ \phi(x)$
$\dfrac{\Gamma \vdash \exists x.\ \phi(x) \quad \Gamma, \phi(x) \vdash \psi}{\Gamma \vdash \psi}\ \exists\mathrm{E}$ where Γ and ψ do not mention x	If one has $\exists x.\ \phi(x)$, and by assuming $\phi(x)$ has ψ, then one can infer ψ

$<, \leq, \ldots$). For the latter two kinds of symbols, each comes with its *arity*, which is the number of arguments it expects. For example, the function symbols $+, \times$ as well as the predicate symbols $=, <$ all have arity 2; constant symbols such as $0, 1$ have arity 0.

The set of *(first-order) terms* is defined by the BNF notation

$$t ::= x \mid f(t, \ldots, t)$$

where x is a variable symbol and f is a function symbol. We require that the number of argument terms t in $f(t, \ldots, t)$ is the arity of f.

Now the set of *(first-order) formulas* is given by the BNF notation

$$\phi ::= P(t, \ldots, t) \mid \top \mid \bot \mid \phi \wedge \phi \mid \phi \vee \phi \mid \neg\phi \mid \phi \rightarrow \phi \mid \forall x.\ \phi \mid \exists x.\ \phi$$

where P is a predicate symbol, the number of argument terms t in $P(t, \ldots, t)$ is the arity of P, and x is a variable.

The set of proof rules for first-order logic extends that for propositional logic, with the proof rules presented in Table 9.3. In the table, $\phi(x)$ represents a formula with a variable x, and $\phi(s)$ denotes the formula where the occurrences of x are replaced[1] by a term s.

Much like in propositional logic, our proof rules for first-order logic are known to be sound and complete, although completeness requires some remarks. Completeness for first-order logic is proved in famous *Gödel's completeness theorem*, which states that formulas that are true in *any possible semantics* are also symbolically derivable using the rules. This variety of semantics, however, can include those which are not intended. For example, for

[1] To be precise, only *free* occurrences of x are replaced, so that occurrences of x in formulas $\forall x.\ \psi$ or $\exists x.\ \psi$ are intact.

the famous grammar of *Peano arithmetic* for natural numbers, besides the *standard model* that interprets variables over natural numbers in the intended manner, there are other *non-standard models*. When we focus on the truth in this standard model (instead of any possible semantics), *Gödel's incompleteness theorem* states that there is no set of mechanisable proof rules that is complete.

For practitioners who wish to use formal logic for system correctness guarantee, completeness of proof rules is usually not very important: even if the proof rules are complete, the complexity for proof search can be so big that a proof may not be found (although its existence is theoretically guaranteed). In contrast, soundness of proof rules is of utmost importance: without it, the theorem "proved" by the proof rules may be semantically false, offering little value to the goal of ensuring system correctness.

9.3 Interactive Theorem Proving

Formal logic enables one to mechanically and automatically check the validity of formal proofs (or *candidates* for formal proofs, to be more precise). These proof candidates must be fed to computers, either manually or automatically. The manual approach, where a human work on a computer to write a proof candidate, is called *interactive theorem proving*. Software for doing so is called *proof assistants*. In contrast, the automatic approach is called *automated theorem proving*, where a proof candidate is searched by some algorithm.

9.3.1 (Incomplete) History of Proof Assistants

The earliest known attempt to mechanically verify mathematics is Automath (automating mathematics) developed by de Bruijn in 1968 [16].

The early 1970s saw the births of several proof assistants whose descendants are still in use today. The *Boyer–Moore theorem prover* NQTHM, developed in 1972, is based on the first-order logic and implemented using Common LISP [13]. Its successor ACL2 has gained wide use in industry, including the verification of processor units of, e.g., AMD [41], IBM [49], Centaur [34], Rockwell Collins [59].

Mizar, initiated by Trybulec in 1973, is now based on the Tarski—Grothendieck set theory and has focus on human-readable proofs [42]. One reason that makes Mizar stand out among other ITP systems is The Mizar Mathematical Library, also known as the *Journal of Formalized Mathematics* in a publication form, consisting of 1290 articles (as of 2018) contributed by the worldwide user base of the Mizar system [5].

The LCF (logic of computable functions) [27] family initiated by Milner in 1973 implements inferences on Scott's formal logic for functional programming (written in 1969 but published only in 1993 [51]). LCF had focus on verifying functional programs. Probably the most well-known spin-off of LCF is the ML programming languages including Standard ML [45] and OCaml [39], but LCF also evolved into various major proof assistants of broader scope.

Gordon evolved LCF to HOL in 1986, shifting the base formal logic into the higher-order logic [28]. Among the well-known applications of the HOL family are the formalisation of IEEE floating-point operations in HOL Light [30], and the formal proof of the Kepler Conjecture (in combination of Isabelle/HOL and HOL Light) [29].

Paulson developed Isabelle in 1983 as a generic inference engine from LCF, but mostly known for it's HOL instance, Isabelle/HOL, emerged in 1991 [46]. Isabelle is still actively maintained and a large body of mathematical libraries, the Archive of Formal Proofs, are publicly available.

Coq [8], developed by Huet and Coquand in 1984, is based on the formal logic called the Calculus of (Inductive) Constructions. Coq is among the most actively used theorem provers of today, probably best known for the formal proof of the four-colour theorem [26] and the formally verified C compiler [38].

9.3.2 Isabelle/HOL

We demonstrate the use of proof assistants by employing *Isabelle/HOL* as an example. A file collecting Isabelle proofs is called a *theory file*. An example theory file `TheoremProvingChapter.thy` would start like this:

```
1. theory TheoremProvingChapter
2.   imports Main
3. begin
```

The second line of the file directs Isabelle to import the standard theory file `HOL/Main.thy`, which declares basic axioms of (a dialect of) the formal logic HOL and natural numbers, and then definitions and lemmas for derived notions, such as sets, lists, and integers. A vast amount of further notions are available in the standard library and external *Archive of Formal Proofs*.

After importing basic concepts, we begin our own definitions, claims and proofs on top of them. As a quick example, let us define a predicate `my_sorted`, which takes a list of integers (represented by type `int list`) and tells if the list is sorted.

```
4. definition my_sorted :: "int list ⇒ bool" where
5.   "my_sorted xs =
6.     (∀ i j. i < j ⟶ j < length xs ⟶ xs!i ≤ xs!j)"
```

The definition would look like the usual mathematical definition, except that in Isabelle (and the LCF family), function application is written "$f\,x$" instead of "$f(x)$". Hence `xs` is the argument of predicate `my_sorted`. The function `length` gives the length of the list, and infix `!` operator takes the *i*th element of the list. These functions, as well as other symbols, are defined in `Main`.

To conveniently use `my_sorted` afterwards, let us prove a (trivial) lemma.

```
7.  lemma my_sortedI:
8.    assumes my_asm: "∀ i j. i < j ⟶ j < length xs ⟶ xs!i ≤ xs!j"
9.    shows "my_sorted xs"
```

In Isabelle, lemmas can be used as inference rules. In notation of Section 9.2, the above lemma can be used as the inference rule

$$\frac{\forall i.\ \forall j.\ i < j \rightarrow j < \mathtt{length}(xs) \rightarrow xs!i \leq xs!j}{\mathtt{my\_sorted}(xs)}\ \text{MY\_SORTEDI}$$

after it is proved. Proving the lemma means proving the judgment

$$\Gamma, (\forall i.\ \forall j.\ i < j \rightarrow j < \mathtt{length}(xs) \rightarrow xs!i \leq xs!j)\ \vdash\ \mathtt{my\_sorted}(xs)$$

where Γ represents the all available *facts*, that is, axioms, definitions, lemmas, and theorems. To prove this goal, we would like to unfold the definition of `my_sorted`. Instead of applying primitive inference rules, Isabelle provides a convenient way of unfolding equations:

```
10.   unfolding my_sorted_def
```

where `my_sorted_def` refers to the equation of line 5 (put in Γ). Now the goal `my_sorted xs` (the right-hand side of $\vdash$) is replaced by

```
"∀ i j. i < j ⟶ j < length xs ⟶ xs!i ≤ xs!j"
```

which is precisely in the premise (the left-hand side of $\vdash$). Thus the assumption would conclude the proof, but Γ, the collection of all known facts, is so big that one should specify which fact one uses. There are several ways to do this, among which we use the following:

```
11.   by (rule my_asm)
```

where `my_asm` refers to the formula of line 8 as the name is explicitly given there. At this point the proof of the lemma `my_sortedI` is done.

Now let us prove the following easy statement:

```
12. theorem my_sorted_123: "my_sorted [1,2,3]"
```

It would look trivial to human readers, but unfortunately it is not to Isabelle. We first tell Isabelle to apply `my_sortedI` as an inference rule to this goal as follows.

```
13.   apply (rule my_sortedI)
```

This yields the following new goal:

```
∀ i j. i < j ⟶ j < length [1,2,3] ⟶ [1,2,3] ! i ≤ [1,2,3] ! j
```

Proving this requires to consider all possibilities of `i` and `j` and requires many proof steps if one is restricted to primitive application of inference rules. Luckily, most proof assistants have *automatic* proof methods that build a proof tree following some strategies. For Isabelle, we need to guide it to do case splitting according to `j < length [1,2,3]`, that is, `j` is either `0`, `1`, or `2`. It needs some training to find how, but this is achievable by the following equation `less_Suc_eq` available in Γ:

```
(m < Suc n) = (m < n ∨ m = n)
```

Here, `Suc` is the internal representation of the successor of natural numbers (i.e., `Suc` n represents $n+1$). We conclude the proof by instructing Isabelle's automatic proof procedure to add `less_Suc_eq` in *simplification* rules:

```
14.  by (auto simp: less_Suc_eq)
```

9.4 Automated Theorem Proving

As we have seen in the previous section, building a formal proof purely in the manner of interactive theorem proving needs some training and is often tedious. There is another line of research directed towards *automating* the theorem proving task: *automated theorem proving (ATP)*.

9.4.1 General-Purpose ATP

The history of automated theorem proving (ATP) is longer than that of interactive theorem proving (ITP); the earliest ATP system (over propositional logic) is *The Logic Theory Machine (LT)* [43] developed by Newell and Simon in 1956, way before the dawn of ITP. Afterwards several ATP systems appeared (see, e.g. [9] for early history), leading to *The CADE ATP System Competitions* [54] held since 1996 to evaluate general-purpose ATP systems on the first-order and higher-order logics. Only a few of remarkable participants of the competitions are: Prover9 (in a lineage since the early 1960s [35]), the E Prover [50], and Vampire [48].

To do justice to ITP, the level of trust ATP offers should be said lower than ITP. This is because the correctness of an ATP tool relies on the implementation of the entire proof search algorithms, while ITP relies only on the implementation of small number of inference rules. In fact, one of the specialities of Isabelle/HOL is *Sledgehammer* [12], a command to delegate proof obligations to automated theorem provers (ATPs), see Fig. 9.2. ATPs are seen as untrusted black-boxes that suggest proofs, and Isabelle tries to reconstruct fully formal proofs following the suggestions.

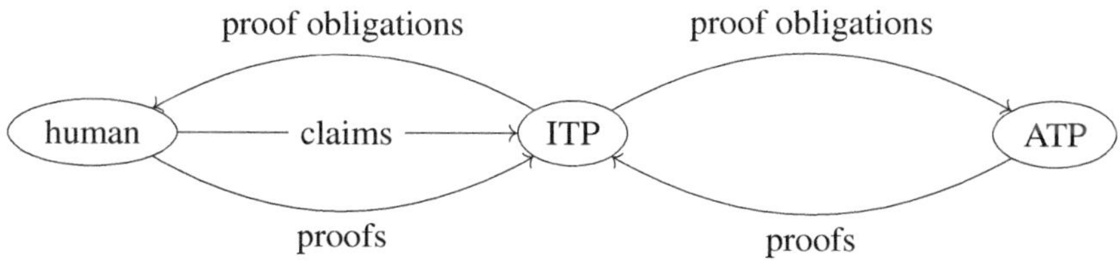

Figure 9.2: Theorem proving by combining interaction and automation. Here, the "proofs" generated by ATP are thought of as "proof candidates" in ITP, and their validity are checked again, using proof assistants for ITP that are considered more trustable than ATP tools.

9.4.2 SAT Solving

It is not expected that a general-purpose ATP system will solve a real-world problem fully automatically. One way of making full ATP possible is to restrict the logic to those for which an efficient way of solving is known. A well-known example of this line is *SAT* (satisfiability) solvers.

Satisfiability of a formula in a formal logic asks if it is possible to instantiate variables in the formula so that the resulting formula evaluates to true ($\top$) in the logic. For example, in propositional logic, the formula $p \wedge (\neg p \vee \neg q)$ is satisfiable, since by instantiating $p \mapsto \top$ and $q \mapsto \bot$ yields $\top \wedge (\bot \vee \top)$, which is simplified into $\top$. In contrast, $p \wedge (\neg p \vee \neg q) \wedge (\neg p \vee q)$ is *unsatisfiable*, since the formula is false under any assignment on p and q. Satisfiability is closely related to theorem proving: ϕ is a theorem if and only if $\neg\phi$ is unsatisfiable.[2]

SAT solvers are concerned with solving satisfiability in propositional logic. Propositional logic is very restrictive, in the sense that the value of each variable has only two choices (either $\top$ or $\bot$). It is however well suited for verifying electronic circuits [20], where each wire must have either *high* ($\top$) or *low* ($\bot$) voltage. Typical SAT solvers further restrict formulas to *conjunctive normal forms (CNFs)*, but there is an efficient transformation of any propositional formula into CNF with the same satisfiability [56]. Moreover, any problem with finite search space can be encoded into a SAT problem, and often solved faster by reducing to SAT solving than other means [11], thanks to the impressive performance of recent fine-tuned SAT solvers.

Example 33 *A typical class of faults in electronic circuit manufacturing is the* stuck-at *faults. For instance, consider a* good, *fault-free circuit in Fig. 9.3a and a faulty one in Fig. 9.3b. Notice that in the faulty circuit, the second input to the AND gate in the center is stuck at $\bot$. To detect such a fault by testing, one wants to find an input such that the outputs of the good and faulty circuits differ.*

[2]This is true since the proof rules employed by SAT solvers are complete in the sense of Section 9.2.1.

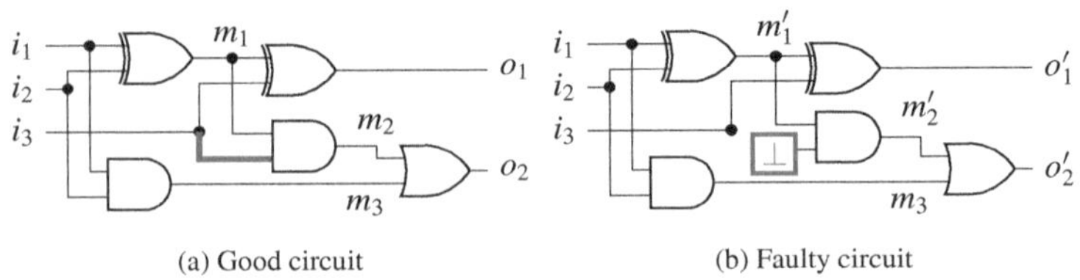

Figure 9.3: Detecting a stuck-at fault.

A propositional formula encoding the good circuit is as follows, where each wire is represented by a propositional variable of the same name:

$$(i_1 \oplus i_2 \leftrightarrow m_1) \wedge (m_1 \wedge \underline{i_3} \leftrightarrow m_2) \wedge (i_2 \wedge i_3 \leftrightarrow m_3) \wedge \\ (m_1 \oplus i_3 \leftrightarrow o_1) \wedge (m_2 \vee m_3 \leftrightarrow o_2) \quad (9.3)$$

where $x \oplus y$ represents the exclusive or: $(x \wedge \neg y) \vee (\neg x \wedge y)$. Similarly, the faulty circuit is encoded as follows:

$$(i_1 \oplus i_2 \leftrightarrow m'_1) \wedge (m'_1 \wedge \underline{\bot} \leftrightarrow m'_2) \wedge (i_2 \wedge i_3 \leftrightarrow m'_3) \wedge \\ (m'_1 \oplus i_3 \leftrightarrow o'_1) \wedge (m'_2 \vee m'_3 \leftrightarrow o'_2) \quad (9.4)$$

In order to detect such faulty circuits by testing, we want an assignment on the input $\{i_1, i_2, i_3\}$ such that the output $\{o_1, o_2\}$ of the good circuit and $\{o'_1, o'_2\}$ of the faulty one differ. This requirement is expressed as a propositional formula as follows:

$$(9.3) \wedge (9.4) \wedge (o_1 \oplus o'_1 \vee o_2 \oplus o'_2)$$

After CNF translation, any reasonable SAT solver will immediately provide a solution to the formula; for instance the SAT solver lingeling [10] finds the following solution (also called a model*):*

$$\begin{array}{lllll} i_1 \mapsto \top & i_2 \mapsto \bot & i_3 \mapsto \bot & & \\ m_1 \mapsto \top & m_2 \mapsto \bot & m_3 \mapsto \bot & o_1 \mapsto \top & o_2 \mapsto \bot \\ m'_1 \mapsto \top & m'_2 \mapsto \top & m'_3 \mapsto \bot & o'_1 \mapsto \top & o'_2 \mapsto \top \end{array}$$

The first line indicates the desired test pattern for detecting such a stuck-at fault.

There are a large number of SAT solvers and they are evaluated annually in the SAT competitions [23]. As the participants and winners dynamically change by years, interested readers are invited to check recent competition results.

9.4.3 Satisfiability Modulo Theory (SMT)

Propositional logic is not the only logic for which efficient mathematical techniques are known. Other examples include integer and real arithmetics and regular expressions. Moreover, other formal logics, such as bit-vector arithmetics or arrays, are highly useful when verifying software.

Satisfiability modulo theories (SMT) solvers [7] are tools that decide whether a formula is satisfiable in such a formal logic other than propositional logic.

Example 34 *Consider the following C program that computes the gravity at time t between two particles A and B, initially at position* $(9,4)$ *and* $(3,14)$ *and moving with velocity* $(0.4,0.8)$ *and* $(0.7,0.3)$*, respectively:*

```
1 double gravity_between_A_and_B(double t) {
2         double Ax, Ay, Bx, By, x, y;
3         Ax = 9.0 + 0.4 * t; Ay = 4.0 + 0.8 * t;
4         Bx = 3.0 + 0.7 * t; By = 14.0 + 0.3 * t;
5         x = Ax - Bx; y = Ay - By;
6         x = x * x; y = y * y;
7         return 1.0 / (x + y);
8 }
```

The program is not safe: it can cause a division-by-zero error. However, finding the error by testing is not a trivial task.

One can find such errors by encoding the problem of finding an erroneous input as an SMT problem. The problem can be encoded as follows:

$$\begin{gathered} Ax = 9.0 + 0.4t \wedge Ay = 4.0 + 0.8t \wedge \\ Bx = 3.0 + 0.7t \wedge By = 14.0 + 0.3t \wedge \\ x_1 = Ax - Bx \wedge y_1 = Ay - By \wedge \\ x_2 = x_1^2 \wedge y_2 = y_1^2 \wedge \\ x_2 + y_2 = 0.0 \end{gathered}$$

where all but last lines correspond to lines 3–6 of the program, and the last line represents the situation where the denominator of the division is 0*. The SMT solver Z3 [17] solves the problem, with the following solution:*

$$\begin{array}{lllll} t \mapsto 20.0 & Ax \mapsto 17.0 & Ay \mapsto 20.0 & Bx \mapsto 17.0 & By \mapsto 20.0 \\ x_1 \mapsto 0.0 & x_2 \mapsto 0.0 & y_1 \mapsto 0.0 & y_2 \mapsto 0.0 & \end{array}$$

Besides the aforementioned Z3 [17] developed by Microsoft, various academic SMT solvers compete in the annual SMT competitions [58], most notably the CVC family [6] and Yices [18].

9.4.4 Discussion: Limit of Automation

Full automation is successful for restricted logic such as propositional logic or (linear) arithmetics. In many other logics, whether a statement is provable or not is undecidable (as in nonlinear integer arithmetic), or decidable but in intractable complexity (as in nonlinear real arithmetic). In the next section we aim at verifying cyber-physical systems (CPSs), which requires more involving formal logic that can deal with differential equations. It should be clear that full automation on such a logic is far beyond our reach. Therefore we will step slightly back towards interactive theorem proving, but still a high degree of automation is achieved by the domain-specific knowledge of differential equations.

9.5 Theorem Proving for Continuous and Hybrid Dynamics

Given their nature, CPSs such as automotive systems require the verification that a system's software, physical components, and their interactions are all correct. This section discusses the *theorem proving* of such systems by *modelling* and *verifying* them as *hybrid(-dynamical) systems*, that combine discrete and continuous dynamics which represent software (or "cyber") and physical components, respectively. Theorem proving for CPS deserves special attention as it requires reasoning about hybrid dynamics, that is continuous evolutions combined with discrete mode switches. As a result—and in contrast to discrete approaches (e.g., SAT solving)—complete proof automation cannot be expected. Nonetheless, by restricting our attention to hybrid system models, we significantly facilitate their proofs and also enable partial automation.

9.5.1 KeYmaera X: An Interactive Theorem Prover

A well-known tool providing such support is *KeYmaera X* [24], a specialised *interactive theorem prover (ITP)* for hybrid systems. Its specialisation enables significant proof automation, a convenient user interface (UI), and a readable modelling language. Despite the emphasis on user-friendliness, KeYmaera X is based on formal logic, providing proofs and confidence in CPS correctness.

9.5.1.1 Hybrid Systems Modelling

Hybrid systems (HSs) is a common *modelling formalism*, which we use as a stepping stone to expressing CPS in KeYmaera X. HSs enable the precise expression of a system's behaviour and enable reasoning about its evolution in an unambiguous way.

In KeYmaera X, HSs are modelled as *hybrid programs (HPs)*, consisting of executable statements representing the CPS's continuous and discrete evolutions. Note that HPs relate closely to *hybrid automata (HAs)* [4], a common and expressivity-wise equivalent modelling formalism. HPs thereby add a *textual syntax* that is well-suited for verification and closely resembles theorem proving intuitions.

The rest of this section uses a one-dimensional vehicle example to discuss CPS theorem proving and HPs. The HP is presented in Fig. 9.4 and an equivalent HA is presented in Fig. 9.5 for the sake of comparison. The example vehicle's aim is to safely drive up towards a stop sign and brake to a complete halt at the sign. As the *controller* is time-triggered, the time between its executions is at most some reaction time *react*. Furthermore, the maximum acceleration A and fixed brake rate B restrict potential evolution, such that the actual acceleration *acc* is in $[-B, A]$. The controller stops, once the velocity v is zero or below. Figure 9.4 shows this behaviour as a HP.

Note how the controller features separate modes for accelerating and braking, whose switching is decided by *staySafe*. The condition *staySafe* is true if and only if, even when the maximum acceleration A is maintained until the next time the controller runs (*react*), the vehicle could still safely brake to a stop (at rate $-B$) without reaching the stop sign (position *sign*). Formally, *staySafe* follows from the standard equations for Newtonian motion:

$$staySafe \quad \leftrightarrow \quad \frac{(vel + A \cdot react)^2}{2 \cdot B} \leq sign - \left(pos + vel \cdot react + A \cdot \frac{react^2}{2}\right)$$

The left-hand side of the comparison $\leq$ represents the distance needed to brake (after the worst-case acceleration). The right-hand side represents the available distance between the stop sign and vehicle (after the worst-case acceleration).

In Fig. 9.4, if the check of *staySafe* succeeds, the vehicle accelerates at any rate ($acc := *$) subject to the condition $acc \in [-B, A]$. Otherwise, it brakes at the maximum rate. We express this behaviour as HP in Fig. 9.4, such that the `while` loop repeats while the vehicle is moving ($vel > 0$). Variable *vel* represents the 1-dimensional velocity of the vehicle, where positive values of *vel* mean forward motion and negative values mean reverse motion. The velocity *vel* and all other variables in the HP are real-valued $\mathbb{R}$. The `if` statement models the mode switching, as described above. The rest of the HP code models the *physical model* (i.e., the "*plant*") and initialises the *timer* for the time-based trigger, and the system of *ordinary differential equations (ODEs)*.

ODEs are the characteristic feature of *hybrid programs*, which distinguish them from discrete programs. Every ODE statement contains a comma-separated list of equations and optionally contains a formula called an *evolution domain constraint*, which is separated from the equations by an ampersand (&). The evolution domain constraint determines how long the ODE evolves. Specifically, the duration of the ODEs is *non-deterministic*, but it can only be used until the evolution domain constraint (ever) becomes false. In the example, the equations mean that the derivative of position is velocity, the derivative of velocity is acceleration, and the derivative of the timer is, by convention, 1. In the example, the evolution domain constraint means that the timer must never exceed the reaction time and the velocity must never be negative. It is important that the ODE ends by *react*, because the control condition *staySafe* is only designed to guarantee safety if the vehicle accelerates for time *react* and is allowed to brake afterwards if needed.

$car \equiv$
`while`$(vel > 0)\{$
  `if`$(staySafe)$
    $\{acc := *; ?acc \in [-B, A];\}$
  `else`
    $\{acc := -B;\}$
  $timer := 0;$
  $\{pos' = vel, vel' = acc, timer' = 1$
    $\&(timer \leq react \wedge vel \geq 0)\};$
$\}$

Figure 9.4: Time-triggered HP of the 1D vehicle.

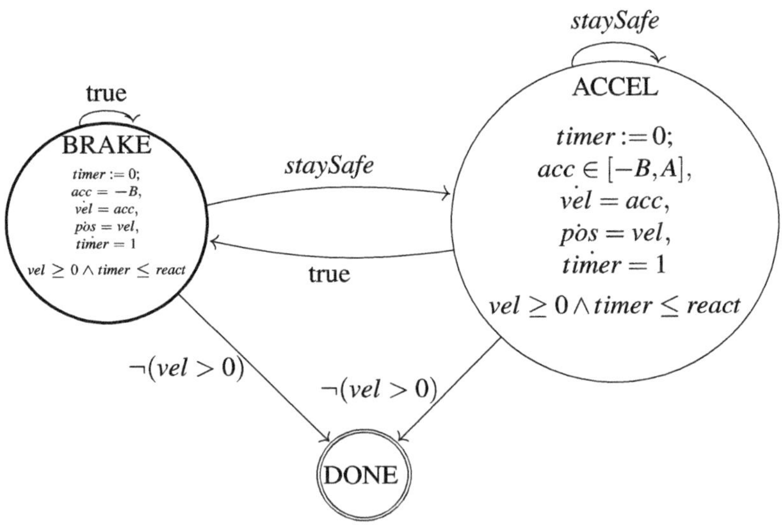

Figure 9.5: A time-triggered model of a 1D-vehicle model expressed as hybrid automaton.

Based on this vehicle model, we use *differential dynamic logic (*dL*)* [47] to state a theorem that we prove in *KeYmaera X*. Using dL it is possible to define that one or all executions of a (possibly non-deterministic) HP are valid. For instance, in the above example, $acc := *$ can take arbitrary values in $[-B, A]$.

Thus, to ensure safety, we want to express that given some starting conditions, *all* executions lead to safe postconditions. By specifying $[car]safe$, we define that the car always satisfies a postcondition $safe$. The standard safety result for our vehicle is that, under appropriate initial conditions, it does not drive through the stop sign, i.e., $pre \rightarrow [car](pos \leq sign)$ where pre is shorthand for a safe initial condition. Next, we discuss the proof UI.

9.5.1.2 Strong Logical Foundations

The design and implementation of KeYmaera X are based on the rigorous underlying theory of dL to increase our trust in the tool. KeYmaera X is in fact sound, meaning that if the tool labels a formula as true, the formula is in fact true. Specifically, if a *safety formula* is true according to the KeYmaera X tool, it is also true according to the theory of dL, in which case we call it *valid.*

The theory of dL defines the truth of formulas by formalising the notion of runs from Section 9.5.1.1. Proof techniques are formalised using *inference rules* which assume that *premises* are true and conclude that a *conclusion* is true. The inference rules are shown to be sound, meaning that only true conclusions are proved. As shown in Fig. 9.6, proof rules necessarily perform small proof steps, so that their soundness can be justified more easily. One of KeYmaera X's major contributions is that it builds complex automation from the *soundness* of these small steps.

$$\text{[if]}\ \frac{Q \rightarrow [\alpha]P \qquad \neg Q \rightarrow [\beta]P}{[\texttt{if}(Q)\{\alpha\}\texttt{else}\{\beta\}]P} \qquad [;]\ \frac{[\alpha][\beta]P}{[\alpha;\beta]P} \qquad [?]\ \frac{Q \rightarrow P}{[?Q]P}$$

$$\text{DI}\ \frac{P \qquad [x' = f\,\&\,Q](P)'}{[x' = f\,\&\,Q]P} \qquad \text{DC}\ \frac{[x' = f\,\&\,Q]C \qquad [x' = f\,\&\,Q \wedge C]P}{[x' = f\,\&\,Q]P}$$

Figure 9.6: Example dL rules.

As shown in the first row of example proof rules in Fig. 9.6, many HP statements can be proved by unfolding their structure, as automated by KeYmaera X. The [`if`]-rule is proved by considering each branch. Sequential statements (rule [;]) require consideration of each statement in a sequence and tests (rule [?]) are inferred by assuming the test condition and showing the postcondition.

The second row shows that greater insight is often required for invariant-style proofs, in this case about ODEs. In rule *dI*, formula P is proved to be an invariant by showing that it holds initially and that its *differential formula* $(P)'$ holds throughout. Although $(P)'$ can be automatically determined [47], rule *dI* still produces two premises which must be proved. Rule *dC* requires more insight: the user chooses a differential *cut* formula C, proves that it holds throughout the ODE, then adds it to the domain constraint to help prove the postcondition P.

In summary, the automation provided by KeYmaera X both helps automate simple steps (first row) even when hard steps (bottom) are performed manually and provides best-effort automatic proof attempts for hard steps (bottom row).

9.5.1.3 High-Level Proof UI

KeYmaera X offers an intuitive and interactive browser-based UI, depicted in Fig. 9.7. For space reasons, we only discuss key aspects here and refer to the literature [40] for additional details. As mentioned above, some dL proofs cannot be fully automated, traditionally requiring significant user input and steep learning curves. Thus, KeYmaera X's UI focuses on the crucial reduction of these efforts.

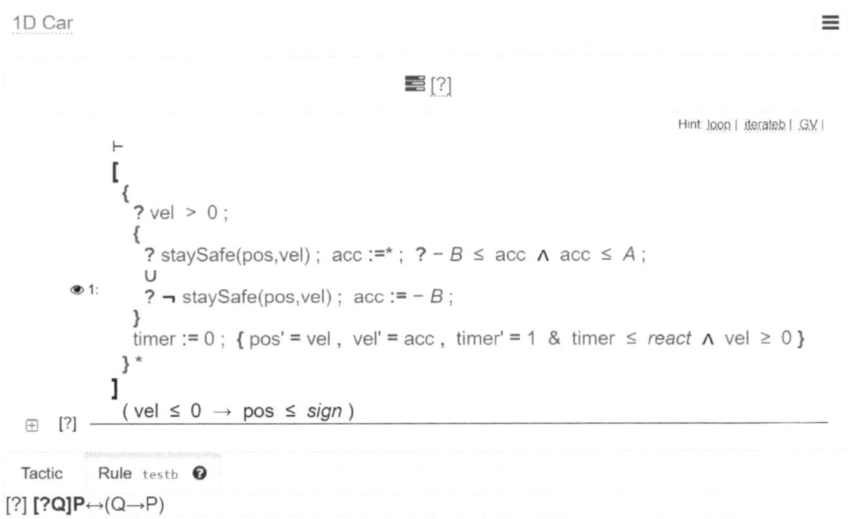

Figure 9.7: KeYmaera X Proof UI.

The tool thus encourages iterative development, and even allows reuse of proof attempts on initial (typically incomplete or faulty) models for later, corrected versions of those models. Further, the *undo* feature helps correcting proof attempt mistakes without the need to restart from scratch. Finally, even though full automation is infeasible, KeYmaera X supports *partial automation* of significant portions of the proof, so that the user may focus on the actually tricky parts, while the rest is handled by the tool. For instance, a user might have to identify invariant formulas for loops or ODEs, then use automation to attempt remaining proof tasks, avoiding unnecessary manual proof effort.

KeYmaera X supports novice users by providing proof suggestions, helping identify *proof mistakes* and automatically identifying which rules are currently relevant. This reduces the learning effort of both the tool and *theorem proving* in general. Finally, for both new and expert users, it is often unclear how to proceed when a proof attempt fails. KeYmaera X provides additional insight by searching for counter examples, i.e., showing specific configurations where the formula is false.

9.5.2 Related Work

Next to KeYmaera X, various other tools and approaches can be used for CPS verification. These "competitor methods" can be organised in terms of the fundamental tradeoff between expressive power and automation. In general, verification of "more powerful" properties will require more manual effort by the user. We compare approaches along this spectrum.

- Given their general-purpose nature, many *ITPs* such as *Isabelle/HOL* [44] aim to cover many problems and domains. However, given the lack of HS automation, CPS proofs require significant manual effort. Recent efforts try to add such automation after-the-fact, though. See [21].
- Hybrid systems model checkers such as *SpaceEx* [22] and *Flow** [14] automatically explore all configurations of a HS to prove or falsify a property, usually a safety property. Even though specific features vary according to their expressive power, common limitations restrict the classes of supported ODEs, impose finite time horizons for safety guarantees, or safety on restricted (compact) sets of configurations. These limitations however enable that model checkers achieve a high degree of automation, sometimes with a high degree of scalability.
- dReach [37] deserves mentioning as it is a special HS *model checker*. Though restricted to *bounded* model-checking, its dReal [25] SMT solver-basis simplifies checking of real-number formulas using δ-robustness, i.e., allowing for small perturbations. dReal excels at automating difficult arithmetic functions, but it shares the limitations of other *SMT solvers*.
- *Simulations*, such as those provided in Stateflow/Simulink [1], provide strong automation and scalability, but typically only consider individual starting configurations, or small sets thereof. The inherent problems of e.g., naive floating-point arithmetics and numerical solvers result in lacks of soundness.

In summary, KeYmaera X excels at proving general-case correctness properties about hybrid systems, which are expressive models of CPSs such as *automotive systems*. The tradeoff is that many KeYmaera X proofs are interactive and that interactive proving takes time to learn. Thus, KeYmaera X is appropriate when general-case correctness is desired, but more automatic tools are appropriate when automation is more important than generality.

9.6 Theorem Proving with Model Refinement

The theorem proving approaches described in Sections 9.3–9.5 are *bottom-up*: they assume a formal description of the target system at the bottom and build up correctness proofs on top of it. However, realistic systems often have black-box components or overwhelming complexity that prevent us from providing formal descriptions, i.e., they have *uncertainties*.

In this section, we introduce a *top-down* approach: we start from the requirements of the target system at the top, and gradually design details of the system so that they satisfy the requirements. This approach is a powerful countermeasure against the challenge of uncertainties — uncertainties are gradually *confined* or *eliminated one by one* as we gradually provide design details.

In the rest of this section, we introduce Event-B [2],[3] a method for top-down theorem proving.

9.6.1 *Top-Down Theorem Proving in Event-B*

Event-B has a mechanism called *stepwise refinement* that enables top-down theorem proving. Using the mechanism, we start with constructing a model of *abstract* aspects of the target system, such as safety requirements, and gradually evolve the model to include *concrete* aspects, such as behavioural details.

In each step of stepwise refinement, we add a new component called **machine** to a model. A machine presents the target system as a state transition system. The primary interest in a refinement step is whether the behaviour of the machine introduced in model version n conforms with that of the machine introduced in model version $n-1$. We call them a *concrete machine* and an *abstract machine*, respectively.

A machine refers to the following constructs:

a variable: A variable whose value is changed by the system's behaviour.

an invariant: A property that must always hold (a safety property).

an event: A set of state transitions regarding the system's behaviour.

[3]See also Chapter 8, where we focus on tackling the complexity of software systems using Event-B.

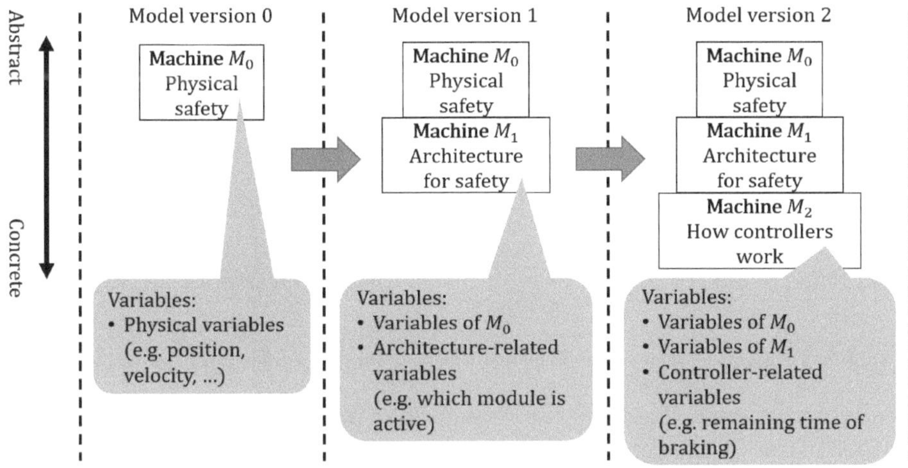

Figure 9.8: Stepwise refinement of train system model.

When we check the conformance of the concrete machine's behaviour with the abstract machine's, we use terms such as *abstract events* and *concrete events* in the same manner as machines.

For instance, a train system with a controller can be gradually modelled as shown in Fig. 9.8. In version 0 of the model, we construct a machine M_0, which focuses on the modelling of the physical safety of the train without going into details of the controller. In version 1, we add a machine M_1, which describes details of the safety and behaviour of the controller's architecture in addition to M_0's focuses. In version 2, we add a machine M_2, which describes details of the concrete behaviour of controller modules in addition to M_1's focuses.

The safety of the final system is proved in a stepwise manner. For example, the train system is verified as follows:

1. We prove that M_0's behaviour satisfies M_0's invariants. The latter express the train's physical safety.
2. We prove that M_1's behaviour satisfies M_1's invariants. The latter express desirable properties that are specific to the structures additionally introduced in M_1. For example, they are about the controller's architecture; they do not refer to the train's physical safety since it is already covered by M_0's invariants.
3. We prove that M_1's behaviour conforms with that of M_0.
4. We prove that M_2's behaviour satisfies M_2's invariants. Much like for M_1, M_2's invariants are concerned about desired properties of M_2 which are not yet covered in M_1.
5. We prove that M_2's behaviour conforms with that of M_1.

Figure 9.9: Train example.

In what follows, we demonstrate how an Event-B model is gradually constructed while verified, by using the train system (Fig. 9.8) as an example.[4]

9.6.2 Constructing and Verifying Abstract Event-B Machine

In this section, we explain how the most abstract machine can be constructed and verified using the abstract machine M_0 of the train system as an example.

9.6.2.1 Requirements of Abstract Machine M_0

Typically, we first construct a machine that focuses on *what* should be satisfied (basic requirements) and abstracts away *how* the system satisfies those requirements.

For instance, assume that the train system has the following requirement for the train's physical safety (Fig. 9.9):

SAF-0 The train must always be able to stop before reaching the destination x_{dst} by activating the emergency brake (deceleration rate: b_{max}). Formally, the following must always hold:

$$x + stopDist(v) \leq x_{dst},$$

where $stopDist(v) = v^2/2b_{max}$ is the stopping distance, b_{max} is the emergency brake rate, and x_{dst} is the position of the destination.

There is another requirement regarding a physical constraint on the train hardware:

EQP-0 The train's maximum acceleration rate is a_{max}.

9.6.2.2 Specifying M_0

In the most abstract machine M_0, we focus on the requirement **SAF-0**. The specification of the controller's behaviour simply says it *somehow* satisfies the requirement; the detailed behaviour such as how the architecture and controller modules work to satisfy the requirements are deferred to concrete machines.

We assume that the controller works once in a cycle.

The machine M_0 is as follows:

[4]For the sake of simplicity, we do not explain the "full" Event-B. It is introduced in [2].

```
machine M0
variables x,v
invariants @inv_stop: x+stopDist(v) ≤ x_dst
events
  event run
    where @grd_whenever: true
    then  @act_xv: x,v :| x′+stopDist(v′) ≤ x_dst
                          ∧ x′ ≤ nextMaxX(x,v) ∧ v′ ≤ nextMaxV(v)
  end
end
```

Variables x and v denote the position and velocity of the train. **Invariant** `@inv_stop` expresses requirement **SAF-0**. **Event** `run` describes the behaviour of the train controller in a cycle.

An **event** is composed of *guards* specified in the **where** clause, which define when the event is allowed to occur, and *actions* specified in the **then** clause, which describe state transitions due to the event.

The event `run` can occur in any situation (guard `@grd_whenever`). The action `@act_xv` updates the values of variables x and v so that their new values (x' and v') satisfy the formula specified in the **then** clause:

$$\begin{aligned} & x' + stopDist(v') \leq x_{dst} \\ & \wedge\ x' \leq nextMaxX(x,v) \wedge v' \leq nextMaxV(v), \end{aligned}$$

Some readers may be interested in definitions of $nextMaxX(x,v)$ and $nextMaxV(v)$. They denote the maximum possible position and velocity at the next cycle following **EQP-0**, respectively, i.e., the position and velocity if the train accelerates with a_{max} for a cycle:

$$nextMaxX(x,v) = x + \int_{t=0}^{1} (v + a_{max}t)dt,$$

$$nextMaxV(v) = v + \int_{t=0}^{1} a_{max}dt.$$

9.6.2.3 Verifying Safety of M_0

After constructing a machine, we need to verify that invariants are satisfied after any possible events, given that they are satisfied beforehand. We call this property *invariant preservation*.

The following judgment expresses the invariant preservation in M_0.

$$
\begin{array}{ll}
x + stopDist(v) \leq x_{dst}, & (\texttt{@inv\_stop} \text{ with old values}) \\
true, & (\texttt{@grd\_whenever}) \\
x' + stopDist(v') \leq x_{dst} \wedge & \\
x' \leq nextMaxX(x,v) \wedge v' \leq nextMaxV(v) & (\texttt{@act\_xv}) \\
\vdash & \\
x' + stopDist(v') \leq x_{dst} & (\texttt{@inv\_stop} \text{ with new values})
\end{array}
$$

This judgment is trivially true, since the conclusion $x' + stopDist(v') \leq x_{dst}$ itself appears as part of the assumption. Note here that the occurrence of $x' + stopDist(v') \leq x_{dst}$ in the consequent part (i.e., on the right of $\vdash$) comes from the invariant `@inv_stop` (we have to preserve it), while its occurrence in the assumption (i.e., on the left of $\vdash$) comes from `@act_xv`. Indeed, in the definition of `@act_xv`, it is specified that the controller *somehow* manages to ensure $x' + stopDist(v') \leq x_{dst}$ after the event.

Rodin platform [3], a modelling environment for Event-B, provides support in verifying machines. It generates judgments such as the above after we construct a machine. We call such judgments *proof obligations (POs)*. Rodin also has automatic provers (including SMT solvers) that help prove generated POs. As we construct M_0 in Rodin, it generates and proves the PO above.

9.6.3 Constructing and Verifying Concrete Event-B Machines

In this section, we explain how concrete machines can be constructed and verified, using the intermediate machine M_1 and the concrete machine M_2 of the train system as examples.

9.6.3.1 Requirements of Intermediate Machine M_1

We assume that the train has a black-box controller (e.g., one based on machine learning). Since it is black-box, we cannot formally guarantee that the train can stop before the destination (**SAF-0**).

To guarantee **SAF-0** while using the black-box controller, in M_1, we adopt the Simplex architecture [52] for the train controller, which is shown in Fig. 9.10.

In the Simplex architecture, we refer to the black-box controller as advanced controller (AC). We provide two additional modules: baseline controller (BC) whose behaviour can be formally specified (e.g., rule-based controller) and thus verifiable; and decision module (DM), which switches the active controller between AC and BC according to the current situation.

The architecture is expected to satisfy the following requirements:

SAF-1 AC is used only when **SAF-0** (stoppability) will be guaranteed after a cycle (i.e., the worst-case situation after a cycle (of acceleration with rate a_{max}) is safe).

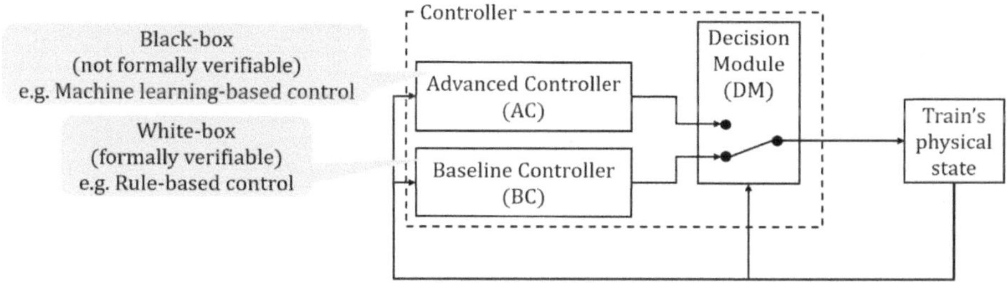

Figure 9.10: Simplex architecture [52] for the train system.

FUN-1 DM switches from BC to AC if **SAF-0** is far from being violated (values of x and v are small enough to be able to use AC at the next cycle) when using BC. Let $ACNext(x,v)$ denote this condition. In contrast, DM switches from AC to BC if **SAF-0** is *about* to be violated ($\neg ACNext(x,v)$) when using AC.

We need to define appropriate $ACNext(x,v)$. This is not a trivial task, because the definition of $ACNext(x,v)$ needs to take into account the *switching delay* — DM should select BC if the current state (current values of x and v) indicates that the next state (values of x and v at the next cycle) may violate **SAF-0**, even if the current state satisfies it.

9.6.3.2 *Specifying* M_1

In M_1, we specify how DM switches between AC and BC according to **FUN-1**. We also specify that **SAF-1** should be satisfied by DM's behaviour. Details of BC's control rules are deferred to M_2.

The overview of M_1 is as follows:

machine M_1 **refines** M_0
variables $x, v, ctrl$
invariants `@inv_ctrl:` $ctrl \in \{AC, BC\}$
`@inv_AC:`
$ctrl = AC \Longrightarrow nextMaxX(x,v) + stopDist(nextMaxV(v)) \leq x_{dst}$
events
...
end

The new **variable** *ctrl* says which of AC and BC is activated. **SAF-1** is specified as **invariant** `@inv_AC` using $nextMaxX(x,v)$ and $nextMaxV(v)$ defined in Section 9.6.2.2. There are four **events** `AC->AC`, `AC->BC`, `BC->AC`, and `BC->BC`, for specifying all switching cases.

We elaborate on events AC->BC and BC->BC below.

```
event AC->BC refines run
  where @grd_AC: ctrl = AC
        @grd_BCNext: ¬ACNext(x,v)
  then @act_xv: x,v :| x' + stopDist(v') ≤ x_dst
                        ∧ x' ≤ nextMaxX(x,v) ∧ v' ≤ nextMaxV(v)
       @act_ctrl: ctrl :| ctrl' = BC
end
event BC->BC refines run
  where @grd_BC: ctrl = BC
        @grd_BCNext: ¬ACNext(x,v)
  then @act_xv: x,v :| x' + stopDist(v') ≤ x_dst
                        ∧ x' ≤ nextMaxX(x,v) ∧ v' ≤ nextMaxV(v)
       @act_ctrl: ctrl :| ctrl' = BC
end
```

They inherit the guard and action from the run event of machine M_0. Newly introduced guards and actions are as follows: In event AC->BC, guard @grd_AC expresses that the event is for the case where the currently active controller is AC; guard @grd_BCNext expresses requirement **FUN-1**; and @act_ctrl expresses that BC will be used from the next cycle. Event BC->BC is the same as AC->BC except that the event is for the case where BC is currently activated (guard @grd_BC).

An important point about M_1 is that *its invariants do not explicitly assert the system level safety*, that is, that the train does not overrun (**SAF-0**). We can establish the safety of M_1, however, by showing that: (1) M_1's behaviour conforms with M_0's, and (2) M_0 is safe.

9.6.3.3 Verifying M_1

The invariant preservation PO of M_1 is generated much like that of M_0. The key to proving it is defining an appropriate $ACNext(x,v)$. We do not present an explicit formula here, but it essentially says @inv_stop is satisfied by a large enough margin. Guided by the PO, we can even derive the definition so that it makes PO provable.

In addition, we need to prove that the behaviour of the concrete machine conforms with that of the abstract machine. *PO* In particular, we need to prove the following two properties.

Guard strengthening: Concrete guards should be stronger than abstract guards. This guarantees that if the concrete event is enabled, then the abstract event is also enabled.

Action simulation: Predicates written in concrete actions should imply those of abstract actions. This guarantees that the concrete state transition is not contradictory to the abstract one.

For example, the following PO is generated for `AC->BC`:

$\ldots$	(M_0's invariants with old values)
$ctrl \in \{AC, BC\}$,	(`@inv_ctrl` with old values)
$ctrl = AC \Longrightarrow$	
$nextMaxX(x, v) +$	
$stopDist(nextMaxV(v)) \leq x_{dst}$,	(`@inv_AC` with old values)
$ctrl = AC$,	(`@grd_AC`)
$\neg ACNext(x, v)$,	(`@grd_BCNext`)
$x' + stopDist(v') \leq x_{dst} \wedge$	
$x' \leq nextMaxX(x, v) \wedge v' \leq nextMaxV(v)$,	(`@act_xv`)
$ctrl' = BC$,	(`@act_ctrl`)
$\vdash$	
$true \wedge x' + stopDist(v') \leq x_{dst} \wedge$	(`@grd_whenever` and `@act_xv`
$x' \leq nextMaxX(x, v) \wedge v' \leq nextMaxV(v)$	of run in M_0)

The assumption of this PO is constructed in the same manner as the invariant preservation, while the consequent part has the abstract guard (`@grd_whenever`) and the abstract action (`@act_xv`) in M_0's `run` event.

Since `AC->BC` inherits the guard and the action of the abstract event `run`, Rodin can easily prove this PO.

9.6.3.4 Requirements of Concrete Machine M_2

In M_2, we specify the deterministic control of BC. BC can be any rule-based controller such that we can guarantee its safety; in this example, it simply brakes with a fixed deceleration rate until it stops.

The following are functional requirements of BC's control:

FUN-2-1 When switching from AC to BC happens, BC calculates T_{brake}, which is how long it should keep braking as follows:

$$T_{brake} = v/b_{max}.$$

FUN-2-2 BC keeps activating the brake with deceleration rate b_{max} for the duration T_{brake}.

FUN-2-3 After T_{brake}, BC releases the brake.

By verifying the conformance of M_2's events with M_1's, we show that the deterministic control of BC satisfies the requirements specified in M_0 and M_1.

9.6.3.5 Specifying M_2

In M_2, we specify BC's control by following **FUN-2-1** – **FUN-2-3**. Since the control changes with time (**FUN-2-2** and **FUN-2-3**), we take care of the remaining time (number of cycles) required for braking.

The overview of machine M_2 is as follows:

```
machine M2 refines M1
variables x,v,ctrl,t_brake
invariants @inv_BC_v: ctrl = BC ⟹ v = b_max t_brake
events
  event AC->AC refines AC->AC
    (same as M1)
  end
  event AC->BC_brake refines AC->BC
    ...
  end
  event BC_brake->AC refines BC->AC
    ...
  end
  event BC_brake->BC_brake refines BC->BC
    ...
  end
end
```

The new **variable** t_{brake} is the remaining time required for braking. **Invariant** @inv_BC_v states that the remaining braking time t_{brake} should be the same as the time required for braking to stop. There are four **events**, each of which is declared as a refined version of an event of M_1. All events except AC->AC are different from M_1's.

Among the four events, we elaborate on AC->BC_brake and BC_brake->BC_brake below.

Event AC->BC_brake inherits predicates of AC->BC but it has additional actions for requirement **FUN-2-1**; in action @act_v_tBrake, the value of t_{brake} is updated with the duration needed to stop ($T_{brake} = v'/b_{max}$).

```
event AC->BC_brake refines AC->BC
  where (same guard as M1)
  then @act_x: x :| x' = x + ∫_{t=0}^{min(1,t_brake)} (v − b_max t) dt
       @act_v_tBrake: v,t_brake :| v' = b_max t'_brake ∧
                                    v' = v − ∫_{t=0}^{min(1,t_brake)} b_max dt
end
```

In event BC_brake->BC_brake, actions are not inherited from its abstract event BC->BC. In particular, actions of updating values x and v (@act_x and @act_v) are *deterministically* specified — recall that they are *non-deterministic* in BC->BC of M_1 (@act_xv). The event also has action @act_tBrake to update the remaining braking time.

```
event BC_brake->BC_brake refines BC->BC
  where @grd_BC: ctrl = BC
        @grd_BCNext: ¬ACNext(x,v)
        @grd_br: 0 < t_brake
  then @act_x: x :| x' = x + ∫_{t=0}^{min(1,t_brake)} (v − b_max t)dt
       @act_v: v :| v' = v − ∫_{t=0}^{min(1,t_brake)} b_max dt
       @act_ctrl: ctrl :| ctrl' = BC
       @act_tBrake: t_brake :| t'_brake = max(0, t_brake − 1)
end
```

9.6.3.6 Verifying M_2

The interesting POs of M_2 are about action simulations between M_1's events and M_2's. They guarantee that the deterministic control specified in M_2 satisfies the requirements specified in M_1.

For instance, the PO of action simulation between `BC_brake->AC` and `BC->AC` is as follows:

$x + stopDist(v) \leq x_{dst},$	(`@inv_stop`)
$\ldots$	(M_0's and M_1's other invariants)
$\ldots$	(M_2's invariants)
$x' = x + \int_{t=0}^{min(1,t_{brake})} (v - b_{max}t)dt,$	(`@act_x`)
$v' = v - \int_{t=0}^{min(1,t_{brake})} b_{max}dt,$	(`@act_v`)
$\ldots$	(other concrete guards/actions)
$\vdash$	
$x' + stopDist(v') \leq x_{dst} \wedge$	
$x' \leq nextMaxX(x,v) \wedge v' \leq nextMaxV(v)$	(`@act_xv` of `BC->AC` in M_1)

For proving the PO, we need to guide Rodin using its interactive proof interface a little (e.g., for case analysis regarding $min(1, t_{brake})$), but we can mostly rely on Rodin's automatic provers.

9.6.4 Refinement Strategy

As described in the previous sections, the refinement in Event-B is done in a *posit-and-prove* style; we construct a machine that is declared as a candidate of a concrete machine of another, and then we verify that its behaviour really conforms with that of the abstract machine.

Therefore, planning which elements of the target system are introduced in each refinement step (*refinement strategy*) is not a trivial task. In fact, engineers often find it difficult.

There is not the "correct" refinement strategy, but good practices of refinement strategies including the following are shared:

In abstract machines	In concrete machines
abstract concepts	concrete concepts
whole system	details of subsystems
requirements	means to satisfy the requirements
what we observe as results	how the system works

Also, our empirical study [36] indicated that the following practices can reduce the complexity of modelling and proving with the stepwise refinement mechanism.

- Having more refinement steps, while trying to introduce, in the same refinement step, those variables which tend to occur together.
- Introducing variables that often appear in invariants early.

9.7 Conclusions

In this chapter, we reviewed about technologies for *theorem proving*—writing mathematical proofs in a formalised, i.e., machine-assisted manner. After reviewing the basics of formal logic (Section 9.2), we covered some general techniques such as interactive theorem proving in a highly expressive logic (Isabelle/HOL, Section 9.3) and fully automated theorem proving in languages with restricted expressivity (SAT/SMT solvers, Section 9.4). We also reviewed two specific lines of development towards automotive applications, namely accommodation of continuous/hybrid dynamics (Section 9.5) and gradual model refinement (Section 9.6). The approaches we covered in this chapter are summarised in Table 9.1.

In general, theorem proving is a methodology that needs careful and strategic planning before its application.

- When it is successfully applied, its benefits are enormous: safety claims as its conclusions come with *absolute guarantees* in the form of mathematical proofs; moreover, the logical reasoning steps explicated in those proofs serve as *explanations* why the conclusions must be true. These benefits should be especially favoured in the automotive domain that is safety critical (strong guarantees are desired in the face of potential loss of human lives) and socially accepted (car manufacturers are under a big pressure for social explainability and accountability).
- However, the cost for these benefits can be enormous as well. The first kind of cost is *modelling cost*: proofs in theorem proving must be established on rigorous system models (i.e., the mathematical *definitions* of target systems, cf. Section 9.1.4). A detailed modelling of a single component of a car can easily take several person month. A comprehensive modelling of a car as a whole requires a fundamental reformation of development processes and decades of time—this is what is being pursued in the aerospace industry. In the automotive industry

where the development processes are always under the pressure from competition and short model cycles, such a fundamental reformation is hardly feasible.

The second kind of cost is the cost of *proving*. Automated theorem proving approaches such as SMT solvers only apply to problems that can be expressed in a restricted language; interactive theorem proving, in contrast, requires a lot of human effort. A single proof costing several person year is not uncommon.

Therefore, theorem proving is in general a *high-cost high-return* methodology. One has to choose a problem setting very carefully and strategically in order to find a good balance between cost and benefit.

One seeming example of such a good balance identified in the automotive domain is the methodology of RSS (mentioned in Section 9.1.4; more details in Section 7.5 and [19, 32, 33, 53]). In RSS, the "frontier" of mathematical modelling is chosen to be some behavioural *safety rules* imposed on each vehicle, and the theorem proving efforts never go into the internal working of vehicles. This makes the modelling efforts (each vehicle can be a black-box) and the proof efforts (proofs are essentially massive case distinctions over high-school physics) feasible. Indeed, in the recent paper [32], a logic similar to the one in Section 9.5 is used to express safety rules and prove their safety. These safety rules play the role of *traffic laws* in the automated driving ecosystem, ensuring the social and industrial relevance of these rules and their safety proofs.

We believe that more targets of theorem proving should be found in the automotive domain. In the coming years where V2V communication is more common, the correctness, security, and privacy of the V2V networks can be a good target of automated theorem proving (Section 9.4), much like the current widespread use of SAT/SMT solvers for the quality assurance of cloud systems [15]. In the quality assurance efforts by automotive manufacturers—e.g. to ensure that their products indeed comply with the safety rules in RSS—theorem proving by iterative model refinement (Section 9.1.4) can serve as a method to decompose system-level requirements into component-level requirements, where the latter can be further passed on to part manufacturers.

References

[1] Stateflow documentation, 2021. Accessed: January 24, 2021.

[2] Abrial, J-R. 2010. Modeling in Event-B: System and Software Engineering. Cambridge University Press.

[3] Abrial, J-R., Butler, M., Hallerstede, S., Hoang, T.S., Mehta, F. and Voisin, L. 2010. Rodin: An open toolset for modelling and reasoning in Event-B. International Journal on Software Tools for Technology Transfer, 12(6): 447–466.

[4] Alur, R., Courcoubetis, C., Henzinger, T.A. and Ho, P-H. 1992. Hybrid automata: An algorithmic approach to the specification and verification of hybrid systems. *In*: Robert L. Grossman, Anil Nerode, Anders P. Ravn and Hans Rischel, (eds). Hybrid Systems, volume 736 of Lecture Notes in Computer Science, pages 209–229. Springer.

[5] Bancerek,G.,Bylinski,C.,Grabowski,A.,Kornilowicz,A.,Matuszewski,R., Naumowicz, A., et al. 2018. The role of the Mizar Mathematical Library for interactive proof development in Mizar. Journal of Autom. Reason., 61(1–4): 9–32.

[6] Barbosa, H., Barrett, C.W., Brain, M., Kremer, G., Lachnitt, H., Mann, M., et al. 2022. cvc5: A versatile and industrial-strength SMT solver. *In*: Dana Fisman and Grigore Rosu, (eds). Tools and Algorithms for the Construction and Analysis of Systems—28th International Conference, TACAS 2022, Held as Part of the European Joint Conferences on Theory and Practice of Software, ETAPS 2022, Munich, Germany, April 2–7, 2022, Proceeding, Part I, volume 13243 of Lecture Notes in Computer Science, pages 415–442. Springer.

[7] Barrett, C.W. and Tinelli, C. 2018. Satisfiability modulo theories. *In*: Edmund M. Clarke, Thomas A. Henzinger, Helmut Veith and Roderick Bloem, (eds). Handbook of Model Checking, pages 305–343. Springer.

[8] Bertot, Y. and Castéran, P. 2004. Interactive Theorem Proving and Program Development—Coq'Art: The Calculus of Inductive Constructions. Texts in Theoretical Computer Science. An EATCS Series. Springer.

[9] Bibel, W. 1982. Automated theorem proving. Vieweg.

[10] Biere, A. 2014. Lingeling essentials, a tutorial on design and implementation aspects of the the SAT solver Lingeling. *In*: Daniel Le Berre, (ed). "2001 by CRC Press LLC POS-14. Fifth Pragmatics of SAT workshop, a workshop of the SAT 2014 conference, part of FLoC 2014 during the Vienna Summer of Logic, July 13, 2014, Vienna, Austria, volume 27 of EPiC Series in Computing, page 88. EasyChair.

[11] Biere, A., Heule, M., van Maaren, H. and Walsh, T. 2009. (eds). Handbook of Satisfiability, volume 185 of Frontiers in Artificial Intelligence and Applications. IOS Press.

[12] Böhme, S. and Nipkow, T. 2010. Sledgehammer: Judgement day. *In*: Jürgen Giesl and Reiner Hähnle, (eds). Automated Reasoning, 5th International Joint Conference, IJCAR 2010, Edinburgh, UK, July 16–19, 2010. Proceeding, volume 6173 of Lecture Notes in Computer Science, pages 107–121. Springer.

[13] Boyer, R.S. and Strother Moore, J. 1980. A computational logic. ACM monograph series. Academic Press.

[14] Chen, X., Ábrahám, E. and Sankaranarayanan, S. 2013. Flow*: An analyzer for non-linear hybrid systems. *In*: Natasha Sharygina and Helmut Veith, (eds). Computer Aided Verification—25th International Conference, CAV 2013, Saint Petersburg, Russia, July 13–19, 2013. Proceeding, volume 8044 of Lecture Notes in Computer Science, pages 258–263. Springer.

[15] Chong, N., Cook, B., Eidelman, J., Kallas, K., Khazem, K., Monteiro, F.R., et al. 2021. Code-level model checking in the software development workflow at amazon web services. Software: Practice and Experience, 51(4): 772–797.

[16] De Bruijn, N.G. 1970. The mathematical language AUTOMATH, its usage, and some of its extensions. *In*: Symposium on Automatic Demonstration, pages 29–61. Springer.

[17] de Moura, L.M. and Bjørner. N.S. 2008. Z3: An efficient SMT solver. *In*: C.R. Ramakrishnan and Jakob Rehof, (eds). Tools and Algorithms for the Construction and Analysis of Systems, 14th International Conference, TACAS 2008, Held as Part of the Joint European Conferences on Theory and Practice of Software, ETAPS 2008, Budapest, Hungary, March 29–April 6, 2008. Proceeding, volume 4963 of Lecture Notes in Computer Science, pages 337–340. Springer.

[18] Dutertre, B. 2014. Yices 2.2. *In*: Armin Biere and Roderick Bloem, (eds). Computer Aided Verification—26th International Conference, CAV 2014, Held as Part of the Vienna Summer of Logic, VSL 2014, Vienna, Austria, ″2001 by CRC Press LLC July 18–22, 2014. Proceeding, volume 8559 of Lecture Notes in Computer Science, pages 737–744. Springer.

[19] Eberhart, C., Dubut, J., Haydon, J. and Hasuo, I. 2023. Formal verification of safety architectures for automated driving. *In*: IEEE Intelligent Vehicles Symposium, IV 2023, Anchorage, AK, USA, June 4–7, 2023, pages 1–8. IEEE, 2023. Preprint available as CoRR abs/2308.10365.

[20] Eggersglüß, S. and Drechsler, R. 2012. High Quality Test Pattern Generation and Boolean Satisfiability. Springer.

[21] Foster, S., Huerta y Munive, J.J. and Struth, G. 2020. Differential hoare logics and refinement calculi for hybrid systems with Isabelle/HOL. *In*: Uli Fahrenberg, Peter Jipsen, and Michael Winter, (eds). Relational and Algebraic Methods in Computer Science—18th International Conference, RAMiCS 2020, Palaiseau, France, April 8–11, 2020, Proceeding [postponed], volume 12062 of Lecture Notes in Computer Science, pages 169–186. Springer.

[22] Frehse, G., Guernic, C.L., Donze, A., Cotton, S., Ray, R., Lebeltel, O., et al. 2011. Spaceex: Scalable verification of hybrid systems. *In*: Ganesh Gopalakrishnan and Shaz Qadeer, (eds). Computer Aided Verification—23rd International Conference, CAV 2011, Snowbird, UT, USA, July 14–20, 2011. Proceeding, volume 6806 of Lecture Notes in Computer Science, pages 379–395. Springer.

[23] Froleyks, N., Heule, M., Iser, M., Järvisalo, M. and Suda, M. 2021. SAT competition 2020. Artificial Intelligence, 301: 103572.

[24] Fulton, N., Mitsch, S., Quesel, J-D., Völp, M. and Platzer, A. 2015. KeYmaera X: An axiomatic tactical theorem prover for hybrid systems. *In*: Amy P. Felty and Aart Middeldorp, (eds). CADE, volume 9195 of LNCS, pages 527–538. Springer.

[25] Gao, S., Kong, S. and Clarke, E.M. 2013. Dreal: An SMT solver for nonlinear theories over the reals. *In*: Maria Paola Bonacina, (ed). Automated Deduction— CADE-24—24th International Conference on Automated Deduction, Lake Placid, NY, USA, June 9–14, 2013. Proceeding, volume 7898 of Lecture Notes in Computer Science, pages 208–214. Springer.

[26] Gonthier, G. 2008. Formal proof: The four-color theorem. Notices of the AMS, 55: 1382–1393.

[27] Gordon, M.J.C., Milner, R. and Wadsworth, C.P. 1979. Edinburgh LCF, volume 78 of Lecture Notes in Computer Science. Springer.

[28] Gordon, M. 2000. From LCF to HOL: A short history. *In*: Gordon D. Plotkin, Colin Stirling, and Mads Tofte, (eds). Proof, Language, and Interaction, Essays in Honour of Robin Milner, pages 169–186. The MIT Press.

[29] Hales, T.C., Adams, M., Bauer, G., Dang, D.T., Harrison, J., Hoang, T.L., et al. 2015. A formal proof of the kepler conjecture. CoRR, abs/1501.02155.

[30] Harrison, J. 2000. Floating point verification in HOL Light: The exponential function. Formal Methods in System Design, 16(3): 271–305, 2000.

[31] Hasuo, I. 2022. Introduction to logic and computability. Lecture Notes, available online.

[32] Hasuo, I., Eberhart, C., Haydon, J., Dubut, J., Bohrer, R., Kobayashi, T., et al. 2023. Goal-aware RSS for complex scenarios via program logic.

IEEE Transactions on Intelligent Vehicles, 8(4): 3040–3072, 2023. Preprint available as CoRR abs/2207.02387.

[33] Haydon, J., Bondu, M., Eberhart, C., Dubut, J. and Hasuo, I. 2023. Formal verification of intersection safety for automated driving. CoRR, abs/2308.06785, 2023. To appear in Proceeding ITSC 2023.

[34] Hunt Jr., W.A. and Swords, S. 2009. Centaur technology media unit verification. *In*: Ahmed Bouajjani and Oded Maler, (eds). Computer Aided Verification, 21st International Conference, CAV 2009, Grenoble, France, June 26 –July 2, 2009. Proceeding, volume 5643 of Lecture Notes in Computer Science, pages 353–367. Springer.

[35] Kalman, J.A. 2001. Automated Reasoning with Otter, volume 3. Rinton Press Princeton NJ.

[36] Kobayashi, T. and Ishikawa, F. 2018. Analysis on strategies of superposition refinement of Event-B specifications. *In*: Jing Sun and Meng Sun, (eds). Formal Methods and Software Engineering, pages 357–372, Cham, 2018. Springer International Publishing.

[37] Kong, S., Gao, S., Chen, W. and Clarke, E.M. 2015. Dreach: δ-reachability analysis for hybrid systems. *In*: Christel Baier and Cesare Tinelli, (eds). Tools and Algorithms for the Construction and Analysis of Systems—21st International Conference, TACAS 2015, Held as Part of the European Joint Conferences on Theory and Practice of Software, ETAPS 2015, London, UK, April 11–18, 2015. Proceeding, volume 9035 of Lecture Notes in Computer Science, pages 200–205. Springer.

[38] Leroy, X. 2009. Formal verification of a realistic compiler. Communications of the ACM, 52(7): 107–115.

[39] Minsky, Y., Madhavapeddy, A. and Hickey, J. 2013. Real World OCaml: Functional Programming for the Masses. O'Reilly.

[40] Mitsch, S. and Platzer, A. 2016. The KeYmaera X proof IDE: Concepts on usability in hybrid systems theorem proving. *In*: Catherine Dubois, Paolo Masci, and Dominique Méry, (eds). 3rd Workshop on Formal Integrated Development Environment, volume 240 of EPTCS, pages 67–81.

[41] Strother Moore, J., Lynch, T.W. and Kaufmann, M. 1998. A mechanically checked proof of the amd5k86tm floating point division program. IEEE Transactions on Computers, 47(9): 913–926.

[42] Naumowicz, A. and Kornilowicz, A. 2009. A brief overview of Mizar. *In*: Stefan Berghofer, Tobias Nipkow, Christian Urban, and Makarius Wenzel, (eds). Theorem Proving in Higher Order Logics, 22nd International Conference, TPHOLs 2009, Munich, Germany, August 17–20, 2009. Proceeding, volume 5674 of Lecture Notes in Computer Science, pages 67–72. Springer.

[43] Newell, A. and Simon, H.S. 2002. The logic theory machine: A complex information processing system. IRE Transactions on Information Theory, 2(3): 61–79, 1956.

[44] Nipkow, T., Paulson, L.C. and Wenzel, M. 2002. Isabelle/HOL— A Proof Assistant for Higher-Order Logic, volume 2283 of Lecture Notes in Computer Science. Springer.

[45] Paulson, L.C. 1996. ML for the Working Programmer (2nd. ed.). Cambridge University Press.

[46] Paulson, L.C., Nipkow, T. and Wenzel, M. 2019. From LCF to Isabelle/HOL. Formal Aspects of Computing, 31(6): 675–698.

[47] Platzer, A. 2018. Logical Foundations of Cyber-Physical Systems. Springer, Cham.

[48] Riazanov, A. and Voronkov, A. 2002. The design and implementation of VAMPIRE. AI Communications, 15(2–3): 91–110.

[49] Sawada, J. and Gamboa, R. 2002. Mechanical verification of a square root algorithm using Taylor's theorem. *In*: Mark D. Aagaard and John W. O'Leary, (eds). Formal Methods in Computer-Aided Design, 4th International Conference, FMCAD 2002, Portland, OR, USA, November 6–8, 2002, Proceeding, volume 2517 of Lecture Notes in Computer Science, pages 274–291. Springer.

[50] Schulz, S. 2002. E: A brainiac theorem prover. AI Communications, 15(2–3): 111–126.

[51] Scott, D.S. 1993. A type-theoretical alternative to iswim, cuch, OWHY. Theoretical Computer Science, 121(1&2): 411–440.

[52] Seto, D., Krogh, B., Sha, L. and Chutinan, A. 1998. The simplex architecture for safe online control system upgrades. *In*: Proceeding of the 1998 American Control Conference. ACC (IEEE Cat. No. 98CH36207), volume 6, pages 3504–3508.

[53] Shalev-Shwartz, S., Shammah, S. and Shashua, A. 2017. On a formal model of safe and scalable self-driving cars. CoRR, abs/1708.06374.

[54] Sutcliffe, G. 2016. The CADE ATP System Competition—CASC. AI Magazine, 37(2): 99–101.

[55] Troelstra, A.S. and Schwichtenberg, H. 1996. Basic Proof Theory. Number 43 in Cambridge Tracts in Theoretical Computer Science. Cambridge University Press.

[56] Tseitin, G.S. 1983. On the Complexity of Derivation in Propositional Calculus, pages 466–483. Springer Berlin Heidelberg, Berlin, Heidelberg.

[57] van Dalen, D. 1994. Logic and Structure (3rd. ed.). Universitext. Springer.

[58] Weber, T., Conchon, S., Daharbe, D., Heizmann, M., Niemetz, A. and Reger, G. 2019. The SMT competition 2015–2018. Journal on Satisfiability, Boolean Modeling and Computation, 11(1): 221–259.

[59] Wilding, M., Greve, D.A., Richards, R.J. and Hardin, D.S. 2010. Formal verification of partition management for the AAMP7G microprocessor. *In*: David S. Hardin, (ed.). Design and Verification of Microprocessor Systems for High-Assurance Applications, pages 175–191. Springer.

Chapter 10

Search-Based Analysis and Engineering

Paolo Arcaini
National Institute of Informatics, Tokyo, Japan

Fuyuki Ishikawa
National Institute of Informatics, The University of Electro-Communications and SOKENDAI, Tokyo, Japan

10.1 Introduction

Many problems in Software Engineering (SE) can be translated to optimisation problems [31]. In software testing, for example, typical problems are the maximization of the coverage achieved by a test suite, or the minimization of the test suite size. In requirements engineering, a classical problem consists of maximizing the satisfaction of the requirements, and minimizing the software development costs. Very often, these problems have multiple objectives that are conflicting with each other; for example, still considering software testing, it is often desirable to maximize coverage and minimize test suite size at the same time. It is clear that the two objectives are conflicting, as, in order to obtain more coverage, one needs to add more tests.

Since the early 2000s,[1] it has been proposed to use and adapt optimisation techniques to solve SE problems; this research area has been named Search-Based Software Engineering (SBSE) [30]. Different approaches have been proposed to solve different problems, such as the next release problem (i.e., the selection of next requirement to implement) [83], fault localization [46], automated program repair [28, 44], repair of DNNs [49, 50], etc. We refer to [31] for a survey, and to the repository [82] for a list of SBSE works till 2017.

Most of the works in SBSE are related to testing problems, and this area is known as Search-Based Testing (SBT) [29, 55, 56]. Works have been proposed for test generation [6, 24], test case prioritization [7, 67], mutation testing [68], combinatorial testing [76], etc. Also falsification (presented in Chapter 1) is an instance of SBT in which the objective function is given by the robustness of a formal specification written in some temporal logic.

Due to its ability to handle large and complex search spaces, SBT has been applied also to testing complex systems, such as machine learning systems [14, 63], quantum programs [72–74], control systems [20], and automated driving systems (ADSs) [5, 11, 13, 15, 16, 41, 47, 53]. In this chapter, we are interested in the application of SBT to ADSs, in particular for scenario generation.

We first provide a minimal background on multi-objective optimisation in Section 10.2. Then, in Section 10.3, we present a general framework for SBT of ADSs. The different components of the framework are elaborated in Sections 10.4 to 10.8. Section 10.9 concludes the chapter.

10.2 Preliminaries

In SBSE, software engineering problems are cast as *search problems* of one or more objectives. In the following, we provide background on multi-objective optimisation; single-objective problems are considered a special case of these.

10.2.1 Multi-Objective Optimisation

A *multi-objective optimisation problem* is given by a set of m objectives that need to be either maximized or minimized; in the following, without loss of generality, we consider minimization problems.[2] We formally define a multi-objective optimisation problem as $F(\mathbf{x}) = (f_1(\mathbf{x}), \ldots, f_m(\mathbf{x}))$, where $\mathbf{x} = (x_1, \ldots, x_n)$ is an n-dimensional decision variable vector defined over a universe X, and each $f_i(\mathbf{x})$ an objective function.

[1]There are earlier works using optimisation for SE problems [57]. Please refer to [30] for an historical survey.

[2]Maximization problems can be obtained by simply negating the objective function.

A *solution* is an element of X. Optimization problems can also define *constraints* that must be satisfied by solutions; in this case, they are called *constrained optimisation problems*. We define the space of feasible solutions as $\Omega \subseteq X$.

Given two feasible solutions $A, B \in \Omega$, we say that A *dominates* B ($A \succ B$) iff,

$$(\forall i \in \{1,\dots,m\}.f_i(A) \leq f_i(B)) \wedge (\exists j \in \{1,\dots,m\}.f_j(A) < f_j(B))$$

i.e., A is never worse than B for any objective, and it is better for at least one.

For simplicity, one can also consider the solutions in the objective space, i.e., $A = (f_1(\mathbf{x}),\dots,f_m(\mathbf{x}))$ with $\mathbf{x} \in \Omega$.

A *search algorithm* explores the search space Ω trying to optimise all the objective functions. The algorithm evaluates some possible solutions S, and returns as output a *Pareto front*, i.e., the subset of S of non-dominated solutions:

$$PF(S) = \{s_i \in S \mid (\nexists s_j \in S.s_j \succ s_i)\}$$

Different algorithms have been proposed to solve optimisation problems, mostly based on evolutionary computation. Notable examples are NSGA-II [21], MOCell [59], SMPSO [58], SPEA2 [85], PAES [39], and CellDE [22].

10.2.2 Quality Indicators

In practical contexts, one often needs to choose the best among different Pareto fronts computed by the same algorithm or by different algorithms. However, this task is not trivial, as Pareto fronts are not directly comparable. Therefore, different *quality indicators* have been proposed. A quality indicator assigns a numerical value to a Pareto front, by considering how well the Pareto front achieves quality aspects such as *uniformity*, *spread*, *convergence*, or *cardinality*. In this way, different Pareto fronts can be compared using their quality indicator assessments.

Hypervolume (HV), for example, is a popular quality indicator that measures the volume of the objective space covered by a Pareto front PF [66] w.r.t. a *reference point* w (i.e., the worst possible solution). It is defined as:

$$HV(PF) = volume(\bigcup_{s_i \in PF} hc(s_i))$$

where $hc(s_i)$ is the hypercube defined between diagonal corners s_i and w. Figure 10.1 shows a visual representation of the hypervolume of the Pareto front computed for a two-objective problem (the gray area).

In literature, several quality indicators have been proposed that consider different quality aspects; please refer to [48] for a complete survey. Given this large set of indicators (100 indicators are surveyed in [48]), SBSE practitioners

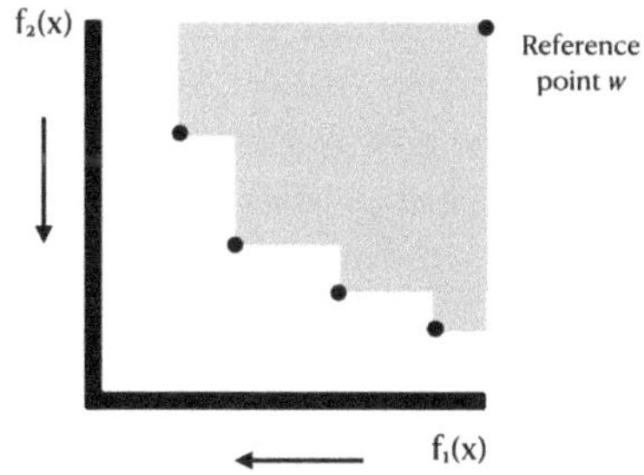

Figure 10.1: Hypervolume (in the figure, f_1 and f_2 need to be minimized).

may have problems in selecting a proper quality indicator for evaluating the solutions produced by the used search algorithm. Therefore, guidelines have been developed that assist the user in selecting a quality indicator (or a set of quality indicators) that is most suitable for the used search algorithm [2]; namely, the guide suggests the minimum set of quality indicators that are representative of the "opinions" of different indicators. In other contexts, instead, the user may know which quality indicator to use (because it reflects some desired properties), but does not know which search algorithm to use; the guide in [3, 77] helps in doing such a selection: given a quality indicator, it suggests the search algorithm that most likely provides solutions that are good for it.

10.3 Search-Based Testing for Autonomous Driving Systems

ADSs can potentially transform the transportation domain, and bring several advantages, such as fewer accidents, and reduced congestion [79]. Being complex and safety-critical systems, the throughout testing of ADSs is mandatory for their acceptance [70]. One type of ADS testing consists of doing on-road driving of the autonomous vehicle. Although such testing is ultimately necessary and it is employed for different industrial ADSs such as Voyage.auto [71] and Waymo [75], it has several limits. First of all, it is costly as the final autonomous vehicle must have been produced, and also potentially very dangerous in case failures occur during testing. Moreover, it has been shown that such testing provides limited guarantees, and it requires to consider a huge amount of driving data; namely, 11 billion miles of drives can provide 95% confidence that the autonomous vehicle has a failure rate lower than that of a human driver [33,54].

A more feasible and pursued approach is *simulation-based* testing, in which the ADS is run in a simulated environment [10,84]. A simulator usually allows to specify the environment in which the ADS is run, in terms of road map, regulatory elements (e.g., traffic lights, traffic signals, etc.), initial position of the autonomous vehicle, initial position and behaviour of other traffic participants (i.e., other vehicles and pedestrians), etc. A tester can specify different test scenarios to exercise the ADS under different driving conditions. Given a test scenario, the ADS is simulated and simulation data is collected. This data can be then analysed to check whether the ADS behaves as expected.

Although simulation-based testing is more flexible than *on-road* testing, it still presents some issues. The main problem is related to the selection of the test scenarios to execute. One may think to use scenarios derived from real test drives or to re-use scenarios developed for other systems. However, Hauer et al. [32] have shown that such an approach is not effective, as these scenarios do not properly challenge the ADS under test, and so some faulty behaviours may not be triggered. They [32], instead, provided evidence that test scenarios should be *generated* for the specific ADS under test.

Assuming that some test goals have been specified, generating test scenarios that cover these goals is challenging. First of all, the search space for scenarios is infinite; moreover, logical approaches such as concolic testing are not applicable to generate such complex inputs. SBT is a promising technique that has demonstrated to be suitable to tackle these issues. Figure 10.2 shows an overview of SBT for ADSs.

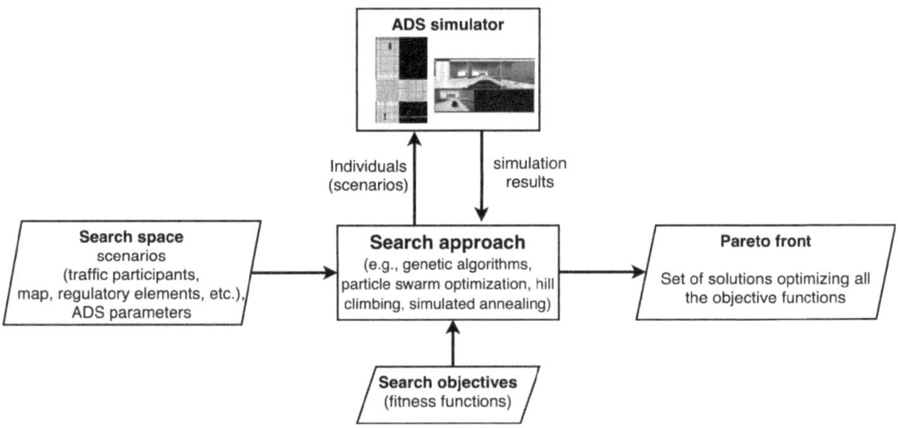

Figure 10.2: Search-based testing for autonomous driving systems.

One of the main components of the approach is the definition of the objective functions, which are implemented as *fitness functions*. These define the *test goals* (or *test requirements*) of the testing technique. Test requirements can be specified in different ways. Some approaches [11, 13, 15, 16, 47] focus on particular performance metrics, such as degree of safety, or comfort, and aim at finding scenarios that minimize them. Other approaches, instead, define specific *coverage criterion*, such as covering the parameters of the ADS [41], covering different types of driving manoeuvres [5], or covering different combinations of requirements violations [53]. We will review different approaches in Section 10.4.

Another component of the approach is the definition of the *search space*, i.e., where to search for scenarios that *cover* the test goals. Search spaces can be defined over different elements of a scenario, such as the behaviour of the traffic participants, the road structure, the weather conditions, etc. Some search approaches also search for different configurations of the ADS that can provide a better driving in some particular situations. An overview of different approaches to specify the search space will be given in Section 10.5.

The computation of the fitness functions requires the simulation of the ADS over the selected scenario, and the collection of the simulation data; Fig. 10.2 shows two examples of ADS simulators: the simulator of an industrial path planner, and WiseSIM [4], the simulator for the Autonomoose ADS [60]. Starting from the simulation data, the values of the fitness functions are computed. There are two main issues in this phase. First of all, running an ADS simulator is computationally expensive, with simulations that can take up to several minutes and more; in order to tackle this problem, some approaches (discussed in Section 10.7) have been proposed to approximate the computation of the fitness function using surrogate models that do not need ADS simulations. The second issue of this phase is that ADS simulations may not be completely deterministic, and so the simulation data is not completely reliable for the computation of the fitness values; some recent approaches, discussed in Section 10.8, have been proposed to tackle this problem.

As the final step of the SBT approach, the user receives as output a set of optimal scenarios. In case of single-objective search, only one scenario is provided. In case of multi-objective search, instead, the output is given by a *Pareto front* of scenarios, i.e., the scenarios that are not strictly dominated by other scenarios.

In the next sections, we will review research works regarding the different phases of the framework as summarised in Table 10.1.

Table 10.1: Research topics in search-based testing for autonomous driving systems

Definition of the search objectives	Definition of the search space	Oracle problem	Scalability	Noisy simulations
Section 10.4	Section 10.5	Section 10.6	Section 10.7	Section 10.8

10.4 Search Objectives

In a search-based approach, the *search objectives* specify the characteristics that the generated scenarios should have. Search objectives can be defined in different ways, depending on which aspects of the ADS need to be tested. We here identify two main categories of search objectives.

Metric-based search objectives aim at finding scenarios that optimise a given metric defined on the simulation of the ADS over the test scenario, such as *Euclidean distance to the other vehicles*, *comfort* (measured by, e.g., maximum acceleration), etc. We will review some of these approaches in Section 10.4.1.

Metric-based objectives are useful to generate particular scenarios possibly showing that the ADS can fail in a given way; however, they do not provide any type of guarantee regarding what ADS behaviours are observed in the generated test scenarios, i.e., they are not interested in the diversity of the ADS behaviours. In testing for classical programs, *coverage criteria* are used to characterize different aspects of the system under test that need to be covered by a test suite; *coverage-based* search objectives follow this approach and, starting from coverage criteria specified for ADSs, use objective functions that aim to obtain full coverage. We will review these approaches in Section 10.4.2. A summary of all the approaches is given in Table 10.2.

Table 10.2: Different types of search objectives in search-based testing for autonomous driving systems

Metric-based search objectives	Coverage-based search objectives
Time to collision [12] (Section 10.4.1.1)	Parameter coverage [40] (Section 10.4.2.1)
Safety distance [47] (Section 10.4.1.2)	Patterns of driving characteristics [5] (Section 10.4.2.2)
Feature interactions [13] (Section 10.4.1.3)	Patterns of requirements violations [53] (Section 10.4.2.3)

10.4.1 Metric-Based Search Objectives

We classify in the *metric-based* category all the approaches that aim at maximizing or minimizing a specific metric defined over the ADS simulation. Usually, the goal is to worsen the metric (e.g., minimizing the minimum Euclidean distance between the ego vehicle and the other vehicles) in order to find specific ADS failures.

Several approaches test the ADS for safety and the typical goal of these approaches is to find dangerous situations in which the autonomous vehicle collides or is close to a collision. We here review some of them.

10.4.1.1 Time to Collision

Ben Abdessalem et al. [12] tested an Automated Emergency Braking (AEB), that is a common component in ADSs. An AEB operates as follows: its senses through sensors the position and speed of other traffic participants (pedestrians, vehicles, etc.) in a given field of view in front of the ego vehicle; if an object is perceived, the *time to collision* (TTC) is computed; if TTC is below a given threshold, the vision component is informed, and this tries to recognize the type of object; on the basis of the recognized object, the system decides whether to apply an emergency braking. AEB can be tested in the *PreScan* simulator that allowed it to specify different scenarios. The authors propose an SBT approach for finding critical test scenarios in which the autonomous vehicle interacts with a pedestrian who is trying to cross the street; specifically, the approach tries to find scenarios in which the autonomous vehicle collides at high speed with the pedestrian. This test goal is specified using two fitness functions:

- One requiring it to minimize the distance between the pedestrian and the vehicle.
- The other one requiring it to maximize the collision speed.

Note that the approach provides a Pareto front of solutions as output, each one showing a different way to satisfy the test goal.

10.4.1.2 Safety Distance

Li et al. [47] propose AV-FUZZER, an SBT approach combined with fuzzing, that has the goal of finding safety violations of an ADS implemented in the Apollo[3] framework. The approach defines a *safety potential* metric as difference $\delta = d_{safe} - d_{stop}$, where d_{safe} identifies the distance that the vehicle can drive without having a collision, and d_{stop} the distance required to stop completely. Of course, the ADS is safe as long as $\delta > 0$. The goal of AV-FUZZER is to find scenarios in which δ is minimized, so possibly exposing the safety violations of the ADS.

10.4.1.3 Feature Interactions

Ben Abdessalem et al. [13] target the safety violations due to feature interactions. An ADS is composed of different units of functionality (also called *features*), such as the Adaptive Cruise Control (ACC), the Automated Emergency Braking (AEB), the Traffic Sign Recognition (TSR), etc. Even if these features work correctly in isolation, they may lead to failures when combined together, in particular if they provide commands to the same actuators: for example, ACC and AEB both provide commands to the vehicle brakes. A feature interaction could occur when the command provided by a

[3] `https://apollo.auto/`

feature is overridden by a command of another feature, e.g., the command of the AEB to stop is overridden by the command of ACC to maintain the speed. The authors propose FITEST (Feature Interaction TESTing), an SBT approach to discover harmful feature interactions; the approach solves a *many-objective problem*, i.e., a problem with more than three objectives. Specifically, for each branch of the code involved in a feature interaction, the approach specifies an objective that aims at covering the branch and, at the same time, violating a safety requirement; in total, there are $k \times n$ objectives, being k the number of branches in the code and n the number of safety requirements. FITEST implements a type of search algorithm, called MOSA [61], that aims at solving each objective individually, rather than trying to optimise all the objectives together.

10.4.2 Coverage-Based Search Objectives

In testing, a *coverage criterion* defines what needs to be *covered* by the test suite in order to consider the system under test sufficiently tested. Specifically, a coverage criterion defines a set of *test goals* (also called *test requirements*); in order for the criterion to be completely fulfilled, each test goal must be covered in at least one test. SBT is an effective technique to generate tests satisfying particular test goals; the satisfaction of a test goal is embedded in the fitness function used to drive the search.

Most of the criteria proposed for ADS testing are *black box*: they are required to cover different types of scenarios in which the ADS may operate, or to observe different types of behaviours (i.e., different types of manoeuvres) of the ADS. For classical programs, different *white box* (also called *structural*) coverage criteria (e.g., statement coverage, branch coverage, MC/DC, etc.) have been proposed and are actively used. However, in [42], we have shown that this type of coverage is rather weak for ADSs, as very simple scenarios can achieve very high structural coverage: this is a problem, as it means that scenarios are not able to discriminate between the execution of different parts of the code and, hence, of possibly different types of ADS behaviours. Recently, some approaches have been proposed that still aim at covering the ADS implementation, but, in order to tackle the problems we identified in [42], they target the specific way ADSs are implemented. In the following, we review these criteria.

10.4.2.1 Parameter Coverage

In [40,41], we proposed a coverage criterion specific for testing the path planner component of the ADS of our industrial partner.[4] The path planner works as follows: At every timestep, it selects a *short-term path* to be followed by the ego

[4]The criterion was later extended in [43] to handle the non-determinism of ADS's simulators.

vehicle in the next few seconds, together with the commands (such as acceleration and angle) to apply to implement it. Figure 10.3 shows an example of simulation of the path planner. In the simulator, the red vehicle is the autonomous vehicle (also called *ego vehicle*) equipped with the path planner, and the black vehicles are other vehicles interacting with the ego vehicle. The coloured lines starting from the ego vehicle are its possible short term paths.

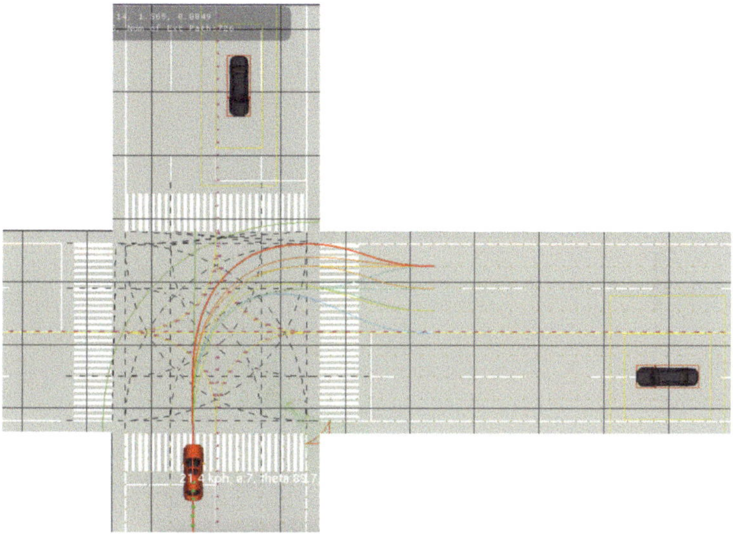

Figure 10.3: Path planner execution.

The next short-term path is decided using an optimisation algorithm. A cost function is used to assign a cost value to all the possible short-term paths: the cost function considers aspects such as *safety of driving*, *compliance to traffic regulations*, and *comfort*. The least costly short-term path is selected to be followed. A more detailed description of the selection of a short-term path is given in Algorithm 2. Given a short-term path *stp*:

- it evaluates the *driving characteristics* that are required in order to follow *stp*. Namely, given a characteristic `dc`, it checks whether the amount of `dc` is non-negligible, using a specific threshold (line 6). The intuition is that, over a given threshold, the driving characteristic `dc` affects some aspect of the drive, such as the safety, comfort, and compliance to traffic rules;
- in case the driving characteristic `dc` is over the threshold, a parameter $p_{\mathtt{dc}}$ is *aggregated* (either summed, or multiplied with some other quantity) to the cost of the path (line 7). Table 10.3 reports some of the driving characteristics of the path planner with their corresponding parameter.

The short-term path with the smallest cost is then followed (line 9).

Algorithm 2 Path planner – selection of the short-term path

```
ShortTermPaths ← retrieveSTPs()
STPcosts ← ∅
for stp ∈ ShortTermPaths do
    cost_stp ← 0
    for dc ∈ {accMax, ..., safeDist} do
        if evaluateDC(stp, dc) then
            cost_stp ← aggregateCost(cost_stp, p_dc)
    STPcosts ← STPcosts ∪ {(stp, cost_stp)}
return getMinCostShortTermPath(STPcosts)
```

In [40], we observed that, in order to thoroughly test the path planner, all the *possible decisions* that can be taken by the path planner should be observed in at least one test. This is particularly challenging, as there is no proper characterization of the path planner's decisions: given two paths computed by the path planner, it is difficult to judge whether they have been taken for the same reason (i.e., the type of decision is the same), or they have been determined for different reasons.

Table 10.3: Path planner – driving characteristics and corresponding parameters

Driving characteristic	Parameter
forward acceleration	$p_{\texttt{accMax}}$
lateral acceleration	$p_{\texttt{latg}}$
speed	$p_{\texttt{spd}}$
deceleration	$p_{\texttt{decMax}}$
safety distance	$p_{\texttt{safeDist}}$
curvature	$p_{\texttt{curvature}}$

To tackle this issue, we proposed to exploit the definition of the path planner's cost function. We claimed that a minimal condition for testing the path planner is that each parameter is shown to be "relevant" in at least one decision in one test; in this case, we will say that the parameter is *covered*. The relevance, and thus the coverage, of a parameter is assessed as follows. A parameter p_i is covered by a scenario S if the path planner configured with a different parameter value p'_i would have selected a path different from the one computed by the original path planner; the intuition is that, if changing the parameter does not lead to a change in followed path, it means that the parameter is irrelevant in the decision taken by the path planner in that scenario.

Since checking all the possible alternative parameter values p'_i is not feasible, we only check a fixed number of values $k \cdot p_i$, with $k \in \{0, 0.5, 0.9, 1.1, 1.5, 2\}$; we call this approximation of the criterion as *mutation-based parameter coverage* and the values k as *mutation levels*.

In [41], we proposed a search-based approach to generate scenarios achieving parameter coverage. We proposed two approaches:

- *Single-parameter approach*: given a parameter p_i, the approach searches for a scenario S that achieves mutation-based parameter coverage for p_i. Specifically, the fitness function prefers the scenarios that achieve the coverage with the smallest mutation level: indeed, if the simulation changes by only slightly changing the parameter p_i, it means that the decisions taken by the path planner in S are highly influenced by p_i. Obviously, the fitness penalizes the scenarios that do not achieve any coverage. Moreover, the fitness also penalizes scenarios in which the vehicle collides (regardless of their coverage) as these are extreme scenarios that do not allow us to properly observe the influence of the parameter in the normal driving condition; such types of scenarios should be considered with other testing approaches, as those presented in Section 10.4.1.
- *Multi-parameter approach*: the single-parameter approach is very effective, as it allows us to find scenarios that cover the parameters with the minimum mutation level. However, it could create test suites that are unnecessarily big, in particular when there are several parameters. Therefore, we proposed a variation of the approach that aims at covering as many parameters as possible with the same scenario. We provided two versions of this approach:
  - *Multiple objectives*: in this version, for each parameter p_i, we specify a fitness function as defined in the single-parameter approach. This version of the approach is therefore multi-objective.
  - *Combined multi-objective*: this version, instead, is single-objective, and all the fitness values for the coverage of the different parameters are aggregated into a single value. The aggregation of the single fitness values for the different parameters follows two principles: (i) covering more parameters is preferred to covering less parameters; (ii) a solution that covers a parameter with mutation level i is worse than a solution in which all the parameters are covered with mutation levels $j < i$.

10.4.2.2 Coverage of Patterns of Driving Characteristics

Parameter coverage simply checks whether some parameter p_i influences the decision of the path planner at some point in time during the execution of scenario S, but it does not impose any constraint on the duration of its influence. It could be that the parameter is relevant for a very short period of time during the scenario execution, making the usefulness of the scenario limited. Therefore, parameter coverage may not guarantee the chance to observe all the possible driving manoeuvres that can be done by the autonomous vehicle.

In [5], we tackled this problem by considering the *driving characteristics* (such as forward acceleration, later acceleration, curvature, etc.; see Table 10.3) that are implemented in a path. During simulation, at each time point, the path planner simulator reports which driving characteristics have been implemented in the path with a non-negligible value (over a given threshold, as shown in Algorithm 2). Our idea is that a driving characteristic is more "relevant" in a scenario execution if it is implemented for a longer period of time. For example, *safety distance* is more relevant if it is violated for a long period, as it may indicate a safety problem. In the same way, if lateral acceleration is applied during a turn, it is more relevant if it is applied for the whole turn rather than for few moments, as it may indicate an abrupt manoeuvre that should be inspected by ADS engineers. In some driving situations, two (or more) driving characteristics could be relevant simultaneously; observing such cases is important as well, as they may show some undesired interactions between the driving characteristics.

To capture these intuitions, in [5], we defined the notion of *patterns of occurrence* for single and pairs of driving characteristics. Namely, given the simulation of the path planner over a scenario S, we define:

- *activation interval* of a given driving characteristic $\mathtt{dc}_i$ as the consecutive time points in which $\mathtt{dc}_i$ was applied; in Fig. 10.4a, for example, there are three activation intervals for $\mathtt{dc}_i$, and the one between t_4 and t_7 is the most relevant one, as it is the longest one;
- *pairs of activation intervals* of two driving characteristics $\mathtt{dc}_i$ and $\mathtt{dc}_j$; the relevance of the pair is determined according to these criteria:
  - overlapping intervals in which the two characteristics are applied together for some time are more relevant than non-overlapping intervals; for example, the pairs of overlapping intervals in Fig. 10.4a are more relevant than the non-overlapping ones in Fig. 10.4b;
  - given different overlapping intervals, those in which the overlapping area is larger are more relevant; for example, in Fig. 10.4a, the pair of intervals between time points t_9 and t_{12} are more relevant because they have the largest overlapping region;[5]
  - given different non-overlapping intervals, those in which the intervals are closer are preferred. For example, in Fig. 10.4b, the

[5]Note that also the pair between time t_0 and t_2 has the same overlapping size, but the other pair is preferred because the average length of the intervals is higher.

pair between time points t_3 and t_8 is preferred, as it is distant of only one time interval.[6]

	t_0	t_1	t_2	t_3	t_4	t_5	t_6	t_7	t_8	t_9	t_{10}	t_{11}	t_{12}	t_{13}	t_{14}
$\texttt{dc}_i$	A	N	N	A	N	N	N	N	A	N	N	N	A	A	A
$\texttt{dc}_j$	N	N	N	A	A	A	A	N	A	A	N	N	N	A	A

(a) Overlapping activation intervals.

	t_0	t_1	t_2	t_3	t_4	t_5	t_6	t_7	t_8	t_9	t_{10}	t_{11}	t_{12}	t_{13}	t_{14}
$\texttt{dc}_i$	A	A	A	N	N	A	A	A	A	A	A	N	N	A	A
$\texttt{dc}_j$	N	N	A	A	A	A	N	N	N	A	A	A	A	A	A

(b) Only close non-overlapping activation intervals

Figure 10.4: Examples of patterns of driving characteristics for the path planner (N: the driving characteristic has been applied (in a non-negligible way); A: the driving characteristic has not been applied).

In [5], we proposed a search-based approach to generate scenarios containing given patterns of driving characteristics. Specifically, we proposed two versions of the approach:

- *Single Driving Characteristic approach* (SDC): given a driving characteristic $\texttt{dc}_i$, it searches for scenarios in which the relevance of $\texttt{dc}_i$ is maximized. The problem is multi-objective. The first objective aims at maximizing the length of the longest activation interval of $\texttt{dc}_i$. The second objective, instead, aims at minimizing the likelihood of collisions; without this objective, the search could find very dangerous scenarios, for which assessing whether the ADS is implementing the correct driving manoeuvres is difficult; the minimization of the objective allows to find the minimum level of danger necessary to obtain the desired driving pattern.
- *Pair Driving Characteristics approach* (PDCs): given a pair of driving characteristics $\texttt{dc}_i$ and $\texttt{dc}_j$, it searches for scenarios in which their interaction is maximized. The problem is still multi-objective. The first objective aims at maximizing the overlapping of the activation intervals of $\texttt{dc}_i$ and $\texttt{dc}_j$ and, if this is not possible, it tries to minimize their distance. As for SDC, the second objective aims at minimizing the likelihood of collisions.

[6]Note that, similarly as before, in case of same distance, the pair with the higher average length is preferred (i.e., t_0 and t_4 is not the most relevant).

Figure 10.5 shows an example of a scenario (snapshots of its simulation) generated for covering the pair of driving characteristics `dec` and `latg`. While

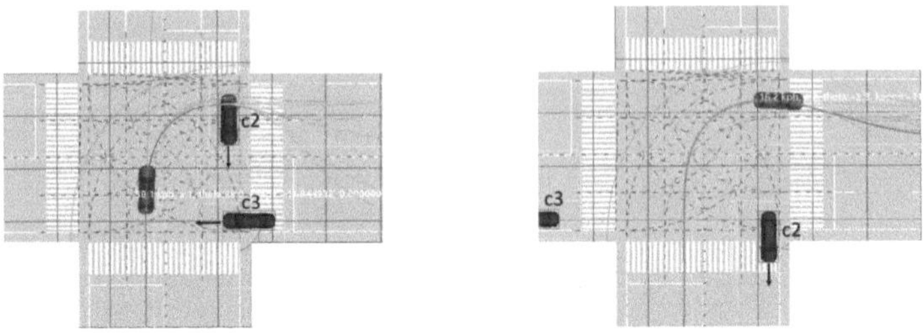

(a) Start of overlapping region between `dec` and `latg`.

(b) End of overlapping region between `dec` and `latg`

Figure 10.5: `PDCs` approach – scenario generated for covering the pair of driving characteristics deceleration (`dec`) and lateral acceleration (`latg`).

the ego vehicle is turning, it must suddenly brake to avoid vehicle c_2 crossing the intersection from top. Figure 10.5a is the simulation moment in which the two driving characteristic related to lateral acceleration (due to the turning) and deceleration start overlapping. Figure 10.5b shows the moment when the ego vehicle does not brake any more, and so the overlapping of the two driving characteristics finishes. You can observe that finding this type of scenario is far from trivial: the search approach managed to find the driving situation in which the ego vehicle decides to turn before vehicle c_3 crosses from right, but then needs to break to let vehicle c_2 pass.

In some cases, finding the overlapping of two driving characteristics is not possible, but it still interesting to find a scenario in which they are activated closely. Figure 10.6 shows such a type of scenario found for `acc` and `dec` (that, by definition, can not be activated simultaneously). The ego vehicle first decelerates for to let vehicle c_3 pass (Fig. 10.6a). Then, as soon as c_3 is passed, it accelerates to do the turn (Fig. 10.6b): there is no time interval between the end of deceleration and the start of acceleration, meaning that the search managed to optimise the coverage of the pair as much as possible. The ego vehicle stops accelerating at the end of the curve, as shown in Fig. 10.6c.

10.4.2.3 Coverage of Patterns of Requirements Violations

The test generation approaches reviewed in Section 10.4.1 usually aimed at worsening a given metric related to some system requirement in order to violate it; examples are the minimization of the minimum Euclidean distance between the ego vehicle and the other vehicles to violate safety requirements, or the maximization of acceleration to violate comfort requirements.

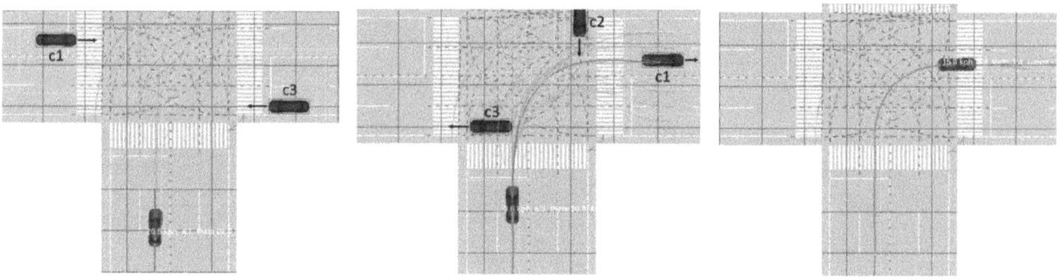

(a) Start of dec (b) End of dec and start of acc (c) End of acc

Figure 10.6: PDCs approach—scenario generated for covering the pair of driving characteristics acceleration (acc) and deceleration (dec).

However, in order to properly test the ADS, the satisfaction/violation of the different requirements should be considered together. Indeed, in some cases, the ADS can not satisfy all the requirements at the same time, and the ADS must trade-off among requirements. We call *requirement violation pattern* a specific combination of satisfied/violated requirements.

From a testing perspective, observing different requirement violation patterns is highly interesting, as this allows to judge whether the ADS gave a correct priority to the different requirements, i.e., it prioritized the satisfaction of the most critical requirements.

Moreover, in case of failure (e.g., accident), different requirement violation patterns could provide different explanations for the failure: the situation in which the ego vehicle collides and runs the red light is different from the one in which the collision happens while the ego vehicle is doing an unsafe lane change.

In [53], we proposed an SBT approach for covering as many requirement violation patterns as possible. Since the number of patterns is very large, trying to cover all of them is not feasible. Therefore, we identify two principles that must be followed: (i) the most critical patterns should be prioritized in the search; (ii) if a pattern is impossible to occur, it should not be searched. The proposed approach EMOOD, based on the previous two principles, is shown in Fig. 10.7.

The central component of the approach is an Evolutionary Many-Objective Optimization (EMOO) that, given a pattern $\mathcal{V}$ taken as input, searches a scenario covering it; namely, for each requirement in $\mathcal{V}$ that needs to be satisfied, an object function is specified that maximizes the requirement satisfaction; instead, for each requirement that must be violated, the corresponding objective function minimizes its satisfaction.

The whole approach works as follows:

1. The *Initial Prioritization* phase (IP) sorts all the possible requirements violation patterns in $\mathcal{V}_c$ according to their criticality, as shown in Fig. 10.8;

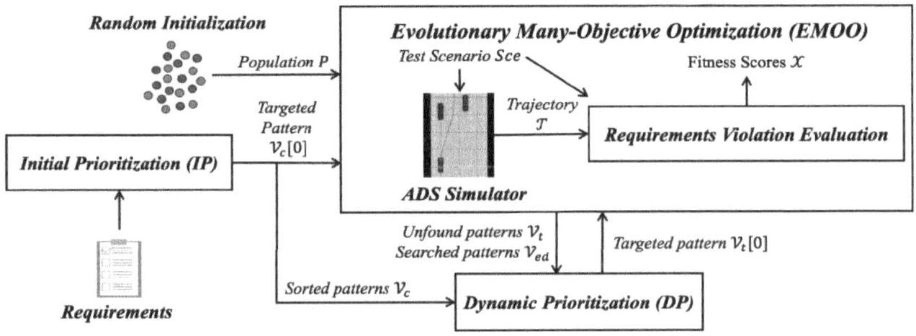

Figure 10.7: EMOOD—search-based testing for requirements violation patterns.

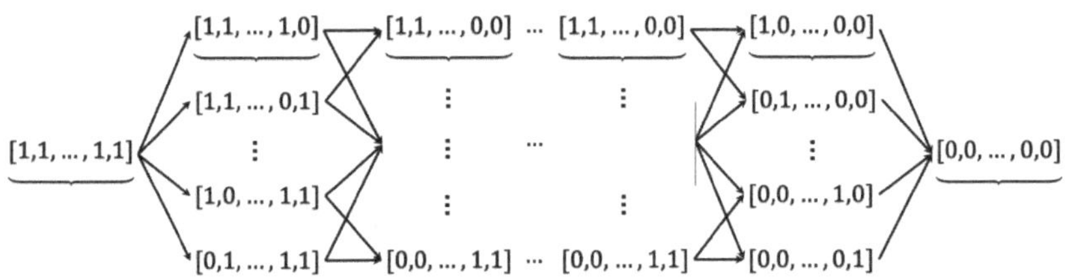

Figure 10.8: Initial relationship of patterns of requirements violations.

2. EMOO is executed using the most critical pattern $\mathcal{V}_c[0]$ as input;
3. Dynamic Prioritization (DP) sorts the patterns in $\mathcal{V}_t$ based on the likelihood of their occurrence. To do this, it uses the information on the patterns that have been previously searched and not found and those that have never been searched;
4. EMOO is executed using as input the most critical pattern $\mathcal{V}_t[0]$ that is likely to occur;
5. If the search budget is not exhausted, the approach performs another iteration starting from Item 3.

10.5 Definition of the Search Space

ADS simulators accept, as input, *scenarios* that describe the environment in which the autonomous vehicle is driving; such scenarios are usually specified using some DSL, such as OpenSCENARIO [8], GeoScenario [64], or Scenic [25]. Depending on the type of simulator, they allow us to specify different elements such as the road structure, the number and behaviour of other traffic participants (i.e., other vehicles and pedestrians), regulatory elements (e.g., traffic lights, traffic signs, etc.), weather conditions (e.g., fog, rain, etc.), and so on.

SBT approaches for ADSs (as those described in the previous sections) usually search for scenarios in which some specific test goals are satisfied. Deciding which scenarios to search for is a difficult task. Indeed, the possible ADS scenarios are infinite and so the search space must be restricted to some particular class of scenarios. A common technique followed by different SBT approaches [5, 11, 13, 15, 16, 41, 53] consists of identifying a particular road structure, a given number of traffic participants, and some regulatory elements, that define an *abstract scenario*; some elements of this initial abstract scenario are parametrized and they constitute the search space of the SBT approach: each individual of the search provides concrete values for the abstract scenario parameters and so identifies a *concrete scenario*.

However, for some particular test goals, the shape of the road could be particularly important; for example, testing the lane keeping component of an ADS requires us to test different types of roads. So, some recent works have proposed road generation. Gambi et al. [27] use *procedural content generation* (a technique used in video games) to generate an initial set of roads that are then evolved using a genetic algorithm, with the aim of finding a road in which the ego vehicle drives off the road.

The generation of roads has been recognized as a particular difficult problem, and so it has been proposed as a challenge in a competition on SBT for ADS at the 14th and 15th International Workshop on Search-Based Software Testing (SBST 2021 and SBST 2022).[7] Five approaches have been submitted to the first edition of the competition [62] and six approaches to the second edition [26], that use different road representations and different search algorithms.

One of the best approaches of the first edition of the competition was our approach Frenetic [18], that adopts a curvature-based representation for roads. The approach was further improved for the second edition of the competition with FreneticV [19], that adopts particular mechanism to generate *valid* roads (e.g., not self-intersecting); also FreneticV obtained good results in the competition. The two tools are integrated in the library Frentic-lib [36]. In both Frenetic and FreneticV, a road is represented by an array of N curvature values $\kappa_0, \dots, \kappa_{N-1} \in \mathbb{R}$. Please refer to Chapter 7 for a complete description of the theoretical framework; we here provide the intuition by means of an example. Figure 10.9a shows the curvature values of a road generated by the approach. A positive value requires a turn to the left, while a negative value requires a turn to the right; given a value κ_i, the higher its absolute value, the sharper the curvature. The value κ_i identifies the curvature between points $i+1$ and $i+2$. Figure 10.9b shows the corresponding generated road. The first two points are fixed, and independent of the concrete κ values; this allows the vehicle to start in a straight segment. Then, between points y_1 and y_3, the road turns left because of the first two positive curvature values κ_0 and κ_1. Then, the three negative

[7] `https://sbst21.github.io/tools/`, `https://sbst22.github.io/tools/`

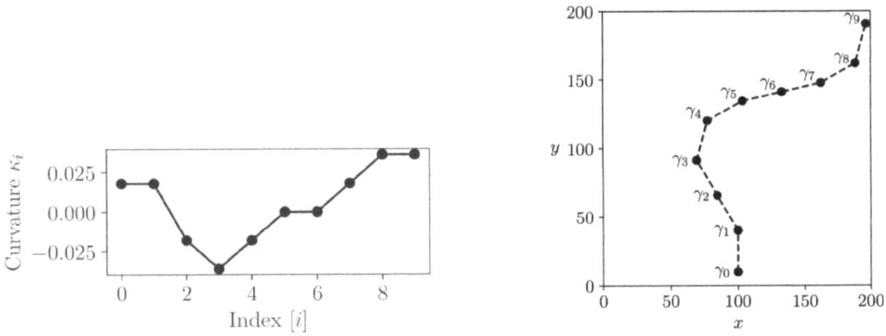

(a) K values (b) Generated road

Figure 10.9: Road representation in Frenetic and FreneticV.

values κ_2, κ_3, and κ_4 make the road turn to the right between points y_3 and y_6. Then, since κ_5 is 0, the road goes straight between the generated points y_6 and y_7. The road construction continues in this way for the remaining κ values.

Other testing works use different road representations. For example, Riccio and Tonella [65] use Catmull-Rom splines, Klück et al. [38] use Bézier curves, and Gambi et al. [27] use parametric straight and curved road segments to construct road networks. Given all these possible representations, one may wonder which is the best representation for their problem. In [17], we performed an empirical study to investigate which is the influence of the road representation on the effectiveness of SBT for ADS. In the study, we considered six road representations: (i) Catmull-Rom splines with Cartesian control nodes, (ii) Bézier curves with Cartesian control nodes, (iii) curvature-based representation, (iv) curvature-based representation with varying segment lengths, (v) orientation-based representation, (vi) orientation-based representation with varying segment lengths. These were used to represent individuals in an SBT approach that searches for failures of an ADS lane-keeping component. The study investigated the following questions: (1) Which representation allows to find more failures? (2) Which representation produces the most number of different roads? (3) Which is the impact of the mutation and crossover operations of the search algorithm on the different representations?

Results of the study show that: (i) curvature-based road representations trigger more failures and generate more diverse roads; (ii) mutation on different positions of an individual have different impacts depending on the adopted representation; (iii) the impact of the crossover operator is similar in the different representations.

10.6 Oracle Problem

One of the main challenges in software testing is the oracle problem [9]. It consists of specifying which is the expected output of each test execution and, therefore, deciding whether the test passes or fails. If complete system requirements are available, the oracle problem can be addressed. However, for complex systems, in some cases it is not clear which should be the expected behaviour. This is particularly true for ADSs. As an example, consider the case in which the autonomous vehicle collides with a fast vehicle that does not give the right of way, although it should: in this case, not only the autonomous vehicle is not responsible for the collision, but also it could be that it has no way to avoid it, i.e., the collision could be effectively *unavoidable*. Specifying whether this is a failure or not is challenging.

Most of the SBT approaches proposed for finding collisions overlook this problem; since their fitness functions aim at maximizing the probability of collisions (e.g., by minimizing the time to collision), they may find collisions that are so severe that they are actually unavoidable. From a testing point of view, these collisions are not interesting, as they do not manifest a failure of the ADS.

Starting from these observations, in [15], we tackled the oracle problem for the path planner of our industry partner. We provided the notion of *avoidable collision* as a collision that could have been avoided by performing some different driving manoeuvre. The core of the idea is that the path planner is configurable with some parameters that affect its decision process, as described in Section 10.4.2; changing the value of these parameters can lead the path planner to take different decisions in some particular scenarios. Starting from this observation, we identify an avoidable collision as the pair $\langle S, \overline{p}' \rangle$, such that:

- the original path planner $PP_{\overline{p}^o}$ (i.e., path planner configured with original parameter values $\overline{p}^o$) leads to a collision if executed over the scenario S;
- the modified path planner $PP_{\overline{p}'}$ (i.e., path planner configured with modified parameter values $\overline{p}'$) does not lead to a collision when executed over S.

In [15], we also proposed a search-based approach to find avoidable collisions. The search space is given by scenarios and by alternative values for the path planner; so an individual of the search is a pair $\langle S, \overline{p}' \rangle$.

The problem is multi-objective. The first objective requires us to find an avoidable collision. It is based on the function $danger(S, \overline{p}')$ that specifies the *danger* of the simulation of scenario S using the path planner $PP_{\overline{p}'}$:

- if there is no collision, it specifies how far the ego vehicle was from colliding;
- in case of collision, it specifies the *impact* of the collision as relative collision speed.

The first objective, that needs to be minimized, is defined as:

$$f_d(\langle S, \overline{p}' \rangle) = (D_{\overline{p}'} - D_{\overline{p}^o}) + penalty(\langle S, \overline{p}' \rangle)$$

where $D_{\overline{p}^o} = danger(S, \overline{p}^o)$ is the danger of the simulation with the original path planner, and $D_{\overline{p}'} = danger(S, \overline{p}')$ the danger of the simulation with the modified path planner. The difference between danger values specifies that we want to find scenarios in which the danger of the original path planner is much higher than that of the modified path planner. However, this term of the sum is not sufficient to specify that we want to find cases in which the original path planner leads to a collision while the modified one does not. To this aim, the second term of the sum (i.e., *penalty*) is used to discriminate among different types of scenarios in which collisions occur or not using the two versions of the path planner:

$$penalty(\langle S, \overline{p}' \rangle) = \begin{cases} 0 & \text{if } C_{\overline{p}^o} \wedge \neg C_{\overline{p}'}, \\ 1000 & \text{if } C_{\overline{p}^o} \wedge C_{\overline{p}'} \vee \neg C_{\overline{p}^o} \wedge \neg C_{\overline{p}'}, \\ 2000 & \text{if } \neg C_{\overline{p}^o} \wedge C_{\overline{p}'}. \end{cases}$$

where $C_{\overline{p}^o}$ and $C_{\overline{p}'}$ specify whether a collision occurred, respectively, using the original and the modified path planner. The intuition is that the most preferred case (i.e., penalty 0) is when there is a collision with the original path planner and not with the modified one, as this is the exact definition of avoidable collision. The second best preferred case (penalty 1000) is when both the original and the modified path planners both lead or do not lead to a collision. The least preferred case (penalty 2000) is when the modified path planner leads to a collision but the original one does not.

The second objective requires that the alternative parameter values are not too different from the original ones, i.e.,

$$f_p(\langle S, \overline{p}' \rangle) = \sqrt{\sum_{i=1}^{n} \left(\frac{p_i - p'_i}{d_i^U - d_i^L} \right)^2}$$

where $[d_i^L, d_i^U]$ is the search interval of ith parameter, and $\overline{p}^o = [p_1, \ldots, p_n]$ and $\overline{p}' = [p_1, \ldots, p_n]$ are the original and the modified parameter values. The motivation is that changing the parameters too much could lead the ADS to behave strangely, e.g., standing still when encountering other vehicles. Moreover, the original parameter values $\overline{p}^o$ have already been extensively tested and, therefore, it makes sense to try not to change them too much.

Figure 10.10 shows an example of avoidable collision found by the approach. The ego vehicle (i.e., the red vehicle) is approaching the crossroad with the aim of turning right. Meanwhile, a vehicle is approaching the intersection from north, and another one from east. Figure 10.10a shows snapshots of the simulation obtained using the original path planner: the ego

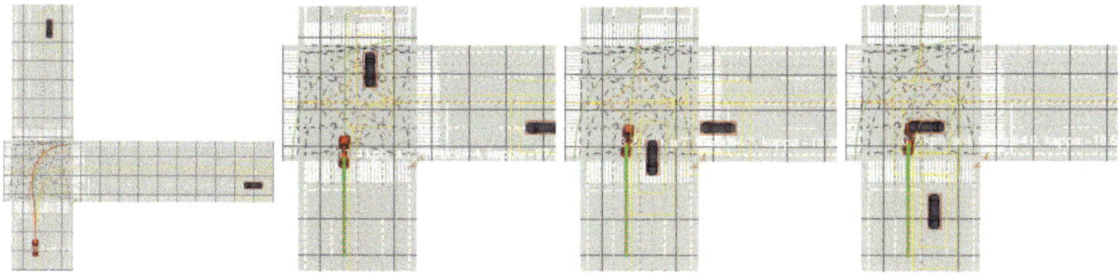

(a) Original path planner $PP_{\overline{p}^o}$– simulation at time points 0, 2.2, 3, and 3.5

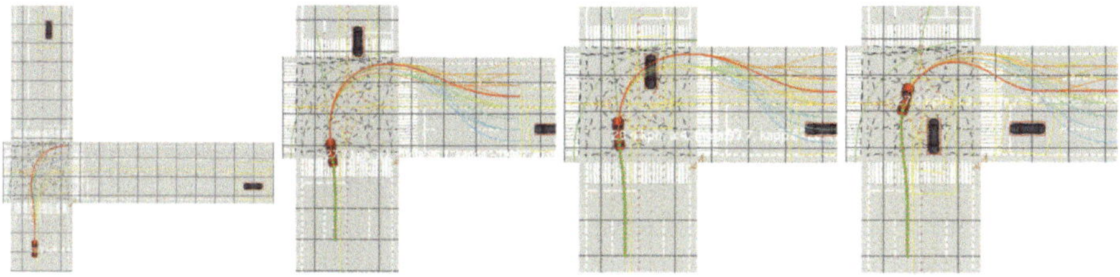

(b) Alternative path planner $PP_{\overline{p}'}$– simulation at time points 0, 1.8, 2.2, and 2.8

Figure 10.10: Avoidable collision found with the approach in [15].

vehicle stops to give the right of way to the vehicle coming from north, but then collides with the vehicle coming from east, as it does not have time to accelerate to avoid the collision. Figure 10.10b shows the behaviour of the alternative path planner: the ego vehicle enters the intersection while giving the right of way to the other vehicle; in this way, it can start turning as soon as the vehicle passes. By looking at the alternative parameter values $\overline{p}'$ found by the search, we observe that high lateral acceleration and high curvature have lower penalties: this allows the path planner to select this more sharp manoeuvre that is needed to avoid the collision.

Experimental results in [15] show that the approach is able to find different types of avoidable collisions and that these require different alternative parameters values in order to be avoided. Such information can be used to do fault localization and debugging [80, 81] (see Chapter 11).

However, re-engineering the path planner starting from these test results is not easy, as it is not clear which parameter values to use. One solution is to build a path planner that at runtime decides which parameter values to use on the basis of the current driving situation; we pursued this approach in [51], where we built a path planner that recognizes whether a driving situation is similar to one observed during testing, and changes the parameter values to those that in testing turned out to lead to safer driving.

A different approach tries to find alternative parameter values that solve different types of collisions. In [16], we extended the approach in [15] by searching different types of collision that are all solved by the same alternative parameter values.

10.7 Scalability

The fitness computation in SBT for ADS requires us to run the generated scenario with the ADS simulator, collect the simulation log, and compute the fitness function from it. The ADS simulation can be computationally very expensive, taking from several seconds to few minutes; for example, all the simulations performed with the industrial path planner described before, take from one to two minutes. This limits the scalability of the approach. In optimisation problems, a typical way to tackle this issue is to use *surrogate models* that approximate the fitness function and so avoid its expensive execution. We refer to [69] for an extensive survey on surrogate models for search approaches.

For example, Ben Abdessalem et al. [11] use surrogate models to accelerate the testing of a Pedestrian Detection Vision (PeVi) system. The system perceives the front of the vehicle with a CCD camera, and uses image processing and object recognition algorithms to decide whether to warn the driver of the presence of pedestrians, in particular when the visibility is reduced due to bad weather conditions. The system can be simulated in the simulator PreScan that allows to simulate the driving environment, including the behaviour of pedestrians. A search-based approach has been implemented to find failures of PeVi; the approach has three objective functions:

- minimization of the distance between the vehicle and the pedestrian;
- minimization of the distance between the pedestrian and the most critical area in front of the vehicle;
- minimization of time to collision.

Since simulating the PreScan simulator is expensive, the authors propose to use surrogate models to speed up the search. Specifically, for each fitness function, a surrogate model using a neural network is created, starting from training data of previous simulations. The evaluation of a scenario S using a surrogate model provides a predicted value $F(S)$ and a confidence value $F_{cl}(S)$, being *cl* the desired confidence level. The approach also computes the *most optimistic value* as $F(S) - F_{cl}(S)$, and the *most pessimistic value* as $F(S) + F_{cl}(S)$. Then, the NSGA-II search algorithm is modified as follows:

- the fitness values of all individuals are computed using the surrogate models; for each fitness, the approach computes the most optimistic and the most pessimistic value;
- it ranks all the individuals in three ways: in *rank* using the approximated fitness values, in $rank^{-}$ using the optimistic fitness values, and in $rank^{+}$ using the pessimistic fitness values;
- then, the approach assumes that the real fitness value of an individual is included between its optimistic and pessimistic approximated value. Starting from this assumption, it identifies all the individuals whose

position in the real ranking would be unchanged (w.r.t. the position in *rank*) if they take any value between its optimistic and pessimistic value;

- for all the individuals identified in the previous step, the computed approximated fitness values are good enough for search purposes (i.e., they do not affect the real ranking). The other individuals, instead, are simulated with the PreScan simulator in order to have precise fitness values.

10.8 Noisy Simulations

The ADS simulators used for executing the generated scenarios can be affected by some noise, i.e., repeated executions of the same scenario may not lead to the exact same simulation results. This is particularly true for simulators based on the Robotic Operating System (ROS), such as WiseSim [4], the simulator of the Autonomoose ADS [60]. Figure 10.11 visualizes the issue. Figure 10.11a shows

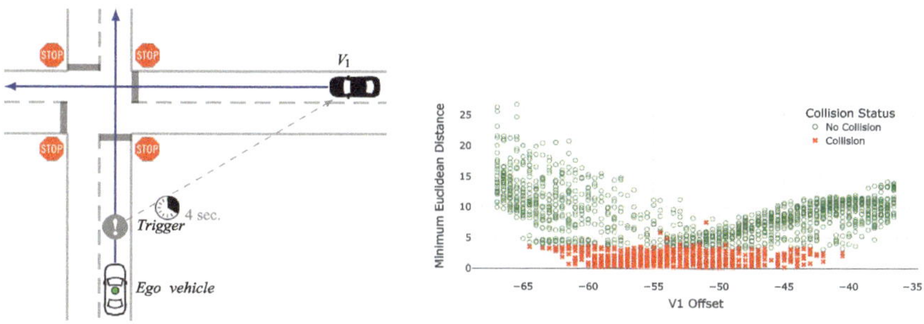

(a) Scenario template

(b) Fitness values of 30 simulations of different scenarios

Figure 10.11: Noisy simulations (adapted from [34]).

the template of a scenario for WiseSim in which the ego vehicle must cross the intersection while there is vehicle $\mathtt{V}_1$ approaching the intersection from right. By modifying the parameters of this template, we can possibly find concrete scenarios in which the two vehicles collide. As an experiment, we modified the $\mathtt{V}_1$'s initial position in the range $[-70, -35]$, and we simulated each scenario 30 times. For each simulation, we computed the minimum Euclidean distance between the two vehicles along the whole simulation, as shown in Fig. 10.11b. We observe that for all the scenarios, the computed metric can vary greatly; for some scenarios, the different repetitions can lead both to collision (red crosses) and non-collision (green circles).

The non-determinism of ROS has been recognized as one of the issues in testing ROS-based systems [1]. Such issue is exacerbated in search-based

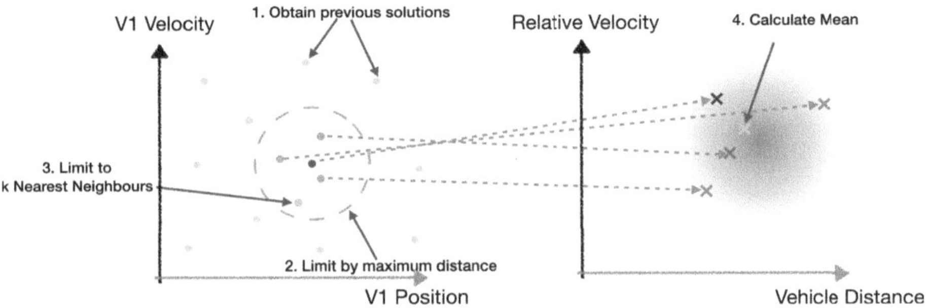

Figure 10.12: kNN-averaging approach.

scenario generation; indeed, noisy simulation values can mislead the search towards solutions that only apparently look good.

A simple solution to tackle the problem is to repeat each scenario several times and compute the fitness value as average of the metric along all the repetitions [23]. Although this approach is very precise, it is also very costly and does not scale for computational expensive systems such as ADSs. Therefore, in [35, 37], we have proposed a more scalable approach to handle noise in multi-objective optimisation; the approach has then been applied to SBT for Autonomoose in [34]. The main idea of the approach is to use the historical data of previous fitness evaluations to compute the fitness of a new individual. Figure 10.12 visualizes the approach. The left side shows the individual in the search space, while the right side shows their computed fitness values in the objective space. Given an individual whose fitness must be computed (e.g., the central point in the circle in the left plot), the approach selects its k nearest neighbours in the history of previously evaluated individuals; only neighbours that are within a given distance are selected. The fitness of the current individual is computed as the mean of the fitness values of the selected neighbours.

10.9 Conclusions

Evolutionary computation has been applied to solve different software engineering problems, in particular related to testing. SBT is an active research area that has targeted the testing of not only classical programs, but also complex systems such as ADSs. In this chapter, we have reviewed different SBT approaches for ADS, by considering different aspects: test objectives, definition of search space, scalability, and handling noisy simulations. The reviewed approaches are all related to the generation of scenarios. However, search approaches have been applied to tackle other testing problems for ADSs, such as test case prioritization [52], and sensitivity analysis of ADS parameters [45, 78].

References

[1] Afzal, A., Le Goues, C., Hilton, M. and Timperley, S.C. 2020. A study on challenges of testing robotic systems. *In*: 2020 IEEE 13th International Conference on Software Testing, Validation and Verification (ICST), pages 96–107.

[2] Ali, S., Arcaini, P., Pradhan, D., Safdar, S.A. and Yue, T. 2020. Quality indicators in search-based software engineering: An empirical evaluation. ACM Transactions on Software Engineering and Methodology, 29(2), March 2020.

[3] Ali, S., Arcaini, P. and Yue, T. 2020. Do quality indicators prefer particular multi-objective search algorithms in search-based software engineering? *In*: Aldeida Aleti and Annibale Panichella, (eds). Search-Based Software Engineering, pages 25–41, Cham, 2020. Springer International Publishing.

[4] Antkiewicz, M., Kahn, M., Ala, M., Czarnecki, K., Wells, P., Acharya, A., et al. 2020. Modes of automated driving system scenario testing: Experience report and recommendations. SAE International Journal of Advances and Current Practices in Mobility, 2(4): 2248–2266, April 2020.

[5] Arcaini, P., Zhang, X-Y. and Ishikawa, F. 2021. Targeting patterns of driving characteristics in testing autonomous driving systems. *In*: 2021 IEEE 14th International Conference on Software Testing, Validation and Verification (ICST), pages 295–305.

[6] Arcuri, A. 2018. EvoMaster: Evolutionary multi-context automated system test generation. *In*: 2018 IEEE 11th International Conference on Software Testing, Verification and Validation (ICST), pages 394–397.

[7] Arrieta, A., Wang, S., Sagardui, G. and Etxeberria. L. 2019. Search based test case prioritization for simulation-based testing of cyber-physical system product lines. Journal of Systems and Software, 149: 1–34.

[8] ASAM. 2022. ASAM OpenSCENARIO. https: //www.asam.net/ standards/ detail/openscenario/, 2021. Last access: January 12, 2022.

[9] Barr, E.T., Harman, M., McMinn, P., Shahbaz, M. and Yoo, S. 2015. The oracle problem in software testing: A survey. IEEE Transactions on Software Engineering, 41(5): 507–525, May 2015.

[10] Belbachir, A., Smal, J-C., Blosseville, J-M. and Gruyer, D. 2012. Simulation-driven validation of advanced driving assistance systems. Procedia-Social and Behavioral Sciences, 48: 1205– 1214.

[11] Abdessalem, R.B., Nejati, S., Briand, L.C. and Stifter, T. 2016. Testing advanced driver assistance systems using multi-objective search and neural

networks. *In*: Proceeding of the 31st IEEE/ACM International Conference on Automated Software Engineering, ASE 2016, pages 63–74, New York, NY, USA, 2016. ACM.

[12] Abdessalem, R.B., Nejati, S., Briand, L.C. and Stifter, T. 2018. Testing vision-based control systems using learnable evolutionary algorithms. *In*: Proceeding of the 40th International Conference on Software Engineering, ICSE 18, pages 1016–1026, New York, NY, USA, ACM.

[13] Abdessalem, R.B., Panichella, A., Nejati, S., Briand, L.C. and Stifter, T. 2018. Testing autonomous cars for feature interaction failures using many-objective search. *In*: Proceeding of the 33rd ACM/IEEE International Conference on Automated Software Engineering, ASE 2018, pages 143–154, New York, NY, USA, ACM.

[14] Ben Braiek, H. and Khomh, F. 2019. DeepEvolution: A search-based testing approach for deep neural networks. *In*: 2019 IEEE International Conference on Software Maintenance and Evolution (ICSME), pages 454–458.

[15] Calò, A., Arcaini, P., Ali, S., Hauer, F. and Ishikawa, F. 2020. Generating avoidable collision scenarios for testing autonomous driving systems. *In*: 2020 IEEE 13th International Conference on Software Testing, Validation and Verification (ICST), pages 375–386.

[16] Calò, A., Arcaini, P., Ali, S., Hauer, F. and Ishikawa, F. 2020. Simultaneously searching and solving multiple avoidable collisions for testing autonomous driving systems. *In*: Proceeding of the 2020 Genetic and Evolutionary Computation Conference, GECCO'20, pages 1055–1063, New York, NY, USA. Association for Computing Machinery.

[17] Castellano, E., Cetinkaya, A. and Arcaini, P. 2021. Analysis of road representations in search-based testing of autonomous driving systems. *In*: 2021 IEEE 21st International Conference on Software Quality, Reliability and Security (QRS), pages 167–178.

[18] Castellano, E., Cetinkaya, A., Ho Thanh, C., Klikovits, S., Zhang, X. and Arcaini, P. 2021. Frenetic at the SBST 2021 tool competition. *In*: Proceeding of the IEEE/ACM 43rd International Conference on Software Engineering Workshops, ICSEW21, pages 36–37.

[19] Castellano, E., Klikovits, S., Cetinkaya, A. and Arcaini, P. 2022. FreneticV at the SBST 2022 tool competition. *In*: 2022 IEEE/ACM ″2001 by CRC Press LLC 15th International Workshop on Search-Based Software Testing (SBST), pages 47–48.

[20] Cetinkaya, A., Arcaini, P., Ishii, H. and Hayakawa, T. 2020. A search-based approach to identifying jamming attacks and defense policies in wireless

networked control. *In*: 2020 59th IEEE Conference on Decision and Control (CDC), pages 5717–5724.

[21] Deb, K., Pratap, A., Agarwal, S. and Meyarivan, T.A.M.T. 2002. A fast and elitist multiobjective genetic algorithm: NSGA-II. IEEE Transactions on Evolutionary Computation, 6(2): 182–197.

[22] Durillo, J.J., Nebro, A.J., Luna, F. and Alba, E. 2008. Solving three-objective optimization problems using a new hybrid cellular genetic algorithm. *In*: Günter Rudolph, Thomas Jansen, Nicola Beume, Simon Lucas and Carlo Poloni, (eds). Parallel Problem Solving from Nature – PPSN X, pages 661–670, Springer, Berlin, Heidelberg.

[23] Fitzpatrick, J.M. and Grefenstette, J.J. 1988. Genetic algorithms in noisy environments. Machine Learning, 3(2–3): 101–120, October 1988.

[24] Fraser, G. and Arcuri, A. 2011. EvoSuite: Automatic test suite generation for object-oriented software. *In*: Proceeding of the 19th ACM SIGSOFT Symposium and the 13th European Conference on Foundations of Software Engineering, ESEC/FSE'11, pages 416–419, New York, NY, USA, 2011. Association for Computing Machinery.

[25] Fremont, D.J., Dreossi, T., Ghosh, S., Yue, X., Sangiovanni-Vincentelli, A.L. and Seshia, S.A. 2019. Scenic: A language for scenario specification and scene generation. *In*: Proceeding of the 40th ACM SIGPLAN Conference on Programming Language Design and Implementation, PLDI 2019, pages 63–78, New York, NY, USA. Association for Computing Machinery.

[26] Gambi, A., Jahangirova, G., Riccio, V. and Zampetti. F. 2022. SBST tool competition 2022. *In*: 2022 IEEE/ACM 15th International Workshop on Search-Based Software Testing (SBST), pages 25–32.

[27] Gambi, A., Mueller, M. and Fraser, G. 2019. Automatically testing self-driving cars with search-based procedural content generation. *In*: Proceeding of the 28th ACM SIGSOFT International Symposium on Software Testing and Analysis, ISSTA 2019, pages 318–328, New York, NY, USA. Association for Computing Machinery.

[28] Gazzola, L., Micucci, D. and Mariani, L. 2019. Automatic software repair: A survey. IEEE Transactions on Software Engineering, 45(1): 34–67, January 2019.

[29] Harman, M., Jia, Y. and Zhang, Y. 2015. Achievements, open problems and challenges for search based software testing. *In*: 2015 IEEE 8th International Conference on Software Testing, Verification and Validation (ICST), pages 1–12.

[30] Harman, M. and Jones, B.F. 2001. Search-based software engineering. Information and Software Technology, 43(14): 833–839.

[31] Harman, M., Mansouri, A.S. and Zhang, Y. 2012. Search-based software engineering: Trends, techniques and applications. ACM Computing Surveys, 45(1), December 2012.

[32] Hauer, F., Pretschner, A. and Holzmüller. B. 2022. Re-using concrete test scenarios generally is a bad idea. *In*: 2020 IEEE Intelligent Vehicles Symposium (IV), pages 1305–1310.

[33] Kalra, N. and Paddock, S.M. 2016. Driving to safety: How many miles of driving would it take to demonstrate autonomous vehicle reliability? Transportation Research Part A: Policy and Practice, 94: 182–193.

[34] Klikovits, S. and Arcaini. P. 2021. Handling noise in search-based scenario generation for autonomous driving systems. *In*: 2021 IEEE 26th Pacific Rim International Symposium on Dependable Computing (PRDC), pages 1–6.

[35] Klikovits, S. and Arcaini. P. 2021. KNN-averaging for noisy multiobjective optimisation. *In*: Ana C.R. Paiva, Ana Rosa Cavalli, Paula Ventura Martins, and Ricardo Pérez-Castillo, (eds). Quality of Information and Communications Technology, pages 503–518, Cham, 2021. Springer International Publishing.

[36] Klikovits, S., Castellano, E., Cetinkaya, A. and Arcaini. P. 2023. Frenetic-lib: An extensible framework for search-based generation of road structures for ADS testing. Science of Computer Programming, 230: 102996.

[37] Klikovits, S., Ho Thanh, C., Cetinkaya, A. and Arcaini. P. 2023. Trust your neighbours: Handling noise in multi-objective optimisation using knn-averaging. Applied Soft Computing, 146: 110631.

[38] Klück, F., Klampfl, L. and Wotawa, F. 2021. GABezier at the SBST 2021 tool competition. *In*: 2021 IEEE/ACM 14th International Workshop on Search-Based Software Testing (SBST), pages 38–39.

[39] Knowles, J.D. and Corne, D.W. 2000. Approximating the nondominated front using the Pareto archived evolution strategy. Evolutionary Computation, 8(2): 149–172, June 2000.

[40] Laurent, T., Arcaini, P., Ishikawa, F. and Ventresque, A. 2019. A mutation-based approach for assessing weight coverage of a path planner. *In*: 2019 26th Asia-Pacific Software Engineering Conference (APSEC), pages 94–101, Dec 2019.

[41] Laurent, T., Arcaini, P., Ishikawa, F. and Ventresque, A. 2020. Achieving weight coverage for an autonomous driving system with search-based test generation. *In*: 2020 25th International Conference on Engineering of Complex Computer Systems (ICECCS), pages 93–102, Oct 2020.

[42] Laurent, T., Arcaini, P., Ishikawa, F. and Ventresque, A. 2021. Shake those system parameters! On the need for parameter coverage for decision systems. *In*: Proceeding of the 36th IEEE/ACM International Conference on Automated Software Engineering, pages 1327–1331.

[43] Laurent, T., Klikovits, S., Arcaini, P., Ishikawa, F. and Ventresque, A. 2022. Parameter coverage for testing of autonomous driving systems under uncertainty. ACM Transactions on Software Engineering and Methodology, Jul 2022.

[44] Le Goues, C., Nguyen, T.V., Forrest, S. and Weimer, W. 2012. GenProg: A generic method for automatic software repair. IEEE Transactions on Software Engineering, 38(1): 54–72, January 2012.

[45] Lee, N-Z., Arcaini, P., Ali, S. and Ishikawa, F. 2019. Stability analysis for safety of automotive multi-product lines: A search-based approach. *In*: Proceeding of the Genetic and Evolutionary Computation Conference, GECCO '19, pages 1241–1249, New York, NY, USA, ACM.

[46] Leitao-Junior, P.S., Freitas, D.M., Vergilio, S.R., Camilo-Junior, C.G. and Harrison, R. 2020. Search-based fault localisation: A systematic mapping study. Information and Software Technology, 123: 106295.

[47] Li, G., Li, Y., Jha, S., Tsai, T., Sullivan, M.B., Hari, S.K.S., et al. 2020. AV-FUZZER: Finding safety violations in autonomous driving systems. *In*: 2020 IEEE 31st International Symposium on Software Reliability Engineering (ISSRE), pages 25–36.

[48] Li, M. and Yao, X. 2019. Quality evaluation of solution sets in multiobjective optimisation: A survey. ACM Computing Surveys, Dec. 2019.

[49] Calsi, D.L., Duran, M., Laurent, T., Zhang, X-Y. Arcaini, P., and Ishikawa, F. 2023. Adaptive search-based repair of deep neural networks. *In*: Proceeding of the Genetic and Evolutionary Computation Conference, GECCO '23, pages 1527–1536, New York, NY, USA. Association for Computing Machinery.

[50] Calsi, D.L., Duran, M., Zhang, X-Y., Arcaini, P. and Ishikawa, F. 2023. Distributed repair of deep neural networks. *In*: 2023 IEEE Conference on Software Testing, Verification and Validation (ICST), pages 83–94.

[51] Liu, K., Zhang, X., Arcaini, P., Ishikawa, F. and Jiao, W. 2020. Leveraging test logs for building a self-adaptive path planner. *In*: Proceeding of the IEEE/ACM 15th International Symposium on Software Engineering for Adaptive and Self-Managing Systems, SEAMS '20, pages 57– 63, New York, NY, USA, 2020. Association for Computing Machinery.

[52] Lu, C., Zhang, H., Yue, T. and Ali, S. 2021. Search-based selection and prioritization of test scenarios for autonomous driving systems.

In: Una-May O'Reilly and Xavier Devroey, (eds). Search-Based Software Engineering, pages 41–55, Cham, 2021. Springer International Publishing.

[53] Luo, Y., Zhang, X-Y., Arcaini, P., Jin, Z., Zhao, H., Ishikawa, F., et al. 2021. Targeting requirements violations of autonomous driving systems by dynamic evolutionary search. *In*: Proceeding of the 36th IEEE/ACM International Conference on Automated Software Engineering, pages 279–291.

[54] Majzik, I., Semerath, O., Hajdu, C., Marussy, K., Zolt.n Szatmari, Z., Micskei, Z., et al. 2019. Towards system-level testing with coverage guarantees for autonomous vehicles. *In*: 2019 ACM/IEEE 22nd International Conference on Model Driven Engineering Languages and Systems (MODELS), pages 89–94. IEEE.

[55] McMinn, P. 2004. Search-based software test data generation: A survey. Software Testing, Verification and Reliability, 14(2): 105–156.

[56] McMinn. 2011. Search-based software testing: Past, present and future. *In*: Proceeding of the 2011 IEEE Fourth International Conference on Software Testing, Verification and Validation Workshops, ICSTW '11, pages 153–163, USA. IEEE Computer Society.

[57] Miller, W. and Spooner, D.L. 1976. Automatic generation of floating-point test data. IEEE Transactions on Software Engineering, SE-2(3): 223–226.

[58] Nebro, A.J., Durillo, J.J., Garcia-Nieto, J., Coello Coello, C.A., Luna, F. and Alba, E. 2009. SMPSO: A new PSO-based metaheuristic for multi-objective optimization. *In*: 2009 IEEE Symposium on Computational Intelligence in Multi-Criteria Decision- Making(MCDM), pages 66–73, March 2009.

[59] Nebro, A.J., Durillo, J.J., Luna, F., Dorronsoro, B. and Alba, E. 2009. MOCell: A cellular genetic algorithm for multiobjective optimization. International Journal of Intelligent Systems, 24(7): 726–746, July 2009.

[60] University of Waterloo. 2022. Autonomoose. https: //www.autonomoose.net/, 2016. Last access: January 12, 2022.

[61] Panichella, A., Kifetew, F.M. and Tonella. P. 2015. Reformulating branch coverage as a many-objective optimization problem. *In*: 2015 IEEE 8th International Conference on Software Testing, Verification and Validation (ICST), pages 1–10.

[62] Panichella, S., Gambi, A., Zampetti, F. and Riccio, V. 2021. SBST tool competition 2021. *In*: 2021 IEEE/ACM 14th International Workshop on Search-Based Software Testing (SBST), pages 20–27.

[63] Pour, M.V., Li, Z., Ma, L. and Hemmati, H. 2021. A search-based testing framework for deep neural networks of source code embedding. *In*: 2021

14th IEEE Conference on Software Testing, Verification and Validation (ICST), pages 36–46.

[64] Queiroz, R., Berger, T. and Czarnecki, K. 2019. GeoScenario: An open DSL for autonomous driving scenario representation. *In*: 2019 IEEE Intelligent Vehicles Symposium (IV), pages 287–294.

[65] Riccio, V. and Tonella, P. 2020. Model-based exploration of the frontier of behaviours for deep learning system testing. *In*: Proceeding of the 28th ACM Joint Meeting on European Software Engineering Conference and Symposium on the Foundations of Software Engineering, ESEC/FSE 2020, pages 876–888, New York, NY, USA. Association for Computing Machinery.

[66] Shang, K., Ishibuchi, H., He, L. and Pang, L.M. 2021. A survey on the hypervolume indicator in evolutionary multiobjective optimization. IEEE Transactions on Evolutionary Computation, 25(1): 1–20.

[67] Shin, S.Y., Nejati, S., Sabetzadeh, M., Briand, L.C. and Zimmer, F. 2018. Test case prioritization for acceptance testing of cyber physical systems: A multi-objective search-based approach. *In*: Proceeding of the 27th ACM SIGSOFT International Symposium on Software Testing and Analysis, ISSTA 2018, pages 49–60, New York, NY, USA. Association for Computing Machinery.

[68] Silva, R.A., de Souza, S.d. R.S. and de Souza, P.S.L. 2017. A systematic review on search based mutation testing. Information and Software Technology, 81: 19–35.

[69] Tong, H., Huang, C., Minku, L.L. and Yao, X. 2021. Surrogate models in evolutionary single-objective optimization: A new taxonomy and experimental study. Information Sciences, 562: 414–437.

[70] Topcu, U., Bliss, N., Cooke, N., Cummings, M.L., Llorens, A., Shrobe, H.E. and Zuck, L. 2020. Assured autonomy: Path toward living with autonomous systems we can trust. CoRR, abs/2010.14443.

[71] Voyage. 2022. Open Autonomous Safety. https: //github.com/voyage/open-autonomous-safety, 2019. Last access: January 12, 2022.

[72] Wang, X., Arcaini, P., Yue, T. and Ali, S. 2021. Generating failing test suites for quantum programs with search. *In*: Una-May O'Reilly and Xavier Devroey, (eds). Search-Based Software Engineering, pages 9–25, Cham, 2021. Springer International Publishing.

[73] Wang, X., Arcaini, P., Yue, T. and Ali, S. 2022. QuSBT: Searchbased testing of quantum programs. *In*: 2022 IEEE/ACM 44th International Conference on Software Engineering: Companion Proceeding (ICSE-Companion), pages 173–177.

[74] Wang, X., Yu, T., Arcaini, P., Yue, T. and Ali, S. 2022. Mutation-based test generation for quantum programs with multi-objective search. GECCO '22, page 1345–1353, New York, NY, USA. Association for Computing Machinery.

[75] Waymo. 2022. On the Road to Fully Self-Driving: Waymo Safety Report. https: //waymo.com/safety/, 2018. Last access: January 12, 2022.

[76] Wu, H. and Nie, C. 2014. An overview of search based combinatorial testing. *In*: Proceeding of the 7th International Workshop on Search-Based Software Testing, SBST 2014, pages 27–30, New York, NY, USA. Association for Computing Machinery.

[77] Wu, J., Arcaini, P., Yue, T., Ali, S. and Zhang, H. 2022. On the preferences of quality indicators for multi-objective search algorithms in search-based software engineering. Empirical Software Engineering, 27(6), Nov 2022.

[78] Yin, K., Arcaini, P., Yue, T. and Ali, S. 2021. Analyzing the impact of product configuration variations on advanced driver assistance systems with search. *In*: Proceeding of the Genetic and Evolutionary Computation Conference, GECCO '21, pages 1106–1114, New York, NY, USA. Association for Computing Machinery.

[79] Yurtsever, E., Lambert, J., Carballo, A. and Takeda, K. 2020. A survey of autonomous driving: Common practices and emerging technologies. IEEE Access, 8: 58443–58469.

[80] Zhang, X-Y., Arcaini, P. and Ishikawa, F. 2022. An incremental approach for understanding collision avoidance of an industrial path planner. IEEE Transactions on Dependable and Secure Computing, pages 1–1.

[81] Zhang, X-Y., Arcaini, P., Ishikawa, F. and Liu, K. 2020. Investigating the configurations of an industrial path planner in terms of collision avoidance. *In*: 2020 IEEE 31st International Symposium on Software Reliability Engineering (ISSRE), pages 301–312.

[82] Zhang, Y., Harman, M. and Mansouri, A. 2022. The SBSE repository: A repository and analysis of authors and research articles on search based software engineering. http: //crestweb.cs.ucl.ac.uk/ resources/sbse_repository/. Last access: January 12, 2022.

[83] Zhang, Y., Harman, M. and Mansouri, S.A. 2007. The multiobjective next release problem. *In*: Proceeding of the 9th Annual Conference on Genetic and Evolutionary Computation, GECCO '07, pages 1129–1137, New York, NY, USA. Association for Computing Machinery.

[84] Zhao, D. and Peng, H. 2017. From the lab to the street: Solving the challenge of accelerating automated vehicle testing. CoRR, abs/1707.04792.

[85] Zitzler, E., Laumanns, M. and Thiele, L. 2001. SPEA2: Improving the strength Pareto evolutionary algorithm for multiobjective optimization. *In*: K.C. Giannakoglou, D.T. Tsahalis, J. Périaux, K.D. Papailiou and T. Fogarty, (eds). Evolutionary Methods for Design Optimization and Control with Applications to Industrial Problems, pages 95–100, Athens, Greece. International Center for Numerical Methods in Engineering.

Chapter 11

Fault Localisation and Understanding

Xiao-Yi Zhang
University of Science and Technology Beijing, Beijing, China

Fuyuki Ishikawa
National Institute of Informatics, The University of Electro-Communications and SOKENDAI, Tokyo, Japan

Toru Takisaka

University of Electronic Science and Technology of China, Chengdu, China

11.1 Introduction

In the industry 4.0 era, software applications, such as cyber-physical systems (CPS), server systems, machine-learning-based systems, database systems, and real-time systems, are playing an increasingly important role in both the industrial world and our daily life. On the other hand, the dependability of these systems is essential, as failures or requirement violations of these systems may lead to disruptive results, even the loss of human lives. Thus, activities for software quality assurance, such as testing and debugging, are becoming increasingly important. However, current software systems tend to be composed of various components and exhibit complex behaviours, making these quality assurance activities extremely resource-consuming. Investigation results show that the process of testing and debugging takes up to 50% – 80% of resources in the classical cycle of software development and maintenance [11,52].

In this chapter, we discuss the activity of *Fault Localisation (FL)*, which is the first and most important step of debugging, aiming to identify the faulty components in the subject program (or system). Let *PG* denote the subject program composed of the set of components $C = \{c_1, c_2, \ldots\}$; here, the components can be program statements, functions, classes, or some high-level modules in a system. Suppose that, during testing, at least one failure was detected. Then, by doing FL, we aim to find a subset of components $C^f \subseteq C$ which are the root causes leading to these detected failures.

Intuitively, if FL can be effectively and efficiently automatised, the whole process of debugging can be accelerated. To date, various studies focusing on the automation of FL have been conducted, accompanied by countless strategies to improve its accuracy and efficiency. According to the survey conducted by Wong et al. [49], these approaches can be categorised as: slice-based techniques [4, 20, 26, 45, 55, 64], program spectrum-based techniques [1, 2, 52, 58], statistics-based techniques [31–33,48], program state-based techniques [19,41,54,62,62], machine learning-based techniques [47,50,51], model-based techniques [35,36, 44], and data mining-based techniques [8, 13, 63].

In this chapter, we take the most widely-used and light-weighted FL technique, called *Spectrum-Based Fault Localisation (SBFL)*, as a representative to show how we can apply SBFL in practice. The implementation of SBFL is introduced in Section 11.2. In Section 11.3, we briefly review the state-of-art studies on the improvement of SBFL. Then, in Section 11.4, we discuss the extension of SBFL to Deep Neural Networks (DNN) and Autonomous Driving Systems (ADS), respectively, which are popular domains in Industrial 4.0. Finally, we summarise this chapter in Section 11.5.

11.2 Spectrum-Based Fault Localisation and Its Applications

In this section, we introduce the most popular and widely applied FL, called Spectrum-Based Fault Localisation (SBFL). Note that in this section, we consider statements as program components under analysis, but other program elements (e.g., functions or classes) could be used as well.

11.2.1 Description of SBFL

Let $S = \{s_1, s_2, \ldots, s_n\}$ be the set of *statements* of the program *PG* to debug. Assume *PG* contains a faulty statement s^f; then, the problem of fault localisation is to find s^f in *S*. SBFL takes as input the test information and predicts the faulty component s^f from the statements' set S. The detailed description of SBFL is presented as follows.

Assume a *test suite* including *m test cases*, i.e., $T = \{t_1, t_2, \ldots, t_m\}$ has already been generated and executed. For each test case t in T, we record its *coverage information*, i.e., which statements are exercised, and its *output correctness*, i.e., whether its output reveals a failure or not. Here, for each $t \in T$, let C^t denote the statements' set that are covered by t during execution; also let $O(t) \in \{Pass, Fail\}$ denote the output correctness of t: $O(t) = Pass$ and $O(t) = Fail$ denote that the output of t is *passed* and *failure-revealing*, respectively. Then, based on T, SBFL calculates the following quantities for each statement s:

- $ef(s)$: the number of failure-revealing test cases that exercise statement s, i.e., $|\{t \in T | s \in C^t, O(t) = Fail\}|$;
- $ep(s)$: the number of passed test cases that exercise statement s, i.e., $|\{t \in T | s \in C^t, O(t) = Pass\}|$;
- $nf(s)$: the number of failure-revealing test cases that do not exercise statement s, i.e., $|\{t \in T | s \notin C^t, O(t) = Fail\}|$;
- $np(s)$: the number of passed test cases that do not exercise statement s, i.e., $|\{t \in T | s \notin C^t, O(t) = Pass\}|$.

Also, let F and P denote the total number of failure-revealing and passed test cases, respectively. Note that, for any statement s, we have $ef(s) = F - nf(s)$ and $ep(s) = P - np(s)$.

SBFL predicts the *suspiciousness* of each statement s using a *metric*, which is a formula involving the above quantities (i.e., *ef*, *ep*, *nf*, *np*, F, and P). Here, let R_X denote the metric with name "X". For example, one of the earliest proposed metrics, *Tarantula* [24], denoted by R_{Tar}, is expressed as follows:

$$R_{Tar}(s) = \frac{ef(s)/F}{ef(s)/F + ep(s)/P} \tag{11.1}$$

As another example, Abreu et al. [1] proposed the *Ochiai* metric R_{Och}, which has been observed to have better performance than R_{Tar}:

$$R_{Och}(s) = \frac{ef(s)}{\sqrt{F(ef(s) + ep(s))}} \tag{11.2}$$

Furthermore, Naish et al. [37] and Xie et al. [52] proposed a metric called *Op*, in which the term *ef* has the major influence on the suspiciousness.

$$R_{Op}(s) = ef(s) - \frac{ep(s)}{np(s) + ep(s) + 1} \tag{11.3}$$

They proved that R_{Op} is the *maximal* metric under the *single-fault* condition. In other words, if the program has only one faulty statement, we cannot find another metric that always performs better than R_{Op}.

To date, hundreds of metrics have been proposed. Although different metrics produce different calculations, they all follow the two heuristics:

1. the larger $ef(s)$ is, the more suspicious statement s is, and
2. the lower $ep(s)$ is, the more suspicious statement s is.

These two heuristics are in accordance with our intuition. Namely, if some statements are exercised by more failure-revealing test cases, they are more likely to be the root cause that triggers the detected failures. Thus, the metric R should be a function that increases with the growth of ef and decreases with the growth of ep.

Finally, as the output, SBFL provides a list of statements, denoted by L, in which the statements are ranked in descending order according to their calculated suspiciousness. Namely, statements with higher suspiciousness values will rank higher than those with lower suspiciousness values. If SBFL is accurate, the faulty statement s^f will have a high ranking. Then, testers can find them earlier by inspecting L.

11.2.2 SBFL for the Function of UAVPP

Here we take a concrete example to show how SBFL can be applied. Consider a function, called *UAVPPSolver*, to solve *the unmanned aerial vehicle path planning (UAVPP) problem*, [72, 73] and its faulty version as shown in Fig. 11.1. In the domain of robotic control, UAVPP describes a situation in which a *Unmanned Aerial Vehicle (UAV)* is performing its task in an environment involving the UAV itself, some obstacles (e.g., walls) and a target (i.e., an objective node); the goal is to plan the actions of the UAV, such that it can reach its target efficiently without collision with the obstacles. The *UAVPPSolver* function realises a path planning approach that relies on three path planning algorithms: the *classical Rapidly-Exploring Random Tree* [25], denoted by *RRT*, the *accelerated RRT*, denoted by *MRRT* [34], and a DNN-based approach (e.g., reinforcement learning based path planning [56]), denoted by *NN*. During the path planning, *UAVPPSolver* takes the status of the current UAV as input, which can be simplified as tuple $\langle objDis, wallDis \rangle$. Here, $objDis$ represents the distance between the subject UAV and the target, and $wallDis$ denotes the distance between the subject UAV and its nearest obstacle. Then, it will assign the next action for the UAV according to the algorithm selected from the candidates (i.e., *RRT*, *MRRT*, and *NN*). Specifically, *UAVPPSolver* assigns the path planning algorithms as follows. If the UAV is far from the obstacles, *NN* will be selected to guide the UAV to the target following a smooth path. Conversely, if the UAV is close to the obstacles ($WallDis < 8$), the $step$ value (indicating its velocity) will decrease to avoid collision. Moreover, methods with strong randomness, i.e., *RRT*, will be selected to let the UAV escape from the obstacles as soon as possible. Finally, if the UAV is very

S	Original Program	Faulty Version
	Input (*wallDis*, *objDis*){	Input (*wallDis*, *objDis*){
	double *step, ranDegree*;	double *step, ranDegree*;
	void (**method*)(double, double);	void (**method*)(double, double);
s_1	*method* = &*MRRT*;	*method* = &*MRRT*;
	\\ The degree of randomness	\\ The degree of randomness
s_2	*ranDegree* =1;	*ranDegree* =1;
s_3	*step* = 2;	*step* = 2;
s_4	if(*wallDis* < 8){	if(*wallDis* < 8){
s_5=s^f	*step* = 1;	***step* = 2;** \\Should be "*step=1*"
s_6	if(*wallDis* < 5){	if(*wallDis* < 5){
s_7	*ranDegree* = 2;	*ranDegree* = 2;
s_8	if(*wallDis* < 1.5)	if(*wallDis* < 1.5)
s_9	*method* = &*RRT*;	*method* = &*RRT*;
	}	}
	}	}
	else{	else{
s_{10}	*method* = &*NN*;}	*method* = &*NN*;}
s_{11}	if(*objDis*<10){	if(*objDis*<10){
s_{12}	*step* = 1;	*step* = 1;
s_{13}	*method* = &*MRRT*;	*method* = &*MRRT*;
s_{14}	*ranDegree* =0.5;	*ranDegree* =0.5;
	}	}
s_{15}	(**method*)(*step*, *ranDegree*);	(**method*)(*step*, *ranDegree*);
s_{16}	return;}	return; }

Figure 11.1: The function of *UAVPPSolver* and its faulty version.

close to the target, *MRRT* whose execution largely relies on the value of *objDis*, will be adopted to let the UAV reach its target soon. Now we assume that *UAVPPSolver* has a faulty statement. Namely, in statement s_5, we set the variable *step* with a wrong value when $wallDis < 8$.

Now we intend to do fault localisation for *UAVPPSolver* with the help of SBFL. First, we do testing. Here, we use five test cases $\{t_1, t_2, t_3, t_4, t_5\}$ as inputs to run *UAVPPSolver* and record the test information, including the coverage and output correctness of each test case. The test information, i.e., the spectrum, is shown in Fig. 11.2.

Here, each column represents the behaviour of a test case and each row represents the spectrum of a statement. Each black point in a cell indicates that a test case exercises a specific statement. For example, we can know that test case t_2 exercises the statement s_4. The *Pass* and *Fail* in the last row denote the test cases that are passed and failure-revealing, respectively. Note that, in *UAVPPSolver*, a test case is considered as failure-revealing if, during execution, the UAV collides with the wall or cannot reach the target before time out. From Fig. 11.2, we can know that there are three failure-revealing test cases (t_1, t_4,

S	t_1 *Input*: 1.5, 50	t_2 *Input*: 10, 5	t_3 *Input*: 20, 45	t_4 *Input*: 4, 45	t_5 *Input*: 7, 60
s_1	•	•	•	•	•
s_2	•	•	•	•	•
s_3	•	•	•	•	•
s_4	•	•	•	•	•
s_5=f	•			•	•
s_6	•			•	•
s_7	•			•	
s_8	•			•	
s_9	•				
s_{10}			•		
s_{11}	•	•	•	•	•
s_{12}		•			
s_{13}		•			
s_{14}		•			
s_{15}	•	•	•	•	•
s_{16}	•	•	•	•	•
o^t	*Fail*	*Pass*	*Pass*	*Fail*	*Fail*

Figure 11.2: The spectra of *UAVPPSolver*.

and t_5) and two passed test cases (t_2 and t_3). Then, based on the test information, we can calculate the quantities *ef*, *ep*, *nf*, and *np* for each statement. For example, for statement s_5, it is exercised by all the three failure-revealing test cases. Thus we have

$$ef(s_5) = 3, nf(s_5) = 0$$

Moreover, s_5 is not exercised by any of the passed test cases. Thus, we have

$$ep(s_5) = 0, np(s_5) = 2$$

Furthermore, we use a metric to calculate the suspiciousness for the statements. Here the metric R_{Op} expressed in Equation (11.3) is applied. Then, the suspiciousness of statement s_5 is calculated as follows:

$$R_{Op}(s_5) = 3.0 - 0/(0 + 2.0 + 1.0) = 3.0$$

Finally, we rank all the statements according to their suspiciousness and get the rank list $L \langle s_5, 3.0\rangle, \langle s_2, 2.3\rangle, \ldots$. Now, we find that statement s_5 is ranked the first and it is indeed the faulty statement in *UAVPPSolver*. This means that if we use SBFL for *UAVPPSolver* we can find the faulty statement immediately.

11.3 Issues in SBFL

In Section 11.2.2, we have shown an example where SBFL is applied successfully in identifying the faulty statement. However, due to the complexity of real-life programs, test strategies, and the potential faults, we may encounter various situations that challenge the effectiveness FL. Figure 11.3 illustrates different activities necessary for performing SBFL, including the execution of test cases (i.e., testing), the collection of test information, and the calculation of statements' suspiciousness (i.e., SBFL metrics). For each activity we can propose a specific technique to handle the constraints/issues that could exist in the implementation of SBFL, such as the budget of testing and the lack of test oracles. In this section, we summarise the potential issues related to SBFL, and then review the related works that, to some extent, can be taken as solutions.

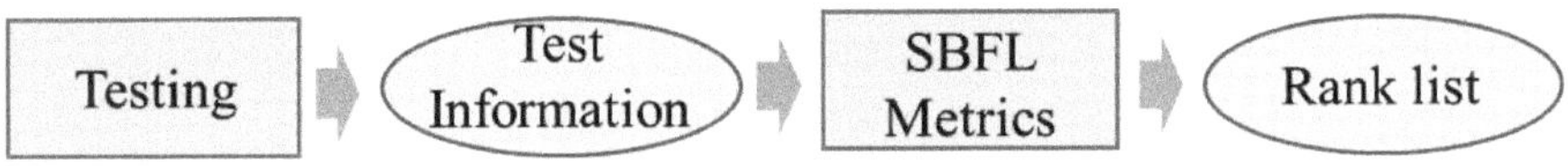

Figure 11.3: The process of conducting SBFL.

Table 11.1 lists the perspectives of research works in terms of *Problems*, *Descriptions*, and *Related Works*. Specifically, the issues related to SBFL include *quality testing information*, *multiple-fault scenarios*, *test oracles*, and *metrics selection*; each of them will be reviewed in this section.

11.3.1 Quality of Test Information

Since testing is the activity preceding fault localisation and provides all the essential information for SBFL, the quality of the information pool derived during testing is important. Yu et al. [59] have observed that fault localisation effectiveness may vary a lot by using different test strategies to generate the test suite. Thus, we expect that the test suite obtained during testing has high quality, such that, by applying SBFL based on this test suite, we can get an accurate localisation of the faulty statement.

We here take an example to show how different test strategies can affect the FL effectiveness. Figure 11.4 shows the program *MEDIAN* that calculates the median value among the inputs u, v, and w. It also shows its faulty version on the right. Specifically, in line 14, we misused v instead of u.

Now we intend to generate some test cases to test *MEDIAN*. Note that each test case is with the form $t = \langle u, v, w \rangle$. Assume that following a given adopted test strategy, two test cases are generated, $t_1 = \langle 3,5,2 \rangle$ and $t_2 = \langle 1,3,0 \rangle$. We execute both t_1 and t_2 and get the test information as shown in Fig. 11.5. Suppose we

Table 11.1: Related works of conducting SBFL under different situations.

Problems	Description	Related Papers
Quality of test information	If the testing resources are limited, how to efficiently generate the test cases with high quality w.r.t. SBFL	Jiang et al. 2012, 2013 [22,23]; Yoo et al. 2013 [57]; Zhang et al. 2015, 2016 [69,70]
Multiple fault scenarios	How to perform SBFL if the subject program contains multiple fault faulty statements	Podgurski et al. 2003 [38]; Abreu et al. 2009 [3]; Gao et al. 2017 [17]; Zakari et al. [61]; Laghari et al. [28]
Test oracles	How to tackle the oracle problem within the context of SBFL	Xie et al. 2013 [53]; Zhang et al. 2018 [72]
Metrics selection	How to select the metrics to calculate the suspiciousness according to benchmarks and test suites	Naish et al. 2011 [37]; Xie et al. 2013 [53]; Le et al. 2013 [29]; Wong et al. 2016; Yoo et al. 2017 [58]; Tang et al. 2017 [43]; Zhang et al. 2019 [71]; Zhang et al. 2021 [68]

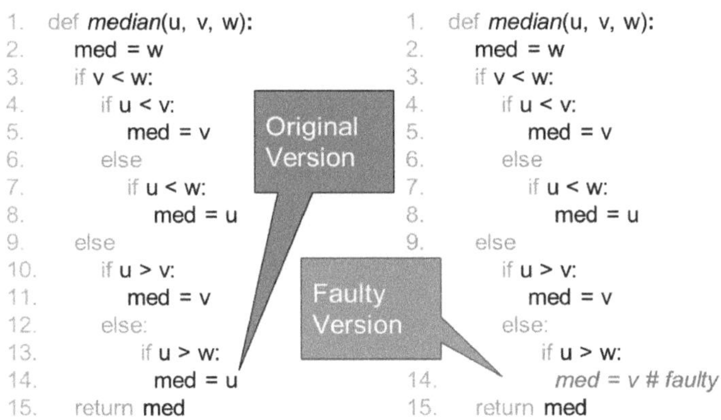

Figure 11.4: Program of calculating the median number and its faulty version.

want to use the coverage of t_1 and t_2 to do fault localisation. Since both t_1 and t_2 are failure-revealing, we can infer that at least one statement covered by t_1 and t_2 is faulty. Unfortunately, although t_1 and t_2 are two failure-revealing test cases, they have the same execution trace, i.e., $w < u < v$, and cover the same statements, s_2, s_3, s_{10}, s_{13}, s_{14}, and s_{15}. Thus, we cannot distinguish the fault

	$t_1=<3,5,2>$	$t_2=<1,3,0>$	$t_3=<3,2,1>$
1.def *median*:			
2.med = w	o	o	o
3.if v < w:	o	o	o
4.if u < v:			
5.med = v			
6.else			
7.if u < w:			
8.med = u			
9.else			
10.if u > v:	o	o	o
11.med = v			o
12.else:			
13.if u > w:	o	o	
14.med = v # *faulty*	o	o	
15.return med	o	o	o
Correctness	*Fail*	*Fail*	*Pass*

Figure 11.5: Coverage of *MEDIAN*.

proneness among these statements by only examining t_1 and t_2. Now we adopt a different test strategy and generate another test suite including t_1 and t_3. Although t_3 is a passed test case and we only have one failure (i.e., t_1), t_3 covers a different branch of program execution $w < v < u$. As a result, we can somehow distinguish s_{14} (i.e., the faulty statement) and the other statements.

To date, various studies focusing on improving the quality of test information have been conducted. *Test strategies* identify the scheduling of generation or execution of each test case. In this context, studies try to achieve high-quality test information to fit the goal of SBFL efficiently. Baudry et al. [5] introduced a metric based on Dynamic Basic Block (DBB) to evaluate the quality of test cases for fault localisation. Campos et al. [7] found that the test suite is suitable for fault localisation if test cases in this test suite cover half of the statements on average. Yoo et al. [57] introduced the fault localisation prioritisation problem and devised a solution method borrowed from the concept of information entropy. Jiang et al. [23] applied several popular test case prioritisation strategies to perform fault localisation. They found that the adequacy criterion and the percentage of a prioritised test suite that is utilised are important factors affecting the effectiveness of SBFL, and they also found that *Random Testing (RT)* and *Adaptive Random Testing (ART)* are effective test strategies. Moreover, they found that SBFL can be more effective if the test suite satisfies the *MC/DC-adequacy* criterion [22].

```
main():                          failure_tackling():
  s1  # Non-faulty                 s5  # Tackling failure
  if input < 0:                    return fail
    s2  # Non-faulty
    return pass

  if 0 < input < 1:
    s3  # Faulty
    failure_tackling()

  if input > 1:
    s4  # Faulty
    failure_tackling()
```

Figure 11.6: Example of multi-fault situation.

11.3.2 Multiple Faults

Another important issue related to SBFL is the presence of "multiple faults" in a program. Whenever we are writing code, we may make errors. Studies have shown that, for an industrial code, the number of bugs per 1000 lines could range between 0.5 and 25 [6, 21]. This means that in a project with a large scale (e.g., one million lines of code), it is likely that multiple faults exist. Furthermore, it has been observed that the interactions among multiple faults can indeed influence the distribution of program spectra, and hence degrade the accuracy of SBFL [12, 14].

Here, we take an example to show the influence of multiple faults. Assume that a program (See Fig. 11.6) has five statements s_1, s_2, s_3, s_4, and s_5, in which both s_3 and s_4 are faulty and assume that, once any of them is exercised, a failure will occur. If we intend to find at least one fault using SBFL, we may expect that one of s_3 and s_4 has a large *ef* value. However, since s_5, although it is a non-faulty statement, belongs to the block of *tackling exceptions*, it will be triggered by the failures brought by both s_3 and s_4. Thus, it will have a large *ef* value, even larger than the *ef* values of s_3 and s_4.

According to the above example, if the program contains multiple faults, the interference among different faulty statements could largely affect the program spectra, and hence degrade the accuracy of SBFL. Currently, various studies have been conducted focusing on alleviating the negative effects caused by multiple faults. A general idea is to identify which failures are caused by which faults. After knowing this, we can eliminate the failures that are caused by non-relevant faults during fault localisation.

For example, Abreu et al. [3] proposed an approach based on Bayesian reasoning to consider the suspiciousness of combinations of statements based on their spectra. Liblit et al. [32] tried to cluster the failure-revealing test cases such that the failures triggering different faults can be isolated [32]. Podgurski et al. [38] proposed a fault identification strategy involving both supervised and unsupervised patterns based on testing reports. Gao et al. [17] proposed a clustering approach called *MSeer*, designed for localising multiple faults in parallel. Although the interactions and interference can have negative effects on SBFL accuracy, it has been observed that SBFL can have similar accuracy for at least one dominant fault, but which fault will be found is unknown [14]. So, in many cases, it is feasible to fix the bugs iteratively: we can focus on fixing one bug at a time, and after each fixing, turn to find other bugs based on the information from regression testing.

11.3.3 Test Oracle

In software testing, the term test oracle identifies the artefact to specify the expected output of each test case and determine whether the observed outputs are correct or not.

However, in most practical cases, the test oracle does not exist or it is incomplete. For example, considering the $sin(\cdot)$ function, although it is easy to know the correctness of the outputs of some special inputs such as $sin(\pi)$, for most test cases (e.g., $sin(1.83)$), the correctness of their outputs cannot be easily identified automatically. In the context of testing, various approaches have been proposed to solve the oracle problem. For example, Chen et al. [9, 10] propose a methodology called Metamorphic Testing (MT). Instead of using a test oracle, MT aims to detect the potential failures by checking some system properties involving multiple inputs, called Metamorphic Relation (MR).

In the context of SBFL, since the calculation of the spectrum (i.e., ef, ep, nf, and np) relies on the information of whether each test case is passed or failure-revealing, in the absence of an oracle, SBFL cannot be applied. To date, to solve the oracle problem in the context of SBFL, various studies have been conducted. For example, Xie et al. [53] adapt MT to the context of SBFL by proposing the concept of "metamorphic slice", which is a trace involving the coverage of multiple test cases. With their approach, we can perform SBFL effectively even if the oracle does not exist. Moreover, Zhang et al. [72] propose an approach to tackle to situation that the test oracle is incomplete, leading to the fact that only some test cases are labelled (i.e., we know their output correctness) while the other test cases are unlabelled (i.e., we do not know whether they are passed or failure-revealing). Specifically, they design a test classifier to assign each unlabelled test case a "predicted label" according to the already labelled ones. In this way, we can use more test cases, including both the labelled and unlabelled ones, and hence the effectiveness of SBFL can be improved.

11.3.4 Suspiciousness Metrics

Finally, as the core part of SBFL, the suspiciousness metric is also an important issue related to the effectiveness of SBFL. In this line of research, studies focus on enriching the theoretical foundations of SBFL. Although all the metrics, such as Equations (11.1) and (11.2), are based on basic intuition of *the larger ef and lower ep that a statement has, the more suspicious it is*, there is no approach or theory to provide an optimal trade off between *ef* and *ep*.

To date, hundreds of metrics have been proposed [1,24,46]. Naish et al. [37] innovatively make a theoretical comparison among various metrics and group them into several equivalent categories. Based on this, Xie et al. [53] explicitly prove the performance relation among these equivalent categories and propose two groups of maximal metrics. Furthermore, Yoo et al. [58] proposed a genetic algorithm to mine useful metrics automatically. Zhang et al. [71] propose a visualisation approach to reproduce the results of [53]. Furthermore, they extend the work to a general framework to review existing SBFL works and make *qualitative* evaluation for the metrics.

11.4 Adaption of Fault Localisation to Other Domains

SBFL is the most popular and widely used technique for automatic fault localisation. The main reason is that SBFL is not only effective but also light-weighted. The implementation of SBFL requires only the statement coverage and the outputs information, and does not need too complex information, such as data dependencies. Thus, for all the systems for which we can define the concepts of "component coverage" and "failures", SBFL can be adapted. Of course, for different artefacts, the distribution of execution spectra could be largely different. Hence, the localisation results may not be as precise as for conventional programs. However, for all the systems, by adapting SBFL, we can get more understanding about the potential system hazard without spending too much computational efforts. In this section, we review some recent works related to SBFL for systems different than conventional programs. Specifically, we mainly introduce two kinds of artefacts, *Deep Neural Networks (DNN)* and *Autonomous Driving Systems (ADS)*.

11.4.1 SBFL for Deep Neural Networks

Recently, the combination of Deep Learning (DL) and fault localisation has been widely studied. The works in [40] and [39] borrow the concept of fault localisation to identify the potential buggy weights in a DNN model. Moreover, Gerasimou et al. [18] and Sun et al. [42] apply fault localisation to identify the important neurons and input elements for the DNN models. According to different tasks (e.g., regression, classification, and decision making), the outputs of a DNN are also different. In this section, we discuss the task of image classification.

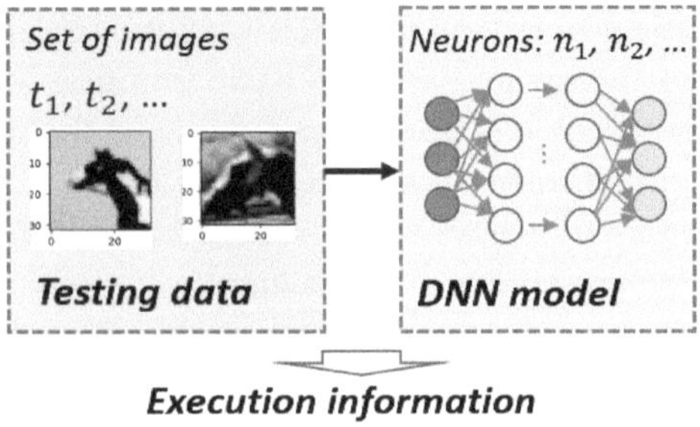

	t_1	t_2	t_3
n_1		activated	
n_2	activated	activated	activated
n_3		activated	
n_4	activated		
n_5			activated
...			
	Correctly Classified	*Correctly Classified*	*Misclassified*

Figure 11.7: Execution information of DNN testing used for fault localisation.

Here, we introduce the work proposed by Eniser et al. [16] that adapts SBFL to localise suspicious neurons. Let $\mathcal{M}$ denote a DNN model containing N neurons $n_1, n_2, \ldots, n_N$. In DNN testing, the test suite is a test dataset. For example, if we use a Convolution Neural Network (CNN) for image classification, the test suite is an image dataset such as *MNIST* [30], and each test case is an individual image. In addition, each test input t is attached with a *label* $C(t)$, indicating the correct classification of t. For example, in the MNIST dataset, we may have $C(t)$ = "7", and in the *CIFAR10* dataset [27], we may have $C(t)$ = "dog". Let $\mathcal{C} = \{l_1, \ldots l_n\}$ denote the set of *labels* (also called *classes*). Now, given a test case t to execute the DNN, the propagation process is triggered, and the neurons are either *activated* or not by their activation functions. Let $\phi(t,n)$ denote the function to calculate the activation status of neuron n under input t. Here, we set $\phi(t,n)$ is 1 when the output of n is above a certain threshold, otherwise it is set to 0. The output is a *predicted class* of the input image, denoted by $PC(\mathcal{M},t)$. Furthermore, if $C(t) = PC(\mathcal{M},t)$, we say the input t is *correctly classified*, denoted by $fl(t,\mathcal{M}) = 0$; otherwise, if $C(t) \neq PC(\mathcal{M},t)$, we say t is *misclassified*, denoted by $fl(t,\mathcal{M}) = 1$.

Figure 11.7 illustrates the execution information of a DNN model that can be used to do fault localisation. Instead of statements as in classical SBFL for programs, we analyse neurons n_1, n_2, ... The concept of "failure" here is the misclassification of an image, e.g., misclassifying "7" to "8", or misclassifying "horse" to "vehicle". The test "coverage" is associated to the activation of neurons. For example, if neuron n is activated by test case t_1, we say t_1 covers neuron n.

According to the test information of the DNN, similarly to conventional SBFL, we can also calculate the values of ef_n, ep_n, nf_n, and np_n, respectively. They denote the spectrum of neuron n after executing all the inputs in T. Specifically, the meaning of ef_n, ep_n, nf_n, and np_n can be summarised as follows:

- ef_n: the number of times that neuron n is *activated* by the *misclassified* inputs, i.e., $|\{t \in T \mid \phi(t,n) = 1 \wedge fl(t,\mathcal{M}) = 1\}|$;
- ep_n: the number of times that neuron n is *activated* by the *correctly classified* inputs, i.e., $|\{t \in T \mid \phi(t,n) = 1 \wedge fl(t,\mathcal{M}) = 0\}|$;
- nf_n: the number of times that neuron n is not *activated* by the *misclassified* inputs, i.e., $|\{t \in T \mid \phi(t,n) = 0 \wedge fl(t,\mathcal{M}) = 1\}|$;
- np_n: the number of times that neuron n is *not activated* by the *correctly classified* inputs, i.e., $|\{t \in T \mid \phi(t,n) = 0 \wedge fl(t,\mathcal{M}) = 0\}|$.

Finally, the suspiciousness metrics, as shown in Equations (11.1) and (11.2), are used to calculate the suspiciousness of each neuron. If a neuron is suspicious, this neuron could be responsible for the misclassifications. The fault localisation results can provide evidence for evaluating DNN's performance and hence predicting the potential hazards when the DNN is applied to safety-critical systems. Furthermore, Duran et al. [15] discuss the potential *masking effect* if we consider all misclassifications for a given class together during the calculation of the values of *ef* and *ep*, and suggest a more fine-grained way (i.e., considering each misclassification individually) when applying SBFL to DNN.

11.4.2 Analysing the Hazard of Autonomous Driving Systems

Nowadays, personal mobility and goods transportation are undergoing a big transformation thanks to autonomous driving. However, errors in Autonomous Driving Systems (ADS) may lead to disruptive economic damages, even the loss of human lives. Thus, it is meaningful to analyse the potential hazard for ADSs once some failures have been exposed during testing.

An ADS system takes as input the signals received by the environment or provided by the system operators. Moreover, the system contains various components performing different tasks. The goal of hazard analysis is to identify which components are the causes of potential system hazard. By knowing this, engineers can use the analysis results to re-engineering the system to improve its safety.

However, since the structure of ADS is complex, it is difficult to do formal verification to prove that the system is safe or dangerous. On the other hand, for ADS, we should execute many test cases before putting them into practice.

Thus, it makes sense to propose a data-driven approach to analyse the system hazard through the derived test information. In this section, we introduce a hazard analysis approach proposed in [65] following the idea of SBFL. Specifically, we consider an ADS whose components are configurable with some *parameters*; some configurations can lead to more(less) safe driving. The goal of the work is to consider the relation between the parameters' values and the potential hazard.

Let $\mathcal{M}$ be a model of ADS under analysis. We assume that the components of $\mathcal{M}$ can be described as a set of parameters $P = \{p_1, \ldots, p_n\}$, in which each parameter p_i is defined within its own domain D_i. $\mathcal{M}$ is configurable. If a concrete value v_i is given to each parameter p_i, a *simulation instance*, denoted by $\mathcal{M}(v_1, \ldots, v_n)$ can be constructed. Then, we can run $\mathcal{M}(v_1, \ldots, v_n)$ and obtain, as output, a signal over time o. Here, we define a *simulation* t as the pair $t = \langle (v_1, \ldots, v_n), o \rangle$. Given system model $\mathcal{M}$ and simulation set T, assume $\mathcal{M}$ is subjected to hazard (i.e., some simulations $t \in T$ can cause system failures). Then, the fundamental problem is:

How to identify the parameters that are the main cause of system hazard?

To do this following the idea of SBFL, we need to know the coverage and result information for each test case. However, the ADS parameters have continuous values at each test execution. Also, due to the complexity of the ADS, it is difficult to have a crisp partition at each parameter. As a result, it is difficult to identify which components are covered and which are not. As a solution, the work in [65] introduces the concept of *fuzzification-based coverage*.

Specifically, to characterise the system parameters, the concept of *fuzzy set* borrowed from fuzzy logic [60] is introduced. Considering a parameter p defined in the domain D, let $\mathcal{F}^p = \{f_1, f_2, \ldots, f_m\}$ denote the set of fuzzy sets that describe certain concepts of p. Each fuzzy set $f \in \mathcal{F}^p$ is defined by a *membership function*. For each concrete value v of p, f assigns a membership degree between 0 and 1, i.e., $f : D \rightarrow [0,1]$, indicating the degree that v belongs to f.

Suppose we have simulation t with parameter values $(v_1, \ldots, v_n)$. We define the coverage of parameter p_k over t as the vector $C^{p_k,t} = [f_1(v_k), \ldots, f_m(v_k)]$, where, for each f_i, we have $f_i \in \mathcal{F}^p$. Furthermore, we define the fuzzification-based coverage of the entire model $\mathcal{M}$ over t as $C^t = \{C^{p,t} | p \in P\}$. For example, we have a parameter p having domain $D = [150, 350]$ and during the execution of a test case t the value of p is 200. Then, we can say that test case t covers both the fuzzy set *Small* and *Slight Small* with 0.5 degree.

Furthermore, considering the correctness of a test case, let φ be a specific safety requirement of $\mathcal{M}$. The satisfaction of φ over a given simulation t can be quantified as $R(t)$ within the interval $[0,1]$, indicating how strongly φ is satisfied. Then, we calculate the *hazard degree* of t, denoted by $h(t)$, as $h(t) = 1 - R(t)$.

After the definition of both the test coverage (i.e., fuzzification-based coverage) and correctness (i.e., hazard degree), we predict the hazard-related parameters following SBFL. That is to say, we finally assign a *hazard impact score* to each parameter p indicating its relation to the hazard. Specifically, we introduce the *scoring metrics* from SBFL and extend them to our context. Given model $\mathcal{M}$ and simulation set T, for each simulation $t \in T$, we know $\mathcal{C}^t$ and $h(t)$. Then, concerning system hazard, we define

$$a_h^{f^{p_i}} = \sum_{t \in T} (f^{p_i}(v_i) \cdot h(t)) \tag{11.4}$$

Similarly, concerning system safety, we define

$$a_s^{f^{p_i}} = \sum_{t \in T} f^{p_i}(v_i) \cdot (1 - h(t)) \tag{11.5}$$

Here, $a_h^{f^{p_i}}$ and $a_s^{f^{p_i}}$ represent the degrees that fuzzy set f^{p_i} of parameter p_i contributes to the hazard and safety, respectively. Specifically, if t covers fuzzy set f^{p_i} at $f^{p_i}(v_i)$ degree, then we consider the term $f^{p_i}(v_i) \cdot h(t)$ as the contribution of f^{p_i} to the hazard at t. Then, we sum $f^{p_i}(v_i) \cdot h(t)$ for each t to calculate $a_h^{f^{p_i}}$, and we calculate $a_s^{f^{p_i}}$ similarly. Finally, we use a metric, involving both $a_s^{f^{p_i}}$ and $a_h^{f^{p_i}}$, to calculate the final hazard impact score, denoted by $\Phi(f^{p_i})$:

$$\Phi(f^{p_i}) = a_h^{f^{p_i}} / (a_h^{f^{p_i}} + a_s^{f^{p_i}}) \tag{11.6}$$

In [67], we adapted the fuzzy-set-based approach to analyse the hazard of a path planner provided by our industrial partner; the approach was later extended in [66], but, for the sake of space, we here only describe the approach in [67]. The path planner can automatically calculate an available path for the ego vehicle at run time. A possible driving scenario for this path planner is shown in Fig. 11.8.

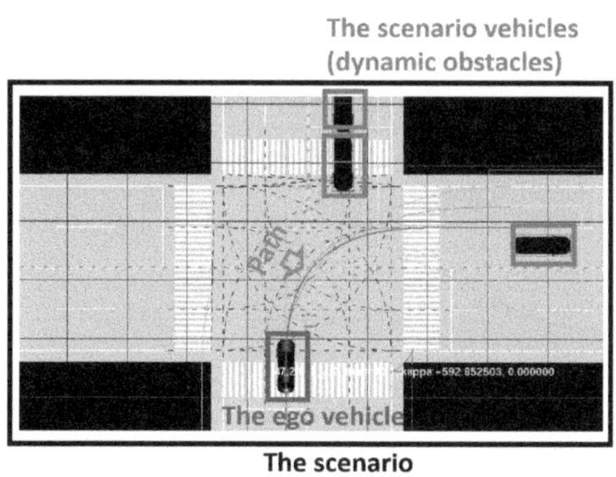

Figure 11.8: The example of an industrial path planner.

During path planning, the path planner decides the action for the ego vehicle, step by step. At each time point τ, it generates some *short-term paths* as candidates, and then selects the least costly one, using a scoring function that employs some parameters *w* to give different importance to *driving characteristics* such as maximum speed, lateral acceleration, curvature, etc. Specifically, for each short-term path, denoted by *stp*, the path planner records some *runtime flags*, indicating whether the driving characteristics corresponding to these runtime flags are high (i.e., exceed the given threshold) or not in *stp*. For example, if the maximal acceleration is high, then the corresponding runtime flag $\texttt{rf}_{\texttt{accMax}}$ is set to 1. Then, during path selection, the cost of *stp* is calculated by aggregating the *parameters* whose corresponding *driving characteristics* have runtime flags set to 1. Generally speaking, setting a parameter `w` to be *high* means that we expect the corresponding driving characteristics being *low*, i.e., the corresponding runtime flag being 0. The parameters analysed here as well as the corresponding runtime flags are listed in Table 11.2.

Table 11.2: Driving characteristics, runtime flags, and parameters

Driving characteristic	Runtime flag	Weight	Description
(forward) acceleration	$\texttt{rf}_{\texttt{accMax}}$	$\texttt{w}_{\texttt{accMax}}$	the parameter is aggregated to the total cost if the maximum acceleration is over a given threshold
deceleration	$\texttt{rf}_{\texttt{decMax}}$	$\texttt{w}_{\texttt{decMax}}$	the parameter is aggregated to the total cost if the maximum deceleration is over a given threshold
lateral acceleration	$\texttt{rf}_{\texttt{latg}}$	$\texttt{w}_{\texttt{latg}}$	the parameter is aggregated to the total cost if the maximum lateral acceleration is over a given threshold
speed	$\texttt{rf}_{\texttt{spd}}$	$\texttt{w}_{\texttt{spd}}$	the parameter is aggregated to the total cost if the speed is over a given speed limit
curvature	$\texttt{rf}_{\texttt{curvature}}$	$\texttt{w}_{\texttt{curvature}}$	the parameter is aggregated to the total cost if the curvature of the short-term path is over a given threshold
safety distance	$\texttt{rf}_{\texttt{safeDist}}$	$\texttt{w}_{\texttt{safeDist}}$	the parameter is aggregated to the total cost if the vehicle violates the safety distance with other scenario vehicles

Now we intend to adapt FL to analyse the system hazard of both parameters and runtime flags. Since the parameters have continuous values, we can calculate their hazard impact scores using fuzzification-based coverage presented before. Instead, since the runtime flags have binary values, we can simply collect their coverage and then calculate their hazard impact scores as in classical SBFL.

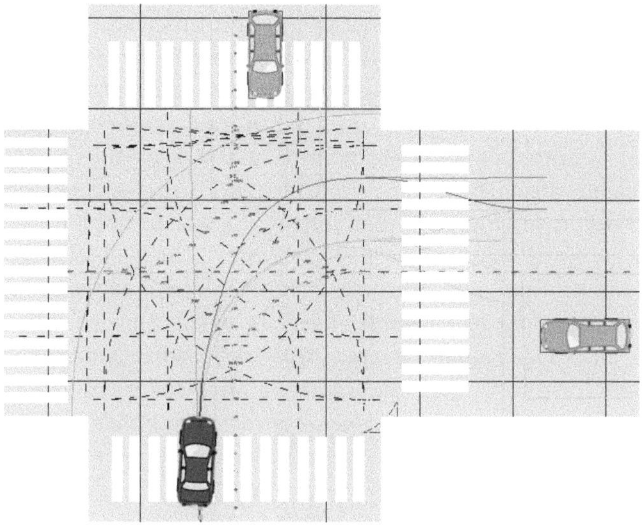

Figure 11.9: The *RightTurn* scenario.

The outputs of such hazard analysis are two rank lists indicating, respectively, the correlations (1) between the parameters and the hazard, and (2) between the runtime flags and the hazard. Finally, through integration of the two lists, we can get an overall explanation about the potential system hazard, such as "if we set the parameters to these values, we obtain a specific activation of the runtime flags, which may be the cause of a potential collision".

For example, we analyse a *RightTurn* scenario, in which, at the intersection, the ego vehicle must turn right, while another vehicle is crossing from right and another vehicle is crossing from the opposite direction, as shown in Fig. 11.9. The analysis result is shown in Table 11.3a. From the results, we can see that, in

Table 11.3: Results for scenario *RightTurn*

(a) Rank of fuzzy sets

Rank	**Parameter** $\mathtt{w}_i$	π_j	$\Phi^{\pi_j^{\mathtt{w}_i}}$
1	$\mathtt{w}_{\mathtt{latg}}$	*VL*	0.63
2	$\mathtt{w}_{\mathtt{latg}}$	*L*	0.62
2	$\mathtt{w}_{\mathtt{latg}}$	*SL*	0.62
4	$\mathtt{w}_{\mathtt{latg}}$	*M*	0.43
...	...	...	...
39	$\mathtt{w}_{\mathtt{curvature}}$	*VS*	0.23
40	$\mathtt{w}_{\mathtt{latg}}$	*SS*	0.20
41	$\mathtt{w}_{\mathtt{latg}}$	*S*	0.06
42	$\mathtt{w}_{\mathtt{latg}}$	*VS*	0.03

(b) Rank of runtime flags

Rank	**Runtime flag** $\mathtt{rf}_i$	$\Phi^{\mathtt{rf}_i}$
1	$\mathtt{rf}_{\mathtt{curvature}}$	1.00
2	$\mathtt{rf}_{\mathtt{decMax}}$	0.82
3	$\mathtt{rf}_{\mathtt{accMax}}$	0.15
3	$\mathtt{rf}_{\mathtt{latg}}$	0.15
5	$\mathtt{rf}_{\mathtt{spd}}$	0.00
5	$\mathtt{rf}_{\mathtt{safeDist}}$	0.00

the rank list of parameters, parameter $\texttt{w}_{\texttt{latg}}$ with fuzzy set *VL* is assigned with the highest hazard impact score of 0.63, which means that setting the parameter of *lateral acceleration Very Large* could lead to hazard. Moreover, the corresponding runtime flag $\texttt{rf}_{\texttt{latg}}$ has a low rank with hazard impact score 0.15, which is consistent with the results of the analysis of the parameters; indeed, when the parameter value is high, its corresponding runtime flag is likely not activated (as the path planner prefers not to implement that driving characteristic); this is why a low hazard score is instead associated with the activation of the runtime flag. Thus, we can get the final explanation as:

In this situation, the ego vehicle needs to make a sharp turn to avoid the collision with the other vehicle coming from the opposite side, which requires a high lateral acceleration. However, since we have set the parameter of lateral acceleration very large, during the turning, the ego vehicle prefers the actions having low lateral acceleration. As a result, the hazard of collision in this traffic situation exist.

11.5 Conclusions

In this chapter, we have introduced the basic concept of software fault localisation (FL) and reviewed some representative automated FL techniques. Then, we have taken Spectrum-Based Fault Localisation (SBFL) as an example, and explicitly described how we can apply this technique in practice. With SBFL, we can (semi-)automatically localise the potential faulty statements in a program. Due to the complexity of current systems and software, various factors can influence the accuracy of fault localisation. Thus, we have reviewed some state-of-art progresses aiming to solve various issues related to SBFL, such as the limitation of test information, the test oracle problem, and the selection of SBFL metrics. Finally, we have discussed the possibility of adapting SBFL to analyse more complex models and systems. Particularly, we have presented two approaches that follow the idea of SBFL to analyse the hazard in the models of Deep Neural Network (DNN) and Autonomous Driving Systems (ADS), respectively. These models and systems have been widely adopted in the era of industrial 4.0. The presented approaches provide a possible way to investigate the potential faulty elements (e.g., neurons or system parameters) in these systems. The output of these approaches can provide useful information to improve the safety during system re-engineering and repairing.

References

[1] Abreu, R., Zoeteweij, P. and Gemund, A.J.C. 2006. An evaluation of similarity coefficients for software fault localization. *In*: 2006 12th Pacific Rim International Symposium on Dependable Computing, PRDC '06, pages 39–46. IEEE.

[2] Abreu, R., Zoeteweij, P. and Gemund, A.J.C. 2007.On the accuracy of spectrum-based fault localization. *In*: Testing: Academic and industrial conference practice and research techniques-MUTATION (TAICPARTMUTATION 2007), pages 89–98. IEEE.

[3] Abreu, R., Zoeteweij, P. and Van Gemund, A.J.C. 2009. Spectrum-based multiple fault localization. *In*: 2009 IEEE/ACM International Conference on Automated Software Engineering, pages 88–99. IEEE.

[4] Agrawal, H. and Horgan, J.R. 1990. Dynamic program slicing. ACM SIGPlan Notices, 25(6): 246–256.

[5] Baudry, B., Fleurey, F. and Traon, Y.L. 2006. Improving test suites for efficient fault localization. *In*: Proceeding of the 28th international conference on Software engineering, pages 82–91.

[6] Birrell, N.D. and Ould, M.A. 1988. A practical handbook for software development. Cambridge University Press.

[7] Campos, J., Abreu, R., Fraser, G. and d'Amorim, M. 2013. Entropybased test generation for improved fault localization. *In*: 2013 28th IEEE/ACM International Conference on Automated Software Engineering (ASE), pages 257–267. IEEE.

[8] Cellier, P., Ducass., M., Ferr., S. and Ridoux, O. 2011. Multiple fault localization with data mining. *In*: SEKE, pages 238–243, 2011.

[9] Chen, T.Y., Cheung, S.C. and Yiu. S.M. 2020. Metamorphic testing: a new approach for generating next test cases. arXiv preprint arXiv: 2002.12543.

[10] Chen, T.Y., Kuo, F.-C., Liu, H., Poon, P-L., Towey, D., Tse, T.H., et al. 2018. Metamorphic testing: A review of challenges and opportunities. ACM Computing Surveys (CSUR), 51(1): 1–27.

[11] Collofello, J.S. and Woodfield, S.N. 1989. Evaluating the effectiveness of reliability-assurance techniques. Journal of Systems and Software, 9(3): 191–195.

[12] Debroy, V. and Wong, W.E. 2009. Insights on fault interference for programs with multiple bugs. *In*: 2009 20th International Symposium on Software Reliability Engineering, pages 165–174. IEEE.

[13] Denmat, T., Ducass., M. and Ridoux, O. 2005. Data mining and cross-checking of execution traces: A re-interpretation of jones, harrold and

stasko test information. *In*: Proceeding of the 20th IEEE/ACM International Conference on Automated Software Engineering, pages 396–399.

[14] DiGiuseppe, N. and Jones, J.A. 2015. Fault density, fault types, and spectra-based fault localization. Empirical Software Engineering, 20(4): 928–967.

[15] Duran, M., Zhang, X-Y., Arcaini, P. and Ishikawa, F. 2021. What to blame? on the granularity of fault localization for deep neural networks. *In*: 2021 IEEE 32nd International Symposium on Software Reliability Engineering (ISSRE), pages 264–275.

[16] Eniser, H.F., Gerasimou, S. and Sen, A. 2019. Deepfault: Fault localization for deep neural networks. *In*: International Conference on Fundamental Approaches to Software Engineering, pages 171–191. Springer.

[17] Gao, R. and Wong, W.E. 2017. Mseer—an advanced technique for locating multiple bugs in parallel. IEEE Transactions on Software Engineering, 45(3): 301–318.

[18] Gerasimou, S., Eniser, H.F., Sen, A. and Cakan, A. 2020. Importance-driven deep learning system testing. *In*: 2020 IEEE/ACM 42nd International Conference on Software Engineering (ICSE), pages 702–713. IEEE.

[19] Gupta, N., He, H., Zhang, X. and Gupta, R. 2005. Locating faulty code using failure-inducing chops. *In*: Proceeding of the 20th IEEE/ACM International Conference on Automated Software Engineering, pages 263–272.

[20] Gyimothy, T., Beszedes, A. and Forgacs, I. 1999. An efficient relevant slicing method for debugging. *In*: Software Engineering—ESEC/FSE '99, pages 303–321. Springer, 1999.

[21] Habib, A. and Pradel. M. 2018. How many of all bugs do we find? A study of static bug detectors. *In*: 2018 33rd IEEE/ACM International Conference on Automated Software Engineering (ASE), pages 317–328.

[22] Jiang, B., Zhai, K., Kwong Chan, W., Tse, T.H. and Zhang, Z. 2013. On the adoption of MC/DC and control-flow adequacy for a tight integration of program testing and statistical fault localization. Information and Software Technology, 55(5): 897–917.

[23] Jiang, B., Zhang, Z., Chan, W.K., Tse, T.H. and Chen, T.Y. 2012. How well does test case prioritization integrate with statistical fault localization? Information and Software Technology, 54(7): 739–758.

[24] Jones, J.A., Harrold, M.J. and Stasko, J. 2002. Visualization of test information to assist fault localization. *In*: Proceeding of the 24th International Conference on Software Engineering., ICSE '02, pages 467–477. IEEE.

[25] Karaman, S. and Frazzoli, E. 2011. Sampling-based algorithms for optimal motion planning. The International Journal of Robotics Research, 30(8): 846–894.

[26] Korel, B. and Laski, J. 1990. Dynamic slicing of computer programs. Journal of Systems and Software, 13(3): 187–195.

[27] Krizhevsky, A. and Hinton, G. 2009. Learning multiple layers of features from tiny images.

[28] Laghari, G., Murgia, A. and Demeyer, S. 2016. Fine-tuning spectrum based fault localisation with frequent method item sets. *In*: Proceeding of the 31st IEEE/ACM International Conference on Automated Software Engineering, pages 274–285.

[29] Le, T-D. B., Thung, F. and Lo, D. 2013. Theory and practice, do they match? a case with spectrum-based fault localization. *In*: 2013 IEEE International Conference on Software Maintenance, pages 380–383. IEEE.

[30] LeCun, Y., Bottou, L., Bengio, Y. and Haffner, P. 1998. Gradient based learning applied to document recognition. Proceeding of the IEEE, 86(11): 2278–2324.

[31] Li, W., Zheng, Z., Hao, P., Gao, Y-C., Rao, P-F. and Gong, C. 2013. Predicate execution-sequence based fault localization algorithm. Chinese Journal of Computers, 36(12): 2406–2419.

[32] Liblit, B., Naik, M., Zheng, A.X., Aiken, A. and Jordan, M.I. 2005. Scalable statistical bug isolation. ACM Sigplan Notices, 40(6): 15–26.

[33] Liu, C., Fei, L., Yan, X., Han, J. and Midkiff, S.P. 2006. Statistical debugging: A hypothesis testing-based approach. IEEE Transactions on Software Engineering, 32(10): 831–848.

[34] Liu, W., Zheng, Z., Cai, K.Y. and Zhu, W.L. 2012. QSRRT based motion planning for unmanned aerial vehicles using quick and smooth convergence strategies. Scientia Sinica Informationis, 42(11): 1403–1422.

[35] Mayer, W. and Stumptner, M. 2007. Model-based debugging–state of the art and future challenges. Electronic Notes in Theoretical Computer Science, 174(4): 61–82.

[36] Mayer, W. and Stumptner, M. 2008. Evaluating models for modelbased debugging. *In*: 2008 23rd IEEE/ACM International Conference on Automated Software Engineering, pages 128–137. IEEE.

[37] Naish, L., Lee, H.J. and Ramamohanarao, K. 2011. A model for spectra-based software diagnosis. ACM Transactions on Software Engineering and Methodology (TOSEM), 20(3): 1–32.

[38] Podgurski, A., Leon, D., Francis, P., Masri, W., Minch, M., Sun, J., et al. 2003. Automated support for classifying software failure reports. *In*: 25th International Conference on Software Engineering, 2003. Proceeding., pages 465–475. IEEE.

[39] Ren, X., Yu, B., Qi, H., Juefei-Xu, F., Li, Z., Xue, et al. 2020. Few-shot guided mix for dnn repairing. *In*: 2020 IEEE International Conference on Software Maintenance and Evolution (ICSME), pages 717–721. IEEE.

[40] Sohn, J., Kang, S. and Yoo, S. 2019. Search based repair of deep neural networks. arXiv preprint arXiv: 1912.12463.

[41] Sumner, W.N. and Zhang, X. 2010. Memory indexing: Canonicalizing addresses across executions. *In*: Proceeding of the Eighteenth ACM SIGSOFT International Symposium on Foundations of Software Engineering, pages 217–226.

[42] Sun, Y., Chockler, H., Huang, X. and Kroening, D. 2020. Explaining image classifiers using statistical fault localization. *In*: 2020 European Conference on Computer Vision (ECCV), pages 391–406, Cham, 2020. Springer International Publishing.

[43] Tang, C.M., Chan, W.K., Yu, Y.T. and Zhang, Z. 2017. Accuracy graphs of spectrum-based fault localization formulas. IEEE Transactions on Reliability, 66(2): 403–424.

[44] Voigt, T., Flad, S. and Struss, P. 2015. Model-based fault localization in bottling plants. Advanced Engineering Informatics, 29(1): 101–114.

[45] Weiser, M. 1984. Program slicing. IEEE Transactions on Software Engineering, (4): 352–357.

[46] Wong, W.E., Debroy, V., Gao, R. and Li, Y. 2013. The DStar method for effective software fault localization. IEEE Transactions on Reliability, 63(1): 290–308.

[47] Wong, W.E., Debroy, V., Golden, R., Xu, X. and Thuraisingham, B. 2011. Effective software fault localization using an RBF neural network. IEEE Transactions on Reliability, 61(1): 149–169.

[48] Wong, W.E., Debroy, V. and Xu, D. 2011. Towards better fault localization: A crosstab-based statistical approach. IEEE Transactions on Systems, Man, and Cybernetics, Part C (Applications and Reviews), 42(3): 378–396.

[49] Wong, W.E., Gao, R., Li, Y., Abreu, R. and Wotawa, F. 2016. A survey on software fault localization. IEEE Transactions on Software Engineering, 42(8): 707–740.

[50] Wong, W.E. and Qi, Y. 2009. BP neural network-based effective fault localization. International Journal of Software Engineering and Knowledge Engineering, 19(04): 573–597.

[51] Xia, L., Li, W., Zhang, Y. and Zhang, L. 2019. DeepFL: Integrating multiple fault diagnosis dimensions for deep fault localization. *In*: Proceeding of the 28th ACM SIGSOFT International Symposium on Software Testing and Analysis (ISSTA), pages 169–180. ACM.

[52] Xie, X., Chen, T.Y., Kuo, F.C. and Xu, B. 2013. A theoretical analysis of the risk evaluation formulas for spectrum-based fault localization. ACM Transactions on Software Engineering and Methodology (TOSEM), 22(4): 1–40.

[53] Xie, X., Wong, W.E., Chen, T.Y. and Xu, B. 2013. Metamorphic slice: An application in spectrum-based fault localization. Information and Software Technology, 55(5): 866 –879.

[54] Xin, B., Sumner, W.N. and Zhang, X. 2008. Efficient program execution indexing. ACM SIGPLAN Notices, 43(6): 238–248.

[55] Xu, B., Qian, J., Zhang, X., Wu, Z. and Chen, L. 2005. A brief survey of program slicing. ACM SIGSOFT Software Engineering Notes, 30(2): 1–36.

[56] Yan, C., Xiang, X. and Wang, C. 2020. Towards real-time path planning through deep reinforcement learning for a UAV in dynamic environments. Journal of Intelligent and Robotic Systems, 98(2): 297–309.

[57] Yoo, S., Harman, M. and Clark, D. 2013. Fault localization prioritization: Comparing information-theoretic and coverage-based approaches. ACM Transactions on Software Engineering and Methodology, 22(3), July 2013.

[58] Yoo, S., Xie, X., Kuo, F-C., Chen, T.Y. and Harman, M. 2017. Human competitiveness of genetic programming in spectrum-based fault localisation: Theoretical and empirical analysis. ACM Transactions on Software Engineering and Methodology (TOSEM), 26(3): 1–30.

[59] Yu, Y., Jones, J. and Harrold, M.J. 2008. An empirical study of the effects of test-suite reduction on fault localization. *In*: 2008 ACM/IEEE 30th International Conference on Software Engineering, pages 201–210.

[60] Zadeh, L.A. 1988. Fuzzy logic. Computer, 21(4): 83–93.

[61] Zakari, A., Lee, S.P., Abreu, R., Ahmed, B.H. and Rasheed, R.A. 2020. Multiple fault localization of software programs: A systematic literature review. Information and Software Technology, 124: 106312.

[62] Zeller, A. and Hildebrandt, R. 2002. Simplifying and isolating failure inducing input. IEEE Transactions on Software Engineering, 28(2): 183–200.

[63] Zhang, S. and Zhang, C. 2014. Software bug localization with markov logic. *In*: Companion Proceeding of the 36th International Conference on Software Engineering, pages 424–427.

[64] Zhang, X., Gupta, N. and Gupta, R. 2007. Locating faulty code by multiple points slicing. Software: Practice and Experience, 37(9): 935–961.

[65] Zhang, X-Y., Arcaini, P. and Ishikawa, F. 2019. Assessing the relation between hazards and variability in automotive systems. *In*: 2019 24th International Conference on Engineering of Complex Computer Systems (ICECCS), pages 190–199. IEEE.

[66] Zhang, X-Y., Arcaini, P. and Ishikawa, F. 2022. An incremental approach for understanding collision avoidance of an industrial path planner. IEEE Transactions on Dependable and Secure Computing, pages 1–1.

[67] Zhang, X-Y., Arcaini, P., Ishikawa, F. and Liu, K. 2020. Investigating the configurations of an industrial path planner in terms of collision avoidance. *In*: 2020 IEEE 31st International Symposium on Software Reliability Engineering (ISSRE), pages 301–312. IEEE.

[68] Zhang, X-Y. and Jiang, M. 2021. Spica: A methodology for reviewing and analysing fault localisation techniques. *In*: 2021 IEEE International Conference on Software Maintenance and Evolution (ICSME), pages 366–377.

[69] Zhang, X-Y., Towey, D., Chen, T.Y., Zheng, Z. and Cai, K-Y. 2015. Using partition information to prioritize test cases for fault localization. *In*: 2015 IEEE 39th Annual Computer Software and Applications Conference, volume 2, pages 121–126. IEEE.

[70] Zhang, X-Y., Towey, D., Chen, T.Y., Zheng, Z. and Cai, K-Y. 2016. A random and coverage-based approach for fault localization prioritization. *In*: 2016 Chinese Control and Decision Conference (CCDC), pages 3354–3361. IEEE.

[71] Zhang, X-Y. and Zheng, Z. 2019. A visualization analytical framework for software fault localization metrics. *In*: 2019 IEEE 24th Pacific Rim International Symposium on Dependable Computing (PRDC), pages 148–14809. IEEE.

[72] Zhang, X-Y., Zheng, Z. and Cai, K-Y. 2018. Exploring the usefulness of unlabelled test cases in software fault localization. Journal of Systems and Software, 136: 278–290.

[73] Zheng, Z., Liu, Y. and Zhang, X. 2016. The more obstacle information sharing, the more effective real-time path planning? Knowledge-Based Systems, 114: 36–46.

Index